VOLUME ONE HUNDRED AND ELEVEN

VITAMINS AND HORMONES

Opioid Hormones

VOLUME ONE HUNDRED AND ELEVEN

VITAMINS AND HORMONES

Opioid Hormones

Series Editor

GERALD LITWACK, PhD

Toluca Lake, North Hollywood, California

ELSEVIER

ACADEMIC PRESS

An imprint of Elsevier

Cover photo credit:
X-ray structure of KOP [kappa opioid receptor] in its inactive state (in blue, PDB 4DJH) and in its active state) in green, PDB 6B7S). The nanobody present in the active state is shown in grey, penetrating a pocket created by the displacement of TM6. The antagonist and agonist in the inactive and active states, respectively, are not displayed.
Vitamins and Hormones (2019) Volume 111.

Academic Press is an imprint of Elsevier
50 Hampshire Street, 5th Floor, Cambridge, MA 02139, United States
525 B Street, Suite 1650, San Diego, CA 92101, United States
The Boulevard, Langford Lane, Kidlington, Oxford OX5 1GB, United Kingdom
125 London Wall, London, EC2Y 5AS, United Kingdom

First edition 2019

ISBN: 978-0-12-818858-3
ISSN: 0083-6729

For information on all Academic Press publications
visit our website at https://www.elsevier.com/books-and-journals

Publisher: Zoe Kruze
Acquisition Editor: Ashlie Jackman
Editorial Project Manager: Joanna Collett
Production Project Manager: Denny Mansingh
Cover Designer: Mark Rogers

Typeset by SPi Global, India

Contents

Contributors

Paula Abate
Laboratorio de Psicología Experimental, Instituto de Investigaciones Psicológicas (IIPsi - CONCET - UNC), Facultad de Psicología, Universidad Nacional de Córdoba, Córdoba, Argentina

Olga Alvarez-Ojeda
Universidad Industrial de Santander, Department of Pathology, School of Medicine, Bucaramanga, Colombia

Inmaculada Banegas
Department of Health Sciences, University of Jaén, Jaén, Spain

Bettina Böttcher
Department of Gynecological Endocrinology and Reproductive Medicine, Medical University of Innsbruck, Innsbruck, Austria

José Miguel Cerdá-Reverter
Department of Fish Physiology and Biotechnology, Institute of Aquaculture from Torre la Sal (IATS-CSIC), Castellon, Spain

Rafael Coveñas
Laboratory of Neuroanatomy of the Peptidergic Systems, Institute of Neurosciences of Castilla y León (INCYL), University of Salamanca, Salamanca, Spain

Georges Czaplicki
Institut de Pharmacologie et de Biologie Structurale (IPBS, CNRS UPS), Université de Toulouse, Toulouse, France

Elisabeth D'Costa
Department of Obstetrics and Gynecology, Medical University of Innsbruck, Innsbruck, Austria

Marc de Gasparo
Cardiovascular & Metabolic Syndrome Adviser, Rossemaison, Switzerland

Pascal Demange
Institut de Pharmacologie et de Biologie Structurale (IPBS, CNRS UPS), Université de Toulouse, Toulouse, France

Ewing Duque-Díaz
Universidad de Santander UDES, Laboratory of Neurosciences, School of Medicine, Bucaramanga, Colombia

Guillaume Ferré
Institut de Pharmacologie et de Biologie Structurale (IPBS, CNRS UPS), Université de Toulouse, Toulouse, France

Allison Jane Fulford
Centre for Applied Anatomy, University of Bristol, Bristol, United Kingdom

Louis Gendron
Department of Pharmacology-Physiology, Université de Sherbrooke, Centre de Recherche du CHUS; Quebec Pain Research Network, Sherbrooke, QC, Canada

Patrick M. Giguère
Department of Biochemistry, Microbiology and Immunology; Brain and Mind Research Institute, University of Ottawa, Ottawa, ON, Canada

Alan R. Gintzler
Department of Obstetrics and Gynecology, State University of New York, Downstate Medical Center, Brooklyn, NY, United States

Alejandra Godino-Gimeno
Department of Fish Physiology and Biotechnology, Institute of Aquaculture from Torre la Sal (IATS-CSIC), Castellon, Spain

Karla Hernández-Fonseca
Departamento de Neuroquímica, Subdirección de Investigaciones Clínicas, Instituto Nacional de Psiquiatría Ramón de la Fuente, Ciudad de México, México

Sarah Keskes
Centre for Applied Anatomy, University of Bristol, Bristol; Keele University Medical School, Keele, United Kingdom

Leda Kovatsi
Laboratory of Forensic Medicine and Toxicology, School of Medicine, Aristotle University of Thessaloniki, Thessaloniki, Greece

Nai-Jiang Liu
Department of Obstetrics and Gynecology, State University of New York, Downstate Medical Center, Brooklyn, NY, United States

Magdalena Martínez-Cañamero
Department of Health Sciences, University of Jaén, Jaén, Spain

Feng Mei
Department of Histology and Embryology, Brain Science Key Laboratory of Chongqing Education Commission, Third Military Medical University, Chongqing, China

Milagros Méndez
Departamento de Neuroquímica, Subdirección de Investigaciones Clínicas, Instituto Nacional de Psiquiatría Ramón de la Fuente, Ciudad de México, México

Alain Milon
Institut de Pharmacologie et de Biologie Structurale (IPBS, CNRS UPS), Université de Toulouse, Toulouse, France

Karim Nagi
College of Medicine, Member of QU Health, Qatar University, Doha, Qatar

Kakia Nikolaou
Addiction Department IANOS, General Hospital of Thessaloniki 'G. Papanikolaou'—Psychiatric Hospital of Thessaloniki, Thessaloniki, Greece

Graciela Pineyro
CHU Sainte-Justine research center, Department of Pharmacology, Faculty of Medicine, University of Montreal, Montreal; Quebec Pain Research Network, Sherbrooke, QC, Canada

Isabel Prieto
Department of Health Sciences, University of Jaén, Jaén, Spain

Manuel Ramírez-Sánchez
Department of Health Sciences, University of Jaén, Jaén, Spain

Ana Rocha
Department of Fish Physiology and Biotechnology, Institute of Aquaculture from Torre la Sal (IATS-CSIC), Castellon, Spain

Hildegard M. Schuller
Department of Biomedical & Diagnostic Science, College of Veterinary Medicine, The University of Tennessee, Knoxville, TN, United States

Beata Seeber
Department of Gynecological Endocrinology and Reproductive Medicine, Medical University of Innsbruck, Innsbruck, Austria

Ana-Belén Segarra
Department of Health Sciences, University of Jaén, Jaén, Spain

Emiliya M. Storman
Department of Obstetrics and Gynecology, State University of New York, Downstate Medical Center, Brooklyn, NY, United States

John M. Streicher
Department of Pharmacology, College of Medicine, University of Arizona, Tucson, AZ, United States

Fei Wang
Department of Histology and Embryology, Brain Science Key Laboratory of Chongqing Education Commission, Third Military Medical University, Chongqing, China

Ludwig Wildt
Department of Gynecological Endocrinology and Reproductive Medicine, Medical University of Innsbruck, Innsbruck, Austria

Manel Zeghal
Department of Biochemistry, Microbiology and Immunology; Brain and Mind Research Institute, University of Ottawa, Ottawa, ON, Canada

Preface

Opioid hormones are produced to modulate pain sensations, usually caused by some form of injury, that are transmitted through the spinal cord to the brain. Opioids produced by the body are β-endorphins, enkephalins, and dynorphins. β-Endorphin derives from the precursor peptide, proopiomelanocortin, in the anterior pituitary, and these hormones are found in hypothalamic neurons in addition to the pituitary. Anterior pituitary proopiomelanocortin is enzymatically cleaved to release adrenocorticotropin (ACTH) and β-lipotropin. In the pituitary, amino acids 61–90 of β-lipotropin give rise to β-endorphin. β-Endorphin has a molecular weight of 3465 consisting of 29 amino acids. In addition to the role of β-endorphin in blocking the transmission of pain in the brain stem, it has an inhibitory action on the secretion of gonadotropin (GnRH) and on reproductive functions. β-Endorphin acts as an agonist for mu opioid receptors located primarily in the hypothalamus.

In addition to β-endorphin, there are two low molecular weight opioids produced in the adrenal medulla and in the central nervous system. These are methionine enkephalin (MetENK) and leucine enkephalin (LeuENK). They are the pentapeptides, YGGFM and YGGFL, respectively. Whereas β-endorphin blocks pain in the brain stem, enkephalins block pain signals in the spinal cord.

In addition to the two classes of opioids described above, dynorphins are a class of opioid peptides that arise from the cleavage of the precursor, prodynorphin. Cleavage of prodynorphin by the enzyme proprotein convertase 2 releases dynorphin A, dynorphin B (also called rinorphin), and α- and β-neo-endorphin. Dynorphins A and B have molecular weights of 981 and 1570, respectively, and each consists of 13 amino acids. This class of opioids preferentially binds to kappa opioid receptors and are regulators of many functional pathways in the brain. α-Neo-endorphin is a decapeptide (TGGFLRKTPK) with a molecular weight of 1228. β-Neo-endorphin is a nonapeptide (TGGFLRKTP) with a molecular weight of 1100. Opioids are produced in various parts of the brain including the hypothalamus, hippocampus, and spinal cord and function in nociception, motor activity, response to injury, stress, anxiety, mood, and learning and memory. When dynorphin production is experimentally knocked out, there are a reduction of fat mass and a stimulation of weight loss, probably caused by an increase in fatty acid oxidation and reduced intestinal nutrient absorption, thus indicating roles in addition to modulating pain transmission.

In this volume, there is coverage of many of the basic and clinical aspects of these hormones. In Chapter 1, A. Rocha, A. Godino-Gimeno, and J.M. Cerdá-Reverter discuss the "Evolution of proopiomelanocortin." In Chapter 2, the "Structure and dynamics of dynorphin peptide and its receptor" is introduced by G. Ferré, G. Czaplicki, P. Demange, and A. Milon. Additional basic studies are exemplified by the contribution of L. Gendron, K. Nagi, M. Zeghal, P.M. Giguère, and P. Pineyro: "Molecular aspects of delta opioid receptors." This chapter is followed by "The role of heat shock protein 90 in regulating pain, opioid signaling and opioid antinociception" by J.M. Streicher. "Enkephalinase regulation" is the topic of M. Ramírez-Sánchez, I. Prieto, A.B. Segarra, M. Martínez-Cañamero, I. Banegas, and M. de Gasparo. In Chapter 6, A.J. Fulford and S. Keskes contribute "Limbic circuit connectivity and the stress response: new insights into the mammalian nociception peptide system." Then, E. Duque-Díaz, O. Alvarez-Ojeda, and R. Covenñs deliver on the subject of "Enkephalins and ACTH in the mammalian nervous system." "Opioids and the hormone oxytocin" is reported by L. Kovatsi and K. Nikolaou.

In physiological and clinical connections, A.R. Gintzler, E.M. Storman, and N.-J. Liu first report on "Estrogens as arbiters of sex-specific and reproductive cycle-dependent opioid mechanisms." This is followed by "Opioids and reproduction" by B. Seeber, B. Böttcher, E. D'Costa, and L. Wildt. F. Wang and F. Mei discuss "Kappa opioid receptor and oligodendrocyte remyelination." The "Inhibitory role of G_i-coupled receptors on cAMP-driven cancers with focus on opioid receptors in lung adenocarcinoma and its stem cells" is related by H.M. Schuller. In the final chapter, M. Méndez, K. Hernández-Fonseca, and P. Abate contribute "Prenatal ethanol exposure and enkephalinergic neurotransmission."

The illustration on the book cover is an X-ray crystal structure of the kappa opioid receptor. It is reproduced from Fig. 4 of Chapter 2 by G. Ferré, G. Czaplicki, P. Demange, and A. Milon entitled "Structure and dynamics of dynorphin peptide and its receptor."

I am grateful for the efforts of Joanna Collett of Elsevier in Oxford, UK who organized the initial phases of the volume and to Denny Mansingh for Elsevier in Taramani, Chennai, India for production.

Gerald Litwack
Toluca Lake, CA
May 20, 2019

CHAPTER ONE

Evolution of proopiomelanocortin

Ana Rocha, Alejandra Godino-Gimeno, José Miguel Cerdá-Reverter*
Department of Fish Physiology and Biotechnology, Institute of Aquaculture from Torre la Sal (IATS-CSIC), Castellon, Spain.
*Corresponding author: e-mail address: jm.cerda.reverter@csic.es

Contents

Abstract

Proopiomelanocortin (POMC) belongs to the opioid/orphanin gene family whose peptide precursors include either opioid (YGGF) or the orphanin/nociceptin core sequences (FGGF). In addition to POMC the family includes the proenkephalin (*PENK*), prodynorphin (*PDYN*), and nociceptin/proorphanin (*PNOC*) precursors. The opioid core sequence in POMC is incorporated by the β-endorphin that occupies the C-terminal region but this propeptide also exhibits at least two "alien" melanocortin core sequences (HFRW). An ACTH/MSH fragment merged into the opioid fragment not earlier than the two tetraploidizations of the vertebrate genome. Therefore, POMC exhibit a complex "evolutionary life" since the gene has coevolved together with two different receptor systems, i.e., opioid and melanocortin following a horse trading system. In this article, we summarize the different evolutionary hypotheses proposed for POMC evolution.

1. Introduction

Proopiomelanocortin (*POMC*) gene encodes a propeptide whose posttranslational processing generates several bioactive peptides. In most vertebrates, POMC contains several domains and each integrates a core sequence (HFRW) that, in turn, distinguishes a melanocyte-stimulating hormone (MSH). The N-terminal domain integrates the pro-γ-MSH, the central domain contains the α-MSH as the N-terminal sequence

Vitamins and Hormones, Volume 111
ISSN 0083-6729
https://doi.org/10.1016/bs.vh.2019.05.008

of the adrenocorticotropic hormone (ACTH), and the C-terminal, β-lipotropin that includes the β-MSH (Nakanishi et al., 1979). The C terminal domain also includes β-endorphin that retains an opioid core sequence, YGGF, characteristic of the opioid/orphanin gene family (Fig. 1).

The tissue-specific posttranscriptional processing rules what peptides are released from the precursor and this differential proteolytic cleavage depends on the prohormone convertase battery available in the specific tissue. In the corticotrophs of the anterior pituitary, the proprotein convertase subtilisin/kexin type 1 (Pcsk1) activity generates pro-γ-MSH, ACTH, and β-lipotropin whereas the cleavage by Pcsk1 and Pcsk1 2 produces α-MSH and β-endorphin in the melanotrophs of the pars intermedia (Castro & Morrison, 1997). *POMC* is mainly expressed in the pituitary being one of the most abundant transcripts in this gland (He, Dai, Chen, He, & Yin, 2014). However, two discrete groups of neurons in the hypothalamus and the medulla, the arcuate nucleus (NAc) and the nucleus of the solitary tract (NTS), respectively, also produce POMC (Bagnol et al., 1999). In the rat brain, POMC is mainly processed to α-MSH and β-endorphin (Castro & Morrison, 1997) but some hypothalamic ACTH-IR after RP-HPLC, elutes also at the position of ACTH, suggesting its presence in the rat brain (Smith & Funder, 1988).

The presence of opioid and melanocortin sequences in the POMC indicates that the molecular evolution of the precursor has been under constrains of either systems, coevolving with different receptor systems, i.e., opioid and melanocortin, as strings pulling together in the same or different direction. In the last years, several reviews on the evolution of opioid system (Elphick, Mirabeau, & Larhammar, 2018; Larhammar, Bergqvist, & Sundstrom, 2015; Sundström, Dreborg, & Larhammar, 2010) and melanocortin system (Cerda-Reverter et al., 2011; Cortés et al., 2014; Dores et al., 2014; Dores & Baron, 2011; Dores, Liang, Davis, Thomas, & Petko, 2016; Navarro et al., 2016; Takahashi, Mizusawa, & Amano, 2014) have been published. We summarize here the main hypothesis on the evolutionary processes of this complex multiprecursor that amalgamates potential gene duplications, insertion, reorganization and coevolutionary process.

2. Evolution of opioid peptides and nociceptin/orphanin family

POMC gene belongs to the opioid peptides and nociception/orphanin gene family, which also includes proenkephalin (*PENK*), prodynorphin (*PDYN*), and nociceptin/proorphanin (*PNOC*) (Danielson

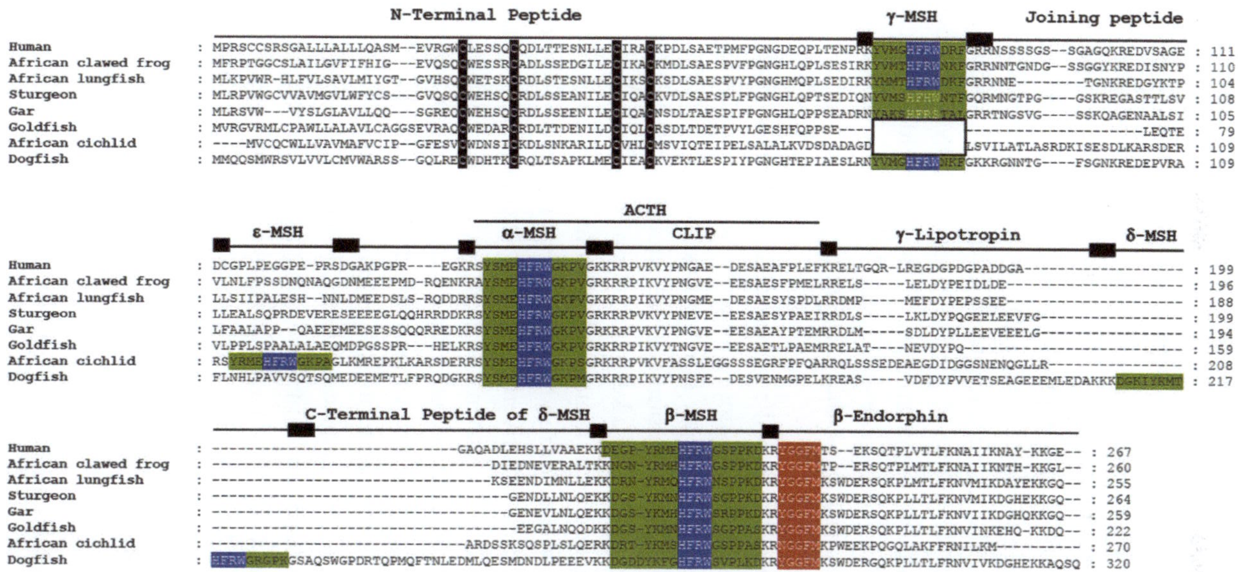

Fig. 1 Alignment of proopiomelanocortin (POMC) sequences from human (*Homo sapiens*). White letters on black background indicate fully conserved cysteine residues in all POMC sequences. Green and red boxes show MSH peptides and endorphin core, respectively. White letters on blue background highlight melanocortin core sequences. Uncoloured boxed area demarcates γ-MSH segment in species lacking γ-MSH. Black boxes indicate endoproteolytic cleavage sites. *Adapted from Navarro, S., Soletto, L., Puchol, S., Rotllant, J., Soengas, J. L., & Cerda-Reverter, J. M. (2016). 60 years of POMC: POMC: An evolutionary perspective.* Journal of Molecular Endocrinology, *56(4), T113–118. https://doi.org/10.1530/JME-15-0288 (web archive link).*

& Dores, 1999). All genes from this family encode at least one opioid "core" sequence, consisting of the peptide Y/FGGF or PNOC (Dores & Lecaude, 2005; Dores, Lecaude, Bauer, & Danielson, 2002). Both PENK and PDYN contain multiple copies of the "core" opioid sequence, varying from three to seven depending on species, whereas POMC has a single core motif (as part of endorphin) and PNOC can have one or two copies depending on the class of vertebrates (Sundström et al., 2010). Outside the opioid core sequence, the members of this peptide family differ substantially among themselves and between species, but they all have retained a set of conserved cysteine residues at the N-terminal region of the molecule. Six residues are found in PENK, PDYN and PNOC, and only four in POMC. In addition, all genes from this family share the same overall structure with a single intron shortly after the region encoding the signal peptide (Larhammar et al., 2015). These unifying signatures suggest that they are all derived from a common ancestral opioid gene.

Gene families comprise several to many genes of similar nucleotide or amino acid sequences that have close functions. Several mechanisms, such as tandem duplications, segmental duplications, or even whole-genome duplications can lead to the expansion of gene families. Since many decades, gene duplication is recognized as a key mechanism of evolution. It can alter gene dosage, rescue gene function, and lead to evolution of new genetic networks, and to the re-wiring or modulation of existing ones. Whether this process is mostly neutral, or driven by natural selection, is controversial (Lespinet, Wolf, Koonin, & Aravind, 2002; Nei, 2013). Notably, during the evolution of chordates, the ancestral deuterostome genome experienced two rounds of whole-genome duplication named 1R and 2R (Fig. 2). These events occurred before the radiation of Gnathostomes (jawed vertebrates) and there are evidences that support that both 1R and 2R happened before the divergence of Cyclostomes (lampreys and hagfishes) and Gnathostomes, approximately 500Mya. However, an alternative scenario maintains that these two lineages might have diverged after the 1R (Smith et al., 2018) (Fig. 2). Even though most duplicated genes were secondarily lost, many evolved new functions, in support of the notion that gene and genome duplications might provide a major mechanism for generating phenotypic diversity in evolution (Ohno, 1970). The common ancestor of teleost fish has experienced an additional whole-genome duplication event approximately 320 Mya, named 3R (Amores et al., 1998; Christoffels et al., 2004; Jaillon et al., 2004; Taylor, Braasch, Frickey, Meyer, & Van de Peer, 2003) (Fig. 2). This event accounts for several divergent features in teleost genomes.

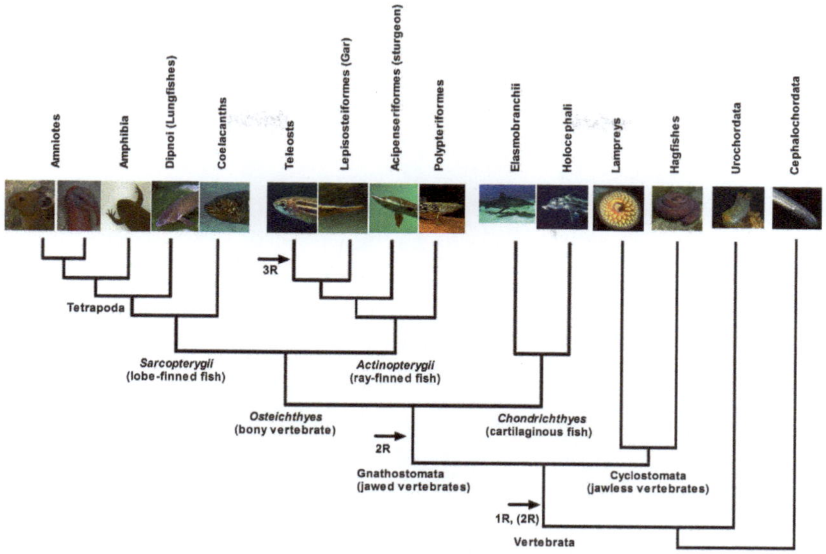

Fig. 2 Simplified phylogenetic tree of Chordate. The phylum Chordata includes three sub-phyla: Cephalochordate, Urochordata and Vertebrata. Vertebrates are the most dominant and successful group of animals on earth, occupying both terrestrial and aquatic habitats. Vertebrates are divided into two broad groups: jawed vertebrates (Gnathostomata) and the jawless vertebrates represented by only two extant lineages, lampreys and hagfishes (Cyclostomata). Gnathostomes, conversely, comprise the vast majority of all vertebrates. There are two main groups within Gnathostomata: cartilaginous fishes (Chondrichthyes), which include sharks, rays, skates and bony vertebrates (Osteichthyes), which include ray-finned fishes, lobe-finned fishes, and tetrapods (land animals). Ray-finned fishes (Actinopterygii) comprise roughly half of all living vertebrates. Two major groups can be recognized in ray-finned fishes: teleosts and basal, nonteleost ray-finned fishes, which comprise four orders: Polypteriformes (e.g., bichir), Acipenseriformes (e.g., sturgeon and paddlefish), Lepisosteiformes (e.g., spotted gar), and Amiiformes (e.g., bowfin). Based on recent findings, it is hypothesized that two rounds of whole-genome duplication occurred, 1R and 2R, although it is still not clear when they took place. The teleost lineage went through an additional whole-genome duplication (3R).

In the human genome, *PDYN* and *POMC* are located on chromosome 20 and 2, respectively (Fig. 3). Chromosome 8 carries both *PENK* and *PNOC* genes. The chicken genome was initially thought to lack *PDYN* but recent genome assembly (Gallus_gallus-5.0) shows this gene localized on chromosome 20, pairing chicken opioid system with that of mammals. However, the chromosomal regions for the four opioid peptide precursor genes are different from those in human genome. *PENK* is found on chromosome 2, whereas both *POMC* and *PNOC* are located to chromosome 3. Therefore, *PNOC* is paired with *POMC* on chromosome 3 and not with

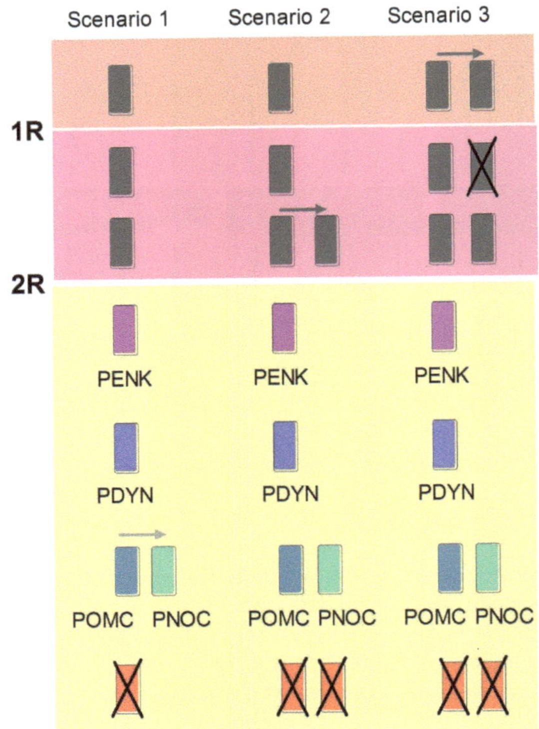

Fig. 3 Anticipated evolutionary history for the opioid peptide and nociceptin/orphanin family by a combination of whole-genome and local duplications. Three different scenarios have been proposed to explain the timing of the duplication events that have occur in the common ancestor leading to POMC and PNOC. Local duplication, indicated by arrows, generating both POMC and PNOC could have taken place after the 2R (scenario 1), after the 1R (scenario 2) or before the 1R (scenario 3). POMC, proopiomelanocortin; PENK, proenkephalin; PDYN, prodynorphin; PNOC, nociceptin/proorphanin. *The figure is an adapted version of the figure published in Sundström, G., Dreborg, S., & Larhammar, D. (2010). Concomitant duplications of opioid peptide and receptor genes before the origin of jawed vertebrates.* PLoS One, *5(5), e10512. https://doi.org/10.1371/journal.pone.0010512 (web archive link); and Navarro, S., Soletto, L., Puchol, S., Rotllant, J., Soengas, J. L., & Cerda-Reverter, J. M. (2016). 60 years of POMC: POMC: An evolutionary perspective.* Journal of Molecular Endocrinology, *56(4), T113–118. https://doi.org/10.1530/JME-15-0288 (web archive link).*

PENK as occurs on human chromosome 8 (Fig. 3). Inferring the distribution pattern of the genes encoding opioid peptide precursors has demonstrate that *PENK*, *PDYN* and *PNOC* are located on three separate chromosomes in the same paralogon (set of paralogous chromosomal regions comprising syntenic genes) of the vertebrate genome (Sundström et al., 2010). The *POMC* gene in many species is locate in the same chromosomal region as the *PNOC*. They share the same chromosome in the opossum (*Monodelphis domestica*) as well as in all species of teleost fish that have been studied (Larhammar et al., 2015; Sundström et al., 2010). In fact, in the soft-shell turtle (*Pelodiscus sinensis*), *POMC* and *PNOC* are just around 1 Mb apart. The spotted gar, *Lepisosteus oculatus*, is a basal nonteleost ray-finned fish, whose lineage diverged from the teleost lineage prior to the teleost whole-genome duplication (Fig. 2). Spotted gar possesses a typical diploid vertebrate genome similar to the human genome (Braasch et al., 2016). Comparative analyses have shown that the gar genome is organized more like chicken and human genomes than like those of teleosts (Amores, Catchen, Ferrara, Fontenot, & Postlethwait, 2011). It has fewer chromosomal rearrangements than both teleost fishes and many mammals including human, converting this ray-finned fish in an exceptional species for studies on vertebrate genome evolution. In the spotted gar *PENK* and *PDYN* are located on chromosomes LG9 and LG18, respectively. *POMC* and *PNOC* are found together on chromosome LG1 suggesting a local gene duplication followed by fusion of an *ACTH/MSH* gene in *POMC*. The different arrangement of these two genes in the human genome could be the result of a gene translocation event, taking place in the ancestor of placental mammals. Therefore, the expansion of the opioid peptide system seems to be the result of two complete genome duplications (1R and 2R) and a single local duplication. There are three possible scenarios for the evolution of the opioid peptides. They differ in the timing of the local duplication that generated *POMC* and *PNOC* that could have taken place before 1R (scenario 1), after 1R but before 2R (scenario 2), or after 2R (scenario 3, Fig. 4) (Navarro et al., 2016). Lampreys, along with hagfishes, constitute the cyclostomes, the sole survivors of a lineage that diverged from the ancestor of the jawed vertebrates more than 500 million years ago (Docker, Hume, & Clemens, 2015). Lampreys have two *POMC* genes, proopiocortin (*POC*) and proopiomelanotropin (*POM*). POC encodes an ACTH sequence, a β-MSH-related sequence and a β-endorphin sequence. *POM* encodes MSH-B (an α-MSH-related peptide), MSH-A (a β-MSH-related peptide) and a β-endorphin sequence (Takahashi et al., 1995; Takahashi, Amemiya, Sarashi, Sower, & Kawauchi, 1995). The presence of *POMC* sequences in the genome of lampreys suggests that the duplication event given rise to *PNOC* and *POMC* probably occurred before the split of cyclostomes from other chordates. If *POMC* emerged before the second genome duplication is impossible to say because both chromosomal and whole-genome duplications have played significant roles in the evolution of ancestral vertebrate lamprey genomes (Smith et al., 2018). In the whole,

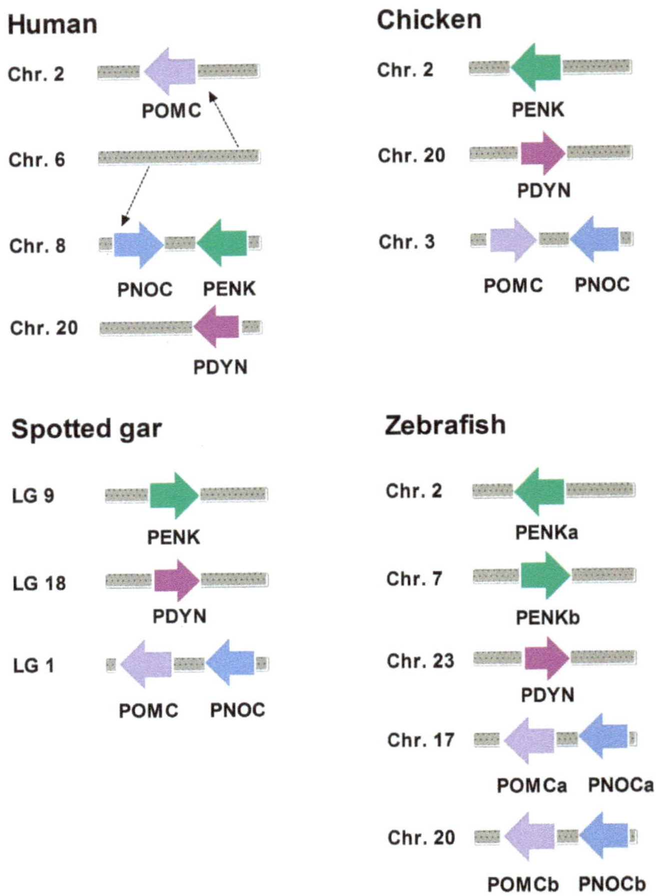

Fig. 4 Chromosomal regions for the four opioid peptide precursor genes in human (*Homo sapiens*), chicken (*Gallus gallus*), spotted gar (*Lepisosteus oculatus*) and zebrafish (*Danio rerio*). Each arrow box represents a gene, and its orientation (forward or reverse) in the respective chromosome (horizontal lines). Dashed arrows symbolize gene translocations. Chr, chromosome; LG, linkage group. *Adapted from Larhammar, D., Bergqvist, C., & Sundstrom, G. (2015). Ancestral vertebrate complexity of the opioid system.* Vitamins and Hormones, *97, 95–122. https://doi.org/10.1016/bs.vh.2014.11.001 (web archive link).*

these considerations point to an ancient vertebrate origin of the opioid system, with all members presently found in humans already established more than 500 Mya.

Grounded in the knowledge of the existence of one opioid peptide precursor gene in the ancestor of the vertebrates prior to the 1R and 2R, efforts have been made to find related sequences also in invertebrates. However, searches on any invertebrate genome project failed to identify opioid propeptides or opioid receptor sequences, not even in deuterostomes invertebrates (reviewed in Larhammar et al., 2015). Also melanocortin receptors are still unknown outside the vertebrates (Larhammar & Bergqvist, 2013).

Mammals, and most other tetrapods, have four opioid peptide precursor genes (Dores et al., 2002). Exceptions include polyploidy frogs that have additional copies. Nociceptin has not yet been reported in a shark but we have identified in the little skate (*Leucoraja erinacea*) draft genome database (LittleSkate-Transcriptome-Contigs-Build2), a sequence of a peptide that contains the opioid YGGF core motif. This peptide is found in the LittleSkate-transcriptB2-contig3929 and is 82% identical to the nociceptin peptide sequence in chicken, gar, Latimeria and zebrafish. Moreover, a kappa-type opioid receptor (oprk1) has been identified in the elephant shark (*Callorhinchus milii*) and therefore Pnoc is expected to be present also in cartilaginous fish (Larhammar et al., 2015) reinforcing the idea of an ancient vertebrate origin of the opioid system. In teleost, the situation is different. The teleost-specific genome duplication is believed to be responsible for the rapid speciation and adaptive radiation of teleost. It is proposed that genome duplication events provide additional genetic raw material on which selection can act and give rise to novel phenotypes. Following genome duplication, there are three common fates for the duplicated genes. The most likely fate is the loss of one of the duplicate loci through accumulation of deleterious mutations, resulting in nonfunctionalization (Lynch & Conery, 2000). This loss of duplicated gene loci can potentially play a role in speciation via the phenomenon of divergent resolution (i.e., differential gene loss) (Semon & Wolfe, 2007; Taylor, Van de Peer, & Meyer, 2001). The other two fates of duplicated genes are subfunctionalization (both copies are retained because of partitioning of ancestral gene functions between the duplicates) and neofunctionalization (whereby one of the copies acquires a novel function) (Force et al., 1999). The retained duplicate pairs often show a difference in overall expression patterns and expression levels (Berthelot et al., 2014). All teleost fish in which the genome has been sequenced have two copies of the *pomc* gene (reviewed in Sundström et al., 2010). In the Tetraodon (*Tetraodon nigroviridis*), *pomc* paralogs have experienced subfunctionalization of both expression and peptide domains during teleost evolution (de Souza, Bumaschny, Low, & Rubinstein, 2005). *pomc*α is expressed in the nucleus lateralis tuberis of the hypothalamus, the homolog of the arcuate nucleus in fish, as well as in the rostral pars distalis and pars intermedia of the pituitary, whereas *pomc*β is expressed in the preoptic area of the brain and weakly in the pituitary pars intermedia. *Pomcβ* genes have a β-endorphin segment that lacks the consensus opioid signal and seems to be under neutral evolution in tetraodontids, whereas *pomcα* genes possess well-conserved peptide regions. Other teleost species, specifically the ones that have experienced a fourth whole-genome duplication event, e.g., rainbow trout, are found to have three copies of the *pomc* gene (Harris, Dijkstra, & Hofmann, 2014; Leder & Silverstein, 2006) but three *pomc* paralogs are also found in the barfin flounder, *Verasper moseri* (Takahashi et al., 2005), Burton's mouthbrooder (*Haplochromis burtoni*) and medaka (*Oryzias latipes*) (Harris et al., 2014). In all cases, presence of three genes should be the consequence of lineage-specific gene duplication events rather than genome duplication (Sundström et al., 2010).

3. POMC bauplan evolution

The presence of "alien" melanocortin sequences in the primary structure of *POMC* converts this gene probably, in the most complex precursor of the opioid family. Suggestively, POMC maintains structural and sequence identities with the three other opioid precursors within the N- and C-terminal region of the precursor thus implying that this complex precursor arose when a fragment of DNA containing one or multiple MSH sequences was inserted into apreproendorphin gene. However, we cannot excluded that the ancestral melanocortin sequence resulted from accumulative mutations in the preproendorphin gene that lead to multiple melanocortin sequences after unequal crossing-over events (Dores & Baron, 2011). When the preproendorphin merged with the MSH sequence(s) is probably one of the key questions in the POMC evolution. Results suggest that a distinct proendorphin gene arose after the two tetraploidizations of the vertebrate genome (Larhammar et al., 2015; Sundström et al., 2010) and therefore POMC emerged, after the fusion of the ACTH/MSH fragment, not earlier than the second genome duplication. A gene encoding ACTH/MSH could have appeared earlier as supported by the fact that the genes encoding melanocortin receptors also arose as the result of the two duplication rounds of the vertebrate genome from a single ancestral gene present before tetraploidizations (Cortés et al., 2014).

The structural plan of the POMC precursors differs between species and these variations apply mainly for the number of MSH sequences/domains (Figs. 1 and 5). All tetrapod species exhibit three MSH core sequences (γ-MSH, α-MSH and β-MSH) suggesting that POMC evolved through

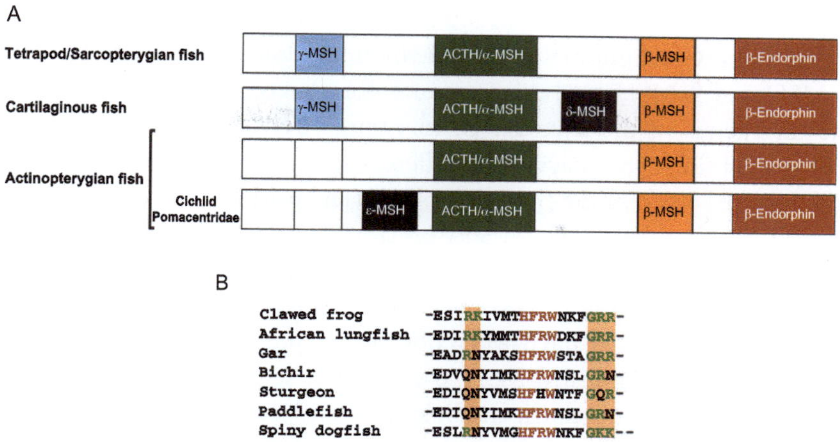

Fig. 5 (A) Structural plans of POMC precursor in main vertebrate groups. Both tetrapod and lobe-finned fish share the same POMC bauplan. Cartilaginous fish exhibit and additional MSH peptide called δ-MSH whereas Actinopterygian fish lack γ-MSH peptide. Some teleost fish have an additional ε-MSH peptide as a putative tandem duplication of the α-MSH-ACTH segment. (B) Comparison of γ-MSH sequences showing the potential degenerative evolutionary process leading to the loss of γ-MSH peptide in teleost fish. The γ-MSH sequence and the flanked proposed endoproteolytic cleavage sites for tetrapod, lobe-finned fish (lungfish), cartilaginous (spiny dogfish) and four ray-finned fish are presented. The N-terminal cleavage site (A) and the C-terminal cleavage and α-amidation site (B) are boxed. An arginine (R)-lysine (K) dibasic pair flanks tetrapod and lobe-finned fish γ-MSH peptide. An arginine (R) residue is still present in gar but lysine is replaced by an asparagine (N) residue, a common substitution in many fish species including elasmobranchs. It is uncertain whether this dibasic pair (RN) can promote γ-MSH cleavage but this motif is fully absent in Polypteriformes and Acipenseriformes precursors, which makes γ-MSH processing questionable. In addition, the C-terminal cleavage motif (RR or KK) is also absent in bichir, sturgeon and paddlefish. Moreover, sturgeon precursor carries a substitution in the MSH core sequence compromising the activity of the peptide. *Adapted from Navarro, S., Soletto, L., Puchol, S., Rotllant, J., Soengas, J. L., & Cerda-Reverter, J. M. (2016). 60 years of POMC: POMC: An evolutionary perspective.* Journal of Molecular Endocrinology, *56(4), T113–118. https://doi.org/10.1530/JME-15-0288 (web archive link).*

intragenic duplication of an ancestral MSH gene (Nakanishi et al., 1979). Lobe-finned or sarcopterygian fish share the same organization displaying identical MSH domains (Amemiya, Takahashi, Meguro, & Kawauchi, 1999; Amemiya, Takahashi, Suzuki, Sasayama, & Kawauchi, 1999; Dores et al., 1999) but elasmobranch fish have an additional MSH domain called δ-MSH (Amemiya, Takahashi, Suzuki, et al., 1999). Some fish like cichlid and Pomacentridae present a novel domain termed ε-MSH due to a potential duplication of the ACTH domain (Harris et al., 2014). The coincidence of the same POMC bauplan in tetrapod and lobe-finned fish suggests that the ancestral Gnathostomes exhibited the three-domain structure (γ-MSH, α-MSH and β-MSH) and this structural plan diverged concomitantly with the divergence of the ancestral Gnathostomes into cartilaginous, ray- and lobe-finned fish. Nowadays, three different structural plans coexist (Fig. 5). The ancestral organization (γ-MSH, α-MSH and β-MSH) has been preserved in lobe-finned fish (lungfishes and coelacanth) and subsequently tetrapod species. The second organization plan is unique to the cartilaginous fish, which has added a fourth MSH peptide (δ-MSH). This fourth domain is placed between ACTH and β-MSH domain and probably evolved as a duplication of the C-terminal β-MSH/β-endorphin segment. The presence of this structure in elasmobranch and holocephalan species suggest that duplication occurred early after the divergence of the cartilaginous lineage.

Finally, the third organization plan is found in ray-finned fish (Actinopterygii) that presently include Polypteriformes (e.g., bichirs), Acipenseriformes (e.g., sturgeons and paddlefish), Lepisosteiformes (e.g., spotted gar), Amiiformes (bowfin) and teleost fish (Fig. 5). Most of ray-finned fish exhibit alterations of the γ-MSH domain and the zenith is reached in the teleost fish where the complete deletion of the domain was undertaken. Probably, the loss of the γ-MSH domain was progressive across the evolutionary process (Dores & Baron, 2011). In fact, basal Actinopterygii still keep the ancestral bauplan, i.e., γ-MSH, α-MSH/ACTH and β-MSH/β-endorphin, but replacements at dibasic pairs flanking the MSH sequence make peptide processing and release improbable. These N- and C-terminal cleavage motifs are clearly absent in Polypteriformes and Acipenseriformes precursors again making peptide processing non-viable. Additionally, in some species as for example the sturgeon, Pomc has mutations in the core sequence HFRW of γ-MSH, which is crucial for binding to melanocortin receptors. All this progressive degeneration of the γ-MSH domain in basal ray-finned fish finally lead to the complete deletion of this domain even affecting the spacer region between γ-MSH and α-MSH/ACTH in teleost fish (Fig. 5). However, the massive deletion in the N-terminal region of the teleost precursors does not affect peptide processing in the pituitary.

Melanocortin peptides signal through a family of G-protein coupled receptors that exhibit different expression patterns and pharmacological properties (Cortés et al., 2014). The γ-MSH peptide presents major affinity biding to the melanocortin 3 receptor (MC3R) or γ-MSH receptor (Roselli-Rehfuss et al., 1993) but physiological levels of γ-MSH are commonly low and full agonist has not been characterized up to now (Hruby et al., 2007).

It fills with suspicion that the precursor could be processed alternatively in the N-terminal region thus generating longer forms including the active melanocortin core much more active at the receptors binding pockets. Suggestively, the genome of most teleost fish lacks a γ-MSH receptor gene, symptomatic of a coevolutionary process of the peptide/receptor system.

Acknowledgments

This research was supported by Grant AGL2016-74857-C3-3-R from the Spanish Research. Agency (AEI) and from the European Fund of Regional Development (FEDER) to JMCR. AG was recipient of a FPI predoctoral fellow from the Spanish AEI.

Conflict of interest

Disclosure summary: The authors have nothing to declare.

References

Amemiya, Y., Takahashi, A., Meguro, H., & Kawauchi, H. (1999). Molecular cloning of lungfish proopiomelanocortin cDNA. *General and Comparative Endocrinology, 115*(3), 415–421. https://doi.org/10.1006/gcen.1999.7327.

Amemiya, Y., Takahashi, A., Suzuki, N., Sasayama, Y., & Kawauchi, H. (1999). A newly characterized melanotropin in proopiomelanocortin in pituitaries of an elasmobranch, squalus acanthias. *General and Comparative Endocrinology, 114*(3), 387–395. https://doi.org/10.1006/gcen.1999.7256.

Amores, A., Catchen, J., Ferrara, A., Fontenot, Q., & Postlethwait, J. H. (2011). Genome evolution and meiotic maps by massively parallel DNA sequencing: Spotted gar, an outgroup for the teleost genome duplication. *Genetics, 188*(4), 799–808. https://doi.org/10.1534/genetics.111.127324.

Amores, A., Force, A., Yan, Y. L., Joly, L., Amemiya, C., Fritz, A., et al. (1998). Zebrafish hox clusters and vertebrate genome evolution. *Science, 282*(5394), 1711–1714.

Bagnol, D., Lu, X. Y., Kaelin, C. B., Day, H. E., Ollmann, M., Gantz, I., et al. (1999). Anatomy of an endogenous antagonist: Relationship between agouti-related protein and proopiomelanocortin in brain. *The Journal of Neuroscience, 19*(18), RC26.

Berthelot, C., Brunet, F., Chalopin, D., Juanchich, A., Bernard, M., Noel, B., et al. (2014). The rainbow trout genome provides novel insights into evolution after whole-genome duplication in vertebrates. *Nature Communications, 5*, 3657. https://doi.org/10.1038/ncomms4657.

Braasch, I., Gehrke, A. R., Smith, J. J., Kawasaki, K., Manousaki, T., Pasquier, J., et al. (2016). The spotted gar genome illuminates vertebrate evolution and facilitates human-teleost comparisons. *Nature Genetics, 48*(4), 427–437. https://doi.org/10.1038/ng.3526.

Castro, M. G., & Morrison, E. (1997). Post-translational processing of proopiomelanocortin in the pituitary and in the brain. *Critical Reviews in Neurobiology, 11*(1), 35–57.

Cerda-Reverter, J. M., Agulleiro, M. J., R, R. G., Sanchez, E., Ceinos, R., & Rotllant, J. (2011). Fish melanocortin system. *European Journal of Pharmacology, 660*(1), 53–60. https://doi.org/10.1016/j.ejphar.2010.10.108.

Christoffels, A., Koh, E. G., Chia, J. M., Brenner, S., Aparicio, S., & Venkatesh, B. (2004). Fugu genome analysis provides evidence for a whole-genome duplication early during the evolution of ray-finned fishes. *Molecular Biology and Evolution, 21*(6), 1146–1151. https://doi.org/10.1093/molbev/msh114.

Cortés, R., Navarro, S., Agulleiro, M. J., Guillot, R., Garcia-Herranz, V., Sanchez, E., et al. (2014). Evolution of the melanocortin system. *General and Comparative Endocrinology, 209*, 3–10. https://doi.org/10.1016/j.ygcen.2014.04.005.

Danielson, P. B., & Dores, R. M. (1999). Molecular evolution of the opioid/orphanin gene family. *General and Comparative Endocrinology, 113*(2), 169–186. https://doi.org/10.1006/gcen.1998.7206.

de Souza, F. S. J., Bumaschny, V. F., Low, M. J., & Rubinstein, M. (2005). Subfunctionalization of expression and peptide domains following the ancient duplication of the proopiomelanocortin gene in teleost fishes. *Molecular Biology and Evolution, 22*(12), 2417–2427. https://doi.org/10.1093/molbev/msi236.

Docker, M. F., Hume, J. B., & Clemens, B. J. (2015). Introduction: A surfeit of lampreys. In M. F. Docker (Ed.), *Vol. 1. Lampreys: Biology, conservation and control* (pp. 1–34). Dordrecht: Springer Netherlands.

Dores, R. M., & Baron, A. J. (2011). Evolution of POMC: Origin, phylogeny, posttranslational processing, and the melanocortins. *Annals of the New York Academy of Sciences, 1220*, 34–48. https://doi.org/10.1111/j.1749-6632.2010.05928.x.

Dores, R. M., & Lecaude, S. (2005). Trends in the evolution of the proopiomelanocortin gene. *General and Comparative Endocrinology, 142*(1–2), 81–93. https://doi.org/10.1016/j.ygcen.2005.02.003.

Dores, R. M., Lecaude, S., Bauer, D., & Danielson, P. B. (2002). Analyzing the evolution of the opioid/orphanin gene family. *Mass Spectrometry Reviews, 21*(4), 220–243. https://doi.org/10.1002/mas.10029.

Dores, R. M., Liang, L., Davis, P., Thomas, A. L., & Petko, B. (2016). 60 years of POMC: Melanocortin receptors: Evolution of ligand selectivity for melanocortin peptides. *Journal of Molecular Endocrinology, 56*(4), T119–T133. https://doi.org/10.1530/JME-15-0292.

Dores, R. M., Londraville, R. L., Prokop, J., Davis, P., Dewey, N., & Lesinski, N. (2014). Molecular evolution of GPCRs: Melanocortin/melanocortin receptors. *Journal of Molecular Endocrinology, 52*(3), T29–T42. https://doi.org/10.1530/JME-14-0050.

Dores, R. M., Sollars, C., Danielson, P., Lee, J., Alrubaian, J., & Joss, J. M. (1999). Cloning of a proopiomelanocortin cDNA from the pituitary of the australian lungfish, neoceratodus forsteri: Analyzing trends in the organization of this prohormone precursor. *General and Comparative Endocrinology, 116*(3), 433–444. https://doi.org/10.1006/gcen.1999.7382.

Elphick, M. R., Mirabeau, O., & Larhammar, D. (2018). Evolution of neuropeptide signalling systems. *The Journal of Experimental Biology, 221*(Pt. 3), jeb151092. https://doi.org/10.1242/jeb.151092.

Force, A., Lynch, M., Pickett, F. B., Amores, A., Yan, Y. L., & Postlethwait, J. (1999). Preservation of duplicate genes by complementary, degenerative mutations. *Genetics, 151*(4), 1531–1545.

Harris, R. M., Dijkstra, P. D., & Hofmann, H. A. (2014). Complex structural and regulatory evolution of the pro-opiomelanocortin gene family. *General and Comparative Endocrinology, 195*, 107–115. https://doi.org/10.1016/j.ygcen.2013.10.007.

He, W., Dai, X., Chen, X., He, J., & Yin, Z. (2014). Zebrafish pituitary gene expression before and after sexual maturation. *The Journal of Endocrinology, 221*(3), 429–440. https://doi.org/10.1530/JOE-13-0488.

Hruby, V. J., Cai, M., Cain, J. P., Mayorov, A. V., Dedek, M. M., & Trivedi, D. (2007). Design, synthesis and biological evaluation of ligands selective for the melanocortin-3 receptor. *Current Topics in Medicinal Chemistry*, 7(11), 1107–1119.

Jaillon, O., Aury, J. M., Brunet, F., Petit, J. L., Stange-Thomann, N., Mauceli, E., et al. (2004). Genome duplication in the teleost fish tetraodon nigroviridis reveals the early vertebrate proto-karyotype. *Nature, 431*(7011), 946–957. https://doi.org/10.1038/nature03025.

Larhammar, D., & Bergqvist, C. A. (2013). Ancient grandeur of the vertebrate neuropeptide Y system shown by the coelacanth latimeria chalumnae. *Frontiers in Neuroscience*, *7*, 27. https://doi.org/10.3389/fnins.2013.00027.

Larhammar, D., Bergqvist, C., & Sundstrom, G. (2015). Ancestral vertebrate complexity of the opioid system. *Vitamins and Hormones*, *97*, 95–122. https://doi.org/10.1016/bs.vh.2014.11.001.

Leder, E. H., & Silverstein, J. T. (2006). The pro-opiomelanocortin genes in rainbow trout (oncorhynchus mykiss): Duplications, splice variants, and differential expression. *The Journal of Endocrinology*, *188*(2), 355–363. https://doi.org/10.1677/joe.1.06283.

Lespinet, O., Wolf, Y. I., Koonin, E. V., & Aravind, L. (2002). The role of lineage-specific gene family expansion in the evolution of eukaryotes. *Genome Research*, *12*(7), 1048–1059. https://doi.org/10.1101/gr.174302.

Lynch, M., & Conery, J. S. (2000). The evolutionary fate and consequences of duplicate genes. *Science*, *290*(5494), 1151–1155.

Nakanishi, S., Inoue, A., Kita, T., Nakamura, M., Chang, A. C., Cohen, S. N., et al. (1979). Nucleotide sequence of cloned cDNA for bovine corticotropin-beta-lipotropin precursor. *Nature*, *278*(5703), 423–427.

Navarro, S., Soletto, L., Puchol, S., Rotllant, J., Soengas, J. L., & Cerda-Reverter, J. M. (2016). 60 years of POMC: POMC: An evolutionary perspective. *Journal of Molecular Endocrinology*, *56*(4), T113–T118. https://doi.org/10.1530/JME-15-0288.

Nei, M. (2013). *Mutation-driven evolution*. Oxford University Press.

Ohno, S. (1970). *Evolution by gene duplication*. Berlin, Heidelberg: Springer.

Roselli-Rehfuss, L., Mountjoy, K. G., Robbins, L. S., Mortrud, M. T., Low, M. J., Tatro, J. B., et al. (1993). Identification of a receptor for gamma melanotropin and other proopiomelanocortin peptides in the hypothalamus and limbic system. *Proceedings of the National Academy of Sciences of the United States of America*, *90*(19), 8856–8860.

Semon, M., & Wolfe, K. H. (2007). Reciprocal gene loss between tetraodon and zebrafish after whole genome duplication in their ancestor. *Trends in Genetics*, *23*(3), 108–112. https://doi.org/10.1016/j.tig.2007.01.003.

Smith, A. I., & Funder, J. W. (1988). Proopiomelanocortin processing in the pituitary, central nervous system, and peripheral tissues. *Endocrine Reviews*, *9*(1), 159–179. https://doi.org/10.1210/edrv-9-1-159.

Smith, J. J., Timoshevskaya, N., Ye, C., Holt, C., Keinath, M. C., Parker, H. J., et al. (2018). The sea lamprey germline genome provides insights into programmed genome rearrangement and vertebrate evolution. *Nature Genetics*, *50*(2), 270–277. https://doi.org/10.1038/s41588-017-0036-1.

Sundström, G., Dreborg, S., & Larhammar, D. (2010). Concomitant duplications of opioid peptide and receptor genes before the origin of jawed vertebrates. *PLoS One*, *5*(5), e10512. https://doi.org/10.1371/journal.pone.0010512.

Takahashi, A., Amano, M., Itoh, T., Yasuda, A., Yamanome, T., Amemiya, Y., et al. (2005). Nucleotide sequence and expression of three subtypes of proopiomelanocortin mRNA in barfin flounder. *General and Comparative Endocrinology*, *141*(3), 291–303. https://doi.org/10.1016/j.ygcen.2005.01.010.

Takahashi, A., Amemiya, Y., Nozaki, M., Sower, S. A., Joss, J., Gorbman, A., et al. (1995). Isolation and characterization of melanotropins from lamprey pituitary glands. *International Journal of Peptide and Protein Research*, *46*(3–4), 197–204.

Takahashi, A., Amemiya, Y., Sarashi, M., Sower, S. A., & Kawauchi, H. (1995). Melanotropin and corticotropin are encoded on two distinct genes in the lamprey, the earliest evolved extant vertebrate. *Biochemical and Biophysical Research Communications*, *213*(2), 490–498. https://doi.org/10.1006/bbrc.1995.2158.

Takahashi, A., Mizusawa, K., & Amano, M. (2014). Multifunctional roles of melanocyte-stimulating hormone and melanin-concentrating hormone in fish: Evolution from classical body color change. *Aqua-BioScience Monographs*, *7*(1), 46. https://doi.org/10.5047/absm.2014.00701.0001.

Taylor, J. S., Braasch, I., Frickey, T., Meyer, A., & Van de Peer, Y. (2003). Genome duplication, a trait shared by 22000 species of ray-finned fish. *Genome Research*, *13*(3), 382–390. https://doi.org/10.1101/gr.640303.

Taylor, J. S., Van de Peer, Y., & Meyer, A. (2001). Genome duplication, divergent resolution and speciation. *Trends in Genetics*, *17*(6), 299–301.

CHAPTER TWO

Structure and dynamics of dynorphin peptide and its receptor

Guillaume Ferré*, Georges Czaplicki, Pascal Demange, Alain Milon*
Institut de Pharmacologie et de Biologie Structurale (IPBS, CNRS UPS),
Université de Toulouse, Toulouse, France
*Corresponding authors: e-mail address: guillaume.ferre@ipbs.fr; alain.milon@ipbs.fr

Contents

Abstract

Dynorphin is a neuropeptide involved in pain, addiction and mood regulation. It exerts its activity by binding to the kappa opioid receptor (KOP) which belongs to the large family of G protein-coupled receptors. The dynorphin peptide was discovered in 1975, while its receptor was cloned in 1993. This review will describe: (a) the activities and physiological functions of dynorphin and its receptor, (b) early structure-activity relationship studies performed before cloning of the receptor (mostly pharmacological and biophysical studies of peptide analogues), (c) structure-activity relationship studies performed after cloning of the receptor via receptor mutagenesis and the development of recombinant receptor expression systems, (d) structural biology of the opiate receptors culminating in X-ray structures of the four opioid receptors in their inactive state and structures of MOP and KOP receptors in their active state. X-ray and EM structures are combined with NMR data, which gives complementary insight into receptor and peptide dynamics. Molecular modeling greatly benefited from the availability of atomic resolution 3D structures of receptor-ligand complexes and an example of the strategy used to model a dynorphin-KOP receptor complex using NMR data will be described. These achievements have led to a better understanding of the complex dynamics of KOP receptor activation and to the development of new ligands and drugs.

Vitamins and Hormones, Volume 111
ISSN 0083-6729
https://doi.org/10.1016/bs.vh.2019.05.006

1. Dynorphin: A neuropeptide involved in pain, addiction and mood regulation

Dynorphin is an endogenous neuropeptide first isolated from porcine pituitary (Cox, Opheim, Teachemacher, & Goldstein, 1975) with a particularly potent opioid activity (Goldstein, Tachibana, Lowney, Hunkapiller, & Hood, 1979). The dynorphin A 1–17 isoform was the first to be fully sequenced (Goldstein, Fischli, Lowney, Hunkapiller, & Hood, 1981) and revealed an amino-terminal sequence identical to leu-enkephalin opioid peptide (YGGFL) with a basic carboxy-terminal extension. Several dynorphin isoforms were further identified: dynorphin A 1–8, dynorphin B 1–13, big dynorphin and leumorphin (Chavkin, 2013) (Fig. 1). All dynorphin isoforms and α- and β-neoendorphin, which are also leu-enkephalin-based opioid peptides, derive from a common precursor named prodynorphin (Chavkin, Bakhit, Weber, & Bloom, 1983; Kakidani et al., 1982; Weber, Evans, & Barchas, 1982; Zamir, Palkovits, Weber, Mezey, & Brownstein, 1984) while other opioids are derived from the precursors proenkephalin and proopiomelanocortin (Chavkin, 2013).

Peptide	Sequence	pK$_i$ KOP	pK$_i$ MOP	pK$_i$ DOP	References
Met-enkephalin	YGGFM	7.3	8.7	9.3	1, 2, 3
Leu-enkephalin	YGGFL	7.0	8.2	9.4	1, 4
Dynorphin A 1-8	YGGFLRRI	9.3	8.6	8.9	4, 5, 6
Dynorphin A 1-13	YGGFLRRIRPKLK	9.6	8.4	8.3	4
Dynorphin A 1-17	YGGFLRRIRPKLKWDNQ	9.4	8.6	8.9	4, 5, 6, 7, 8
Big dynorphin	YGGFLRRIRPKLKWDNQKRYGGFLRRQFKVVT	8.5	8.2	8.5	6
Leumorphin	YGGFLRRQFKVVTRSQQDPNPNAYYGGLFNV	9.3	8.4	8.4	
Dynorphin B 1-13	YGGFLRRQFKVVT	9.3	8.7	8.5	6
α-neo-endorphin	YGGFLRKYPK	9.9	8.9	8.9	3, 5, 7
β-neo-endorphin	YGGFLRKYP	9.3	8.3	8.9	5

Fig. 1 Dynorphin-related opioid peptides. Met-enkephalin, leu-enkephalin, dynorphin A 1–13 and prodynorphin-derived peptides amino acid sequence and affinity for opioid receptors. The common N-terminal leu-enkephalin "message" sequence is colored in green and C-terminal "address" residues conserved with dynorphin A 1–17 in blue. In order to compare affinities in homogenous systems, pK_i values from Mansour, Hoversten, Taylor, Watson, and Akil (1995) are reported where competition binding experiments were conducted against rat opioid receptors transiently expressed in COS-1 cells. Binding properties have been additionally reviewed from the indicated references from IUPHAR/BPS guide to pharmacology (Alexander et al., 2017): 1 (Meng et al., 1993), 2 (Raynor et al., 1994), 3 (Yasuda et al., 1993), 4 (Toll et al., 1998), 5 (Simonin et al., 1995), 6 (Merg et al., 2006), 7 (Zhu et al., 1995), 8 (Zhu, Luo, Li, Chen, & Liu-Chen, 1997). Similar trends are observed despite discrepancies arising from differences in experimental models and conditions.

Prodynorphin and its processing enzyme, prohormone convertase 2 (PC2), are largely expressed in the central nervous system (Berman et al., 2000; Civelli, Douglass, Goldstein, & Herbert, 1985) and dynorphin peptides are present in presynaptic neurosecretory vesicles (Molineaux & Cox, 1982; Pickel, Chan, & Sesack, 1993; Pickel, Chan, Veznedaroglu, & Milner, 1995; Whitnall, Gainer, Cox, & Molineaux, 1983). They can be released by membrane depolarization (Chavkin et al., 1983) and subsequently activate the kappa opioid receptor (KOP) (Wagner, Evans, & Chavkin, 1991) which modulates neurotransmitter release and postsynaptic neural activity (Wagner, Terman, & Chavkin, 1993).

KOP belongs to the opioid receptor family which is composed of several subtypes, originally defined by the pharmacological profiles of the first receptors to be characterized (Dhawan et al., 1996): the μ and δ opioid receptors (MOP and DOP). KOP was initially named from ketocyclazocine opioid activity (Martin, Eades, Thompson, Huppler, & Gilbert, 1976) and highly KOP-specific agonists were synthesized (Dhawan et al., 1996) such as U50488 (Lahti, Von Voigtlander, & Barsuhn, 1982) or U69593 (Lahti, Mickelson, McCall, & Von Voigtlander, 1985). Opioid receptor cloning led to the identification of a fourth member of the family, the nociceptin opioid receptor (NOP), and of its endogenous ligand, nociceptin (Meunier et al., 1995; Mollereau et al., 1994). Several studies reported the cloning of KOP from rodents (Li et al., 1993; Meng et al., 1993; Yasuda et al., 1993) and then from human (Simonin et al., 1995; Zhu et al., 1995), showing that it, together with the other opioid receptors, belongs to the G protein-coupled receptors (GPCRs) superfamily. Opioid receptors are mainly G_i/G_o-coupled (Al-Hasani & Bruchas, 2011; Prather et al., 1995). They display a small basal intracellular signaling activity, at about 10% of the maximal response for KOP, in the absence of any ligand, and this is modulated by extracellular ligand binding (Wang, Sun, & Sadee, 2007).

Opioid receptor signaling controls a multitude of intracellular effectors by both G protein-dependent and G protein-independent pathways (Al-Hasani & Bruchas, 2011; Bruchas & Chavkin, 2010; Law, Wong, & Loh, 2000). G protein-dependent opioid receptor signaling predominantly results in neuronal excitability and synaptic transmission inhibition. G_i/G_o-protein activation promotes G protein-coupled inwardly rectifying potassium channel (GIRK) activity (Barchfeld & Medzihradsky, 1984; Childers & Snyder, 1978; Minneman & Iversen, 1976). It causes neuronal membrane hyperpolarization and thus attenuates the neuron's ability to generate and propagate action potentials (Henry, Grandy, Lester, Davidson, & Chavkin, 1995;

Sadja, Alagem, & Reuveny, 2003; Schneider, Eckert, & Light, 1998). Activation of G_i/G_o-proteins also inhibits voltage-dependent calcium channels, diminishing calcium influx in response to action potentials, and thus prevents neurotransmitter synaptic release (Bourinet, Soong, Stea, & Snutch, 1996; Rhim & Miller, 1994; Rusin, Giovannucci, Stuenkel, & Moises, 1997; Zamponi & Snutch, 1998). Activated G_i/G_o-proteins also inhibit adenylate cyclase thus decreasing intracellular cyclic adenosine monophosphate (cAMP) concentration (Taussig, Iniguez-Lluhi, & Gilman, 1993), which in turn regulates numerous targets. Opioid receptors also signal through G protein-independent pathways. Following agonist stimulation, they can be intracellularly phosphorylated by G protein-coupled receptor kinases (GRKs) leading to receptor interaction with β-arrestins (Al-Hasani & Bruchas, 2011; Law et al., 2000). This interaction is involved in receptor internalization, one of the consequences of which is to prevent exposition to extracellular ligands. It is also responsible for intracellular signaling *per se*, notably through mitogen activated kinase (MAPK) pathways that regulate predominant cellular processes such as gene transcription, proliferation and differentiation (Bruchas & Chavkin, 2010; Raman, Chen, & Cobb, 2007).

Because of their involvement in cellular communication, opioid peptides and their receptors participate in numerous physiological processes (Feng et al., 2012). First, opioid receptor activation results in spinal and supra-spinal pain modulation (Ahlbeck, 2011; Waldhoer, Bartlett, & Whistler, 2004). The afferent nociceptive influx from peripheral tissues is regulated by descending neurons and opioid receptors participate in this system at multiple levels (Vanderah, 2010). As an example, KOP has been found to be present in spinal cord dorsal root ganglia (Attali & Vogel, 1989; Corder, Castro, Bruchas, & Scherrer, 2018; Ji et al., 1995) where it inhibits synaptic transmission between primary and secondary afferent neurons (Vanderah, 2010) in response to agonists released by descending neurons. Opioid peptides and their receptors also regulate monoaminergic systems and notably mesolimbic dopaminergic functions (Lutz & Kieffer, 2013). KOP agonists inhibit dopaminergic neurons release activity by their receptor-specific action both in the *nucleus accumbens* and the ventral tegmental area (Lutz & Kieffer, 2013; Margolis, Hjelmstad, Bonci, & Fields, 2003; Spanagel, Herz, & Shippenberg, 1992), resulting in an aversive effect (Bals-Kubik, Ableitner, Herz, & Shippenberg, 1993). In contrast, MOP agonists cause a rewarding effect (Bals-Kubik et al., 1993) because they indirectly stimulate dopamine release by diminishing inhibitory gamma-aminobutyric acid (GABA) neuron activity in the ventral tegmental area (Lutz & Kieffer, 2013; Spanagel et al., 1992). Opioid receptors thus mediate abusive opioid seeking behavior but are also more generally involved in modulating drug addictions with pronounced subtype discrepancies (Kreek et al., 2012). In addition, the mesolimbic dopamine reward system is linked to mood disorders (Nestler & Carlezon, 2006) and opioid receptors further regulate serotonin and noradrenaline neurons (Lutz & Kieffer, 2013), thought to participate in depression (Krishnan & Nestler, 2010). Consistent with this, opioid receptors and their ligands are involved in depression-like behaviors and KOP antagonists such as JDTic induce antidepressant-like effects (Lutz & Kieffer, 2013). Furthermore, opioid peptides and their receptors, especially dynorphin and KOP, control the hypothalamic–pituitary–adrenal axis (HPA) and are thus involved in stress-related phenomenon with implications in addiction, depression and anxiety disorders (Bruchas, Land, & Chavkin, 2010; Knoll & Carlezon, 2010; Kreek et al., 2012; Ribeiro, Kennedy, Smith, Stohler, & Zubieta, 2005). Beside their predominant role in pain, addiction and mood control, opioid receptors have been linked to a multitude of functions such as immunity, neuroprotection, cell proliferation, neural differentiation, cardiovascular system regulation and feeding (Feng et al., 2012). Of particular importance, opioid receptors signaling stimulation can influence the respiratory system function (Pattinson, 2008) with strong respiratory depression observed upon activation of MOP (but not KOP). Together with its impact on breathing, MOP agonism induces an important inhibition of gastrointestinal transit (Tavani, Petrillo, La Regina, & Sbacchi, 1990), which represents a significant limitation of opioid use in the clinic.

The biological functions of opioid peptides and their receptors renders them key targets to interfere in pain, addiction and mood disorders. There is considerable interest in discovering new opioids with reduced side effects and compounds that target the dynorphin/KOP system are being developed in order to produce analgesic, antidepressant, anxiolytic or anti-addiction drugs (Dogra & Yadav, 2015; Zheng et al., 2017). Because of the biological significance of this system, extensive research is under way to decipher the molecular mechanisms underlying the dynorphin/KOP interaction and the resulting modulation of receptor signaling activity. These studies, which form the subject of this review, could also help in the design of novel KOP ligands with pharmacologically relevant properties such as biased agonism or allosteric modulation.

2. The molecular mechanism of action of dynorphin: Research and hypotheses prior to KOP cloning

The ability to produce dynorphin and analogues by solid-phase peptide synthesis (Goldstein et al., 1979) opened the way to a wide range of structure–activity relationship studies on various peptide structures. These have been reviewed thoroughly elsewhere (Aldrich & McLaughlin, 2009; Lapalu et al., 1997; Naqvi, Haq, & Mathur, 1998; Ramos-Colon et al., 2016). In brief, the 17 amino acid long dynorphin 1–17 may be shortened at its C-terminus to dynorphin 1–13 without affecting its activity (activation of KOP is typically assessed on the guinea pig ileum electrical contraction assay), and further to dynorphin 1–8 with a 50-fold reduction in activity and no loss in affinity (Fig. 1 and Mansour et al., 1995). Further shortening is deleterious for both affinity and activity. Substitution of Gly2 by L-amino acids resulted in a reduction in activity while substitution by D-amino acids resulted in reduced selectivity for KOP versus MOP, since D-amino acids tend to increase activity on MOP and to decrease it on KOP. Substitution of Pro10 by a D-proline led to a marked increase in selectivity for KOP over the other opiate receptors. The further alkylation of the amino group of Tyr1 gave rise to highly KOP-selective peptide ligands such as N–Benzyl[D–Pro10]–Dyn A(1 − 11) (Choi, Murray, DeLander, Caldwell, & Aldrich, 1992).

Peptide synthesis also enabled biophysical studies of dynorphin conformation and dynamics in solution and in membrane biomimetic media (Lancaster et al., 1991; Lind, Graslund, & Maler, 2006; Naito & Nishimura, 2004; Spadaccini, Crescenzi, Picone, Tancredi, & Temussi, 1999). These included circular dichroism, Raman, FT-IR and NMR spectroscopies. In aqueous solution, dynorphin is largely disordered. It may adopt secondary structures in specific environments, such as a type I β-turn involving the first five residues in DMSO (Renugopalakrishnan, Rapaka, Huang, Moore, & Hutson, 1988) or a α-helical conformation from residue Gly3 to Arg9 when bound to dodecylphosphocholine detergent micelles (Kallick, 1993). Limiting the conformational space available to constrained peptide analogues increases receptor binding affinity and specificity (Naqvi et al., 1998). However, without knowing the structure adopted by dynorphin upon KOP binding, it was difficult to derive a clear understanding of the binding mechanism from these results.

A detailed study of dynorphin and dynorphin analogues bound to lipid bilayers was performed in the early 1980s. Using a combination of infrared attenuated total reflection spectroscopy and capacitance minimization, dynorphin was found to bind to POPC lipid bilayers by forming an α-helix from Tyr1 to Pro10, which inserts into the bilayer perpendicular to the bilayer plane (Erne, Sargent, & Schwyzer, 1985). In the same study, the affinity for a neutral bilayer was determined to have a K_d of 11 μM. In a subsequent theoretical estimation of the preferred orientation and binding energy of a series of dynorphin analogues of various lengths (from 1–13 to 1–5), a good correlation between the peptide amphiphilic moment, the affinity for lipid bilayers and the KOP receptor subtype specificity was found. Loss of the C-terminal positively-charged residues converts a KOP-selective peptide dynorphin, into a MOP/DOP-selective peptide Leu-enkephalin (Fig. 1). This, and similar observations on other neuropeptide families, gave rise to two major hypotheses: (a) the *message-address concept*, first introduced in 1977 (Schwyzer, 1977), in which the N-terminal five residues (message) are responsible for specific binding and receptor activation, while the positively charged C-terminus (address) is responsible for KOP receptor selectivity by concentrating the peptide in the vicinity of the receptor; (b) the *membrane compartment concept*, in which the lipid bilayer plays an active role in catalyzing the peptide-receptor interaction (Auge, Bersch, Tropis, & Milon, 2000; Axelrod & Wang, 1994; Bersch, Koehl, Nakatani, Ourisson, & Milon, 1993; Czaplicki & Milon, 1998, 2005; Milon, Miyazawa, & Higashijima, 1990; Sargent & Schwyzer, 1986). One reason for the correlation between KOP receptor subtype specificity and the positively charged C-terminus (besides the potential role of the bilayer itself) became clear when the opiate receptors were cloned in 1992–1994 (Chen, Mestek, Liu, Hurley, & Yu, 1993; Evans, Keith, Morrison, Magendzo, & Edwards, 1992; Kieffer, Befort, Gaveriaux-Ruff, & Hirth, 1992; Mollereau et al., 1994; Yasuda et al., 1993). Indeed, it appeared that a specific feature of KOP and NOP receptors as compared to MOP and DOP receptors is the high negative potential surrounding the extracellular loops 2 (Fig. 2), which is expected to increase the local concentration of highly positively-charged neuropeptides (such as dynorphin and nociceptin for KOP and NOP, respectively). Dynorphin affinity for the extracellular loop 2 was further confirmed experimentally (Bjorneras et al., 2014). The sole effect of an attractive electrostatic potential may account for increased on-rate binding kinetics and receptor binding affinity by at least two orders of magnitude (Fersht, 1999).

```
          ECL1                     ECL2                     ECL3          ECL2
                                                                          charge
             -   +        + +-- -  -      ---    -    -
KOP: LMNSWPFGDVLCKI....GGTKVREDVDVIECSLQFPDDDYSWWD....EALGSTSHST-AALSSY   -6
                 +        + +      -             -    +      -
MOP: LMGTWPFGTILCKI....ATTKYRQ--GSIDCTLTFSHPTW-YWE....KALVTIPETT-FQTVSW    0
       -     -   +        + +-                   -        - -++-
DOP: LMETWPFGELLCKA....AVTRPRD--GAVVCMLQFPSPSW-YWD....WTLVDIDRRDPLVVAAL    0
                 +          --  -- -   -     -                 -
NOP: LLGFWPFGNALCKT....GSAQVED--EEIECLVEIPTPQD-YWG....QGLGVQPSSE-TAVAIL   -7
```

Fig. 2 Primary sequences of the extracellular loops of the opioid receptor subtypes, KOP, MOP, DOP and NOP showing that the extracellular loop 2 (ECL2) is particularly rich in negative charges in KOP and NOP, while it is neutral for MOP and DOP. A similar trend is observed for the entire extracellular surface. This characteristic contributes to KOP specificity of positively charged dynorphin analogues as shown with MOP/KOP chimeric receptors (Wang, Johnson, Wu, Wang, & Uhl, 1994).

3. KOP cloning, site-directed mutagenesis and heterologous expression systems

KOP (Li et al., 1993; Meng et al., 1993; Simonin et al., 1995; Yasuda et al., 1993; Zhu et al., 1995) and the other three opioid receptors DOP, MOP and NOP (Chen et al., 1993; Evans et al., 1992; Kieffer et al., 1992; Mollereau et al., 1994; Yasuda et al., 1993) were cloned in the early 1990s. KOP shares 60% sequence similarity with the other receptors. The highest sequence diversity is located in the N-terminus, C-terminus and extracellular loops (Waldhoer et al., 2004). Moreover, all four opioid receptors possess a remarkable signature, specific to class A GPCRs, which includes a sodium binding pocket (residues numbering corresponds to human KOP; superscripts are given according to Ballesteros-Weinstein numbering (Ballesteros & Weinstein, 1995)): D105$^{2.50}$, N141$^{3.35}$, S145$^{3.39}$; a PIF micro-switch P238$^{5.50}$, I146$^{3.40}$, F283$^{6.44}$; a NPxxY motif N326$^{7.49}$, Y330$^{7.53}$; a DRY motif D155$^{3.49}$, R156$^{3.50}$, Y157$^{3.51}$ and a cysteine disulfide bridge C131$^{3.25}$, C210ECL2 (Fig. 3). Heterologous expression and site-directed mutagenesis experiments were then used intensively to better understand the receptor architecture, ligand recognition specificity and activation mechanisms. It was shown that the extracellular loops play a major role in receptor subtype specificity (Metzger & Ferguson, 1995; Seki et al., 1998). Several key transmembrane residues were identified, such as S187$^{4.54}$ (Claude et al., 1996), E297$^{6.58}$ (Larson, Jones, Hjorth, Schwartz, & Portoghese, 2000; Sharma, Jones, Metzger, Ferguson, & Portoghese, 2001), Y312$^{7.35}$ (Metzger, Paterlini, Ferguson, & Portoghese, 2001) and I316$^{7.39}$ (Owens & Akil, 2002). Using KOP/DOP chimeric receptor, Kong et al. obtained indirect

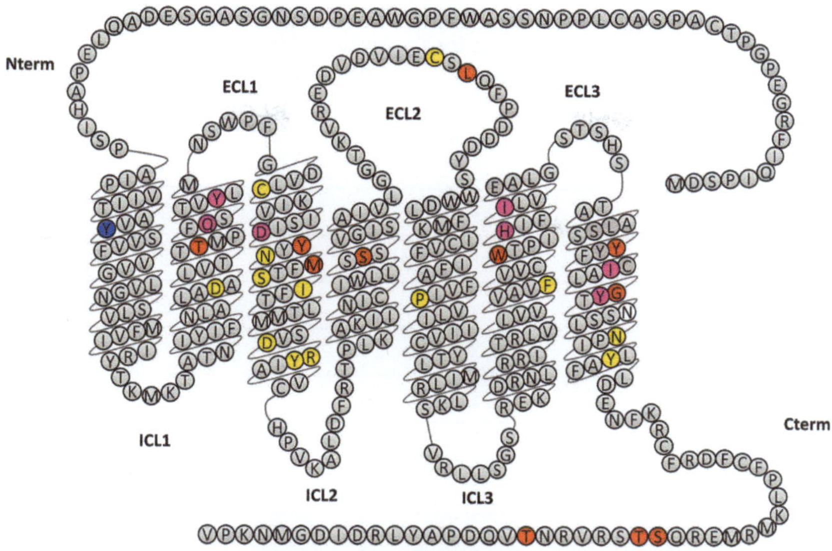

Fig. 3 Primary sequence of human KOP highlighting mutations shown to affect KOP-dynorphin interactions. The limits of the seven α-helices were defined according to the activated KOP 3D structure (Che et al., 2018). In the three-dimensional structure of inactive KOP (Wu et al., 2012) these limits differ slightly for α-helix 5 ($D218^{5.30}$ to $S260^{5.72}$ instead of $W221^{5.33}$ to $S255^{5.67}$), α-helix 6 ($R267^{6.28}$ to $L299^{6.60}$ instead of $R263^{6.24}$ to $G300^{6.61}$) and α-helix 7 ($L309^{7.32}$ to $L333^{7.56}$ instead of $T306^{7.29}$ to $D334^{7.57}$). Mutations affecting dynorphin binding (>10-fold increase in K_d) are colored in blue ($Y66^{1.39}$), those affecting signaling (>10-fold increase in EC_{50} in cAMP inhibition assay) are colored in red ($T111^{2.56}$, $Y139^{3.33}$, $M142^{3.36}$, $S187^{4.54}$, $C210^{ECL2}$, $L212^{ECL2}$, $W287^{6.48}$, $Y312^{7.35}$, $G319^{7.43}$, $S356^{Ct}$, $T357^{Ct}$, $T363^{Ct}$) and those affecting both are colored in magenta ($Q115^{2.60}$, $Y119^{2.64}$, $D138^{3.32}$, $H291^{6.52}$, $I294^{6.55}$, $I316^{7.39}$, $Y320^{7.43}$). Other residues generally considered to play a role in opioid receptor activation (Che et al., 2018; Koehl et al., 2018) are colored in yellow (cysteine disulfide bridge $C131^{3.25}$, $C210^{ECL2}$; sodium binding pocket: $D105^{2.50}$, $N141^{3.35}$, $S145^{3.39}$; PIF microswitch: $P238^{5.50}$, $I146^{3.40}$, $F283^{6.44}$; NPxxY motif: $N326^{7.49}$, $Y330^{7.53}$; DRY motif: $D155^{3.49}$, $R156^{3.50}$, $Y157^{3.51}$). Residues $D^{3.32}$ and $Y^{7.43}$ (conserved in all four opioid receptors) were shown to form direct contacts with DAMGO N-terminus in the DAMGO-MOP-Gi structure (Koehl et al., 2018) and thus presumably with the N-terminus of dynorphin in the case of KOP. Residue $H^{6.52}$ is conserved in MOP, KOP and DOP and forms a water-mediated contact with the phenol group of Tyr1 in the same MOP structure. Interestingly, this residue $H^{6.52}$ is mutated to a glutamine in NOP, whose ligand nociceptin possesses a phenylalanine at position 1 and is thus devoid of the phenol hydroxyl group. It should be noted that for clarity, this figure presents data obtained with dynorphin (mostly from Che et al., 2018, p. 199; Claude et al., 1996; Vardy et al., 2013) and not with other non-peptidic ligands for which other deleterious mutations have been described. Superscripts are given according to Ballesteros-Weinstein numbering (Ballesteros & Weinstein, 1995); residues numbering corresponds to human KOP.

Table 1 Main point mutations of KOP and their consequences on dynorphin binding and functional activity (cAMP inhibition assay).

KOP mutant	Binding	Potency	References
T111$^{2.56}$	↔	↓	3
Q115$^{2.60}$	↓	↓	2
Y119$^{2.64}$	↓	↓	1, 2
D138$^{3.32}$	↓	↓	2, 3
Y139$^{3.33}$	↔	↓	1, 2, 3
M142$^{3.36}$	↔	↓	2
W287$^{6.48}$	↔	↓	3
H291$^{6.52}$	↓	↓	2, 3
I294$^{6.55}$	↓	↓	2
Y312$^{7.34}$	↔	↓	1, 2, 3
Y313$^{7.35}$	↓	↓	1, 3
I316$^{7.39}$	↓	↓	2
G319$^{7.42}$	↔	↓	3
Y320$^{7.43}$	↓	↓	1, 2, 3

References: 1 (Yan et al., 2005); 2 (Vardy et al., 2013); 3 (Che et al., 2018).

evidence that "agonists and antagonists bind to different domains of the cloned kappa opioid receptor" (Kong et al., 1994). However, recent structures of KOP in the presence of antagonist (Wu et al., 2012) and agonist (Che et al., 2018) do not support this statement, illustrating the difficulties of drawing firm conclusions from these approaches in the absence of precise 3D structures. Using the 3D structures of KOP, extensive mutagenesis experiments were performed to characterize the interaction of KOP with dynorphin, as well as with other non-peptidic ligands such as salvinorin, which are summarized, together with previous studies, in Fig. 3 and Table 1.

4. Structure and dynamics of dynorphin and its receptor based on experimental 3D structures

Three-dimensional structures of GPCRs at atomic resolution began to appear with the structure of rhodopsin A (Palczewski et al., 2000). However, rhodopsin was a specific case due to its unusual stability and availability from natural sources, and further efforts were necessary to solve the three-dimensional structures of recombinant GPCRs. Several international consortia developed crucial methodologies in protein expression systems, receptor stabilization by mutagenesis, fusion proteins, the selection of stabilizing ligands, binding to antibodies (particularly nanobodies), the development of better solubilizing and crystallizing media (Caffrey & Cherezov, 2009; Cherezov, 2011; Granier & Kobilka, 2012; Kobilka & Schertler, 2008; Rosenbaum, Rasmussen, & Kobilka, 2009; Stevens et al., 2013; Tate, 2012).

Most GPCR structures were obtained from membrane proteins expressed in eukaryotic insect cells where the flexible N- and C-termini, as well as the intracellular loops (mostly ICL3), were deleted or replaced by exogenous protein domains promoting thermostability and crystallization, such as T4 lysozyme or the thermostabilized apocytochrome b562 RIL (BRIL) (Chun et al., 2012; Lv et al., 2016). Other expression systems have been used such as the methylotrophic yeast *P. pastoris* (Talmont, Sidobre, Demange, Milon, & Emorine, 1996), which allows stable isotope labeling, including perdeuteration (Massou et al., 1999), mostly for NMR experiments (Eddy et al., 2018). *E. coli* is also an interesting host for isotope-labeled GPCR biosynthesis which can be achieved by receptor expression as inclusion bodies followed by in vitro refolding during the protein purification (Baneres et al., 2003; Baneres, Popot, & Mouillac, 2011; Casiraghi et al., 2016). We have shown this strategy to be efficient for KOP: the dynorphin-KOP interaction was measured in our laboratory using KOP expressed in *E. coli* and refolded (unpublished results), and the same results were obtained as with KOP produced and purified from sf9 cell membranes (O'Connor et al., 2015).

These methodological developments allowed the first three-dimensional structure of a recombinant GPCR, the β2-adrenergic receptor, to be solved in 2007 (Rosenbaum et al., 2007). This marked the beginning of a new era of GPCR structural biology: according to the database GPCRdb, 270 structures of receptor-ligand complexes had been solved by September 2018, including 52 unique receptor complexes (http://gpcrdb.org/structure/statistics). In 2011 the first structure of the β2-adrenergic receptor in its active form (that is in the presence of an agonist and a G protein or a nanobody mimicking the G protein) was solved, thus revealing for the first time the atomic details of an activation mechanism of a GPCR (Rasmussen, Choi, et al., 2011; Rasmussen, DeVree, et al., 2011). The field of opioid receptors followed closely this revolution, with structures of the four opioid receptors solved in their inactive state in 2012 (Granier et al., 2012; Manglik et al., 2012; Thompson et al., 2012; Wu et al., 2012), and later in their active states for MOP (Huang et al., 2015; Koehl et al., 2018) and KOP (Che et al., 2018). An overlay of KOP in its inactive state, in complex with the antagonist JDTic (PDB 4DJH), and in its activated state, in complex with the agonist MP1104 and a nanobody mimicking G protein (Nb39) (PDB 6B7S) illustrates the general mechanism of activation (Fig. 4): it is characterized by outward movements of transmembrane helix 6 (TM6) (by 10 Å) and ICL2 and inward movements of TM5 and TM7, leading to the creation of an intracellular pocket into which G proteins can penetrate.

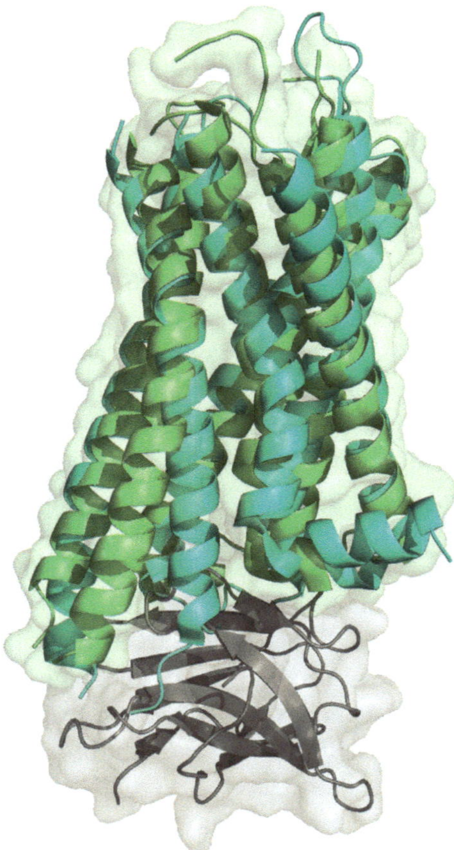

Fig. 4 Superposition of the X-ray structures of KOP in its inactive state (in blue, PDB 4DJH) and in its active state (in green, PDB 6B7S). The nanobody present in the active state is shown in gray, penetrating a pocket created by the displacement of TM6. The antagonist and agonist in the inactive and active states, respectively, are not displayed.

These movements are associated with a contraction (10% reduction in volume of the ligand binding pocket) of the extracellular portion in the active-state KOP, with extracellular loop 2 (ECL2) and TM4 and TM6 moving closer to the receptor core (Che et al., 2018). Both ligands, the antagonist JDTic and the agonist MP1104 bind at a similar location, with conserved contacts, in particular a salt bridge to $D138^{3.32}$ in TM3 and a water-mediated hydrogen bond with the backbone carbonyl oxygen of $K227^{5.39}$. Comparison of the active and inactive states of KOP indicates structural changes involving several residues of TM3, which are thus believed to be critical for coupling ligand-mediated changes in the orthosteric site and the transducer interface. This coupling is in part mediated by changes in a sodium binding pocket (formed by residues $D105^{2.50}$, $N141^{3.35}$ and $S145^{3.39}$) which acts as a negative allosteric modulator at opioid receptors (Fenalti et al., 2014; Katritch et al., 2014; Pasternak, Snowman, & Snyder, 1975).

With these data in hand, it may appear that the structural biology of opioid receptors and the molecular details of their activation mechanism are now well understood. This is not entirely true for several reasons: first, GPCRs in general and opioid receptors in particular must be understood in terms of interactions with other intracellular protein partners such as arrestins (Kang et al., 2015; Zhou et al., 2017), and with phospholipids and cholesterol within membrane domains (Dawaliby et al., 2016; Lagane et al., 2000; Meral et al., 2018; Pucadyil & Chattopadhyay, 2006; Xu et al., 2006), where they can form homo- and hetero-oligomers (Ferre et al., 2014; Jordan & Devi, 1999). Second, another important emerging characteristic of GPCRs is their extremely complex conformational landscape, within which the X-ray structures determined to date should be viewed as specific snapshots (Casiraghi et al., 2016; Deupi & Kobilka, 2010; Hilger, Masureel, & Kobilka, 2018). Third, it is extremely difficult to solve three-dimensional structures of complexes of a GPCR with its natural peptidic agonist. So far, the structure of DOP in complex with a non-natural peptidic *antagonist* has been solved (Fenalti et al., 2015), and the structure of MOP in complex with DAMGO, a highly specific synthetic peptide agonist analogue, and Gi heterotrimeric protein has been solved at 3.5 Å resolution by cryo-electron microscopy (Koehl et al., 2018).

NMR has proven highly efficient in demonstrating the conformational heterogeneity of GPCRs and their ligands (Bokoch et al., 2010; Casiraghi et al., 2016; Didenko, Liu, Horst, Stevens, & Wuthrich, 2013; Nygaard et al., 2013), including opioid receptors (Sounier et al., 2015). In collaboration

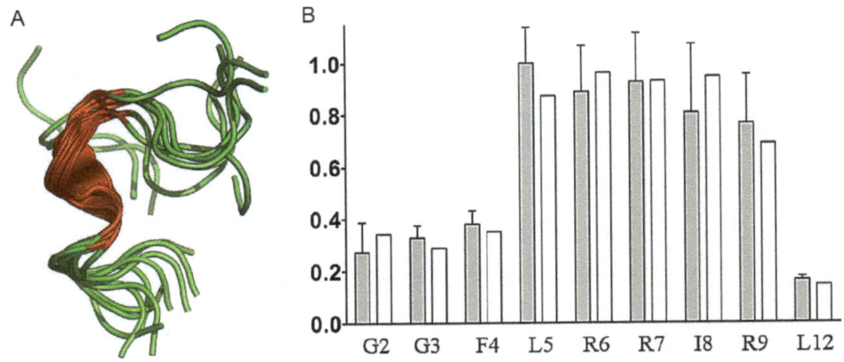

Fig. 5 (A) Receptor-bound conformation of dynorphin 1–13: a α-helical conformation is formed between Phe4 and Arg9; (B) Order parameter profile of dynorphin N—H bonds in the receptor-bound state. Gray: experimental data; White: calculated S^2 profiles from molecular dynamics simulations of dynorphin-receptor complexes. Note that both the N- and C-termini remain flexible in the receptor-bound state.

with Stevens and Wüthrich, we have solved the structure of dynorphin 1–13 bound to KOP in the absence of G proteins by NMR, thus in its low affinity state for agonists (O'Connor et al., 2015). A well-defined α-helical conformation forms from Leu5 to Arg9 upon receptor binding (Fig. 5A). Most interestingly, ^{15}N relaxation measurements indicate that the peptide remains flexible on a nanosecond time scale in its receptor-bound state (Fig. 5B). This was expected for the C-terminus in which non-specific electrostatic interaction contribute to receptor binding (i.e., for the "address" part of dynorphin). It was, however, unexpected for the first four amino acids Tyr1-Gly2-Gly3-Phe4 which form the signature of opioid peptides (the "message") and which cannot be modified without affecting receptor binding and activation (Naqvi et al., 1998). This mobility may be characteristic of an intermediate binding state observed for the G protein uncoupled receptor, and work is in progress to determine whether N-terminal immobilization occurs in the high-affinity ternary complex of peptide, receptor and G protein or the Nb39 nanobody used to stabilize the active conformation (Che et al., 2018).

5. Building 3D models of dynorphin-KOP complexes

Modeling structures of complexes formed by KOP and its agonists or antagonists has been attempted for >20 years (Alonso, Bliznyuk, & Gready, 2006; Bailey & Husbands, 2018; Benyhe, Zador, & Otvos, 2015;

Bera, Marathe, Payghan, & Ghoshal, 2018; Gentilucci, Tolomelli, De Marco, & Artali, 2012; Johnson, 2017; Kane, Svensson, & Ferguson, 2006; Kaserer, Lantero, Schmidhammer, Spetea, & Schuster, 2016; Kolinski & Filipek, 2010; Lavecchia, Greco, Novellino, Vittorio, & Ronsisvalle, 2000; Martinez-Mayorga et al., 2013; Patra, Kumar, Pasha, & Chopra, 2012; Tessmer, Meyer, Hruby, & Kallick, 1997; Wu, Song, Graaf, & Stevens, 2017; Yongye & Martínez-Mayorga, 2012). Some of these models specifically focused on dynorphin (Bjorneras et al., 2014; Chavkin, 2013; Iadanza, Höltje, Ronsisvalle, & Höltje, 2002; Kang et al., 2015; O'Connor et al., 2015; Paterlini, Portoghese, & Ferguson, 1997; Sankararamakrishnan & Weinstein, 2000; Smeets et al., 2016; Vardy et al., 2013). Early studies, performed before any experimentally determined receptor structures were available, were based entirely on modeling (Iadanza et al., 2002; Paterlini et al., 1997; Wan et al., 2000). Dynorphin was positioned within the receptor such that it was in agreement with mutagenesis data. Specifically, the spatial proximity between the N-terminus of dynorphin and residue $D138^{3.32}$ was preserved. Recent progress in obtaining X-ray and electron microscopy structures of the opioid receptors has enabled significant advances, due to the wealth of details for both the receptor structure and the binding modes of associated ligands. A binding mode was proposed for dynorphin 1–8, in which the peptide's N-terminal mimics the orientation of the phenol-like ring of the JDTic antagonist (Vardy et al., 2013). The structure of dynorphin 1–13 in the bound state was determined using transferred NOE experiments and that of the dynorphin-KOP complex was modeled using restraints from NMR, in particular N—H bond order parameters derived from $^{15}N\ R_2$ relaxation rates, as discussed below (O'Connor et al., 2015).

A protocol has been developed for creating three-dimensional structures of dynorphin-KOP complexes (O'Connor et al., 2015). The molecular modeling approach (Fig. 6) starts with the separate preparation of the two molecules. Although a KOP structure in complex with an antagonist was known from X-ray studies (Wu et al., 2012) a MOP structure mutated into KOP was used as starting point to avoid possible structural distortions caused by JDTic binding. Following the addition of missing residues and the prediction of rotamers of their side-chains (Krivov, Shapovalov, & Dunbrack, 2009; Nagata, Randall, & Baldi, 2012), the structure was relaxed and equilibrated in a 100 ns molecular dynamics (MD) run (Salomon-Ferrer, Case, & Walker, 2013; Salomon-Ferrer, Götz, Poole, Le Grand, & Walker, 2013), followed by a clustering procedure which allowed major conformers

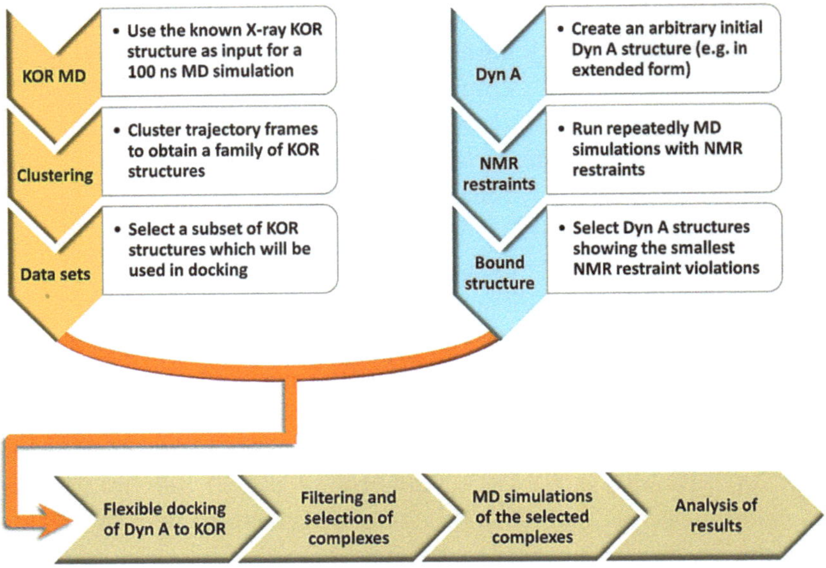

Fig. 6 Molecular modeling protocol leading to three-dimensional structures of dynorphin-KOP complexes. Briefly, plausible KOP structures were obtained from MD simulations using an X-ray structure as input, while dynorphin models were obtained from simulations with NMR constraints. Subsequent docking, filtering and verification of stability of complex structures thus obtained permitted a selection of the optimal result. See the text for a more detailed description of the procedure.

of the receptor to be identified. The use of a cluster radius of 2 Å resulted in eight families of structures, whose representative members were selected as those being closest to the cluster centroids. Six major KOP structures were retained for docking. The modeling of dynorphin involved a typical MD simulation coupled with a simulated annealing protocol (Nilges, Clore, & Gronenborn, 1988), run in the presence of NMR constraints to preserve the peptide structure previously determined by NMR. The results indicated the existence of an α-helical turn involving residues Leu5-Ile8, and a disordered peptide elsewhere. In the subsequent docking procedure, the backbone of residues Leu5-Ile8 was therefore held fixed, while the rest of the molecule (backbone and side-chains) remained flexible. The KOP molecule was mostly held rigid, except for 16 residues whose side-chains were allowed to be flexible as they were considered likely to interact with the peptide in the binding pocket. Flexible docking was performed with the AutoDock Vina program (Trott & Olson, 2009), by launching multiple runs on each of the six retained KOP structures. The results were filtered to keep the 10 best poses per each KOP structure. This produced a set of 60 structures of the dynorphin-KOP complex, which was reduced to a set of 22 structures after selecting the best representatives from each family, characterized by lowest energies. The stability of the complexes was verified by running further MD simulations for times ranging from 50 to 100 ns.

The crucial final step consisted of comparing the values of the order parameters calculated from the generated structures with those obtained from NMR experiments. The calculations were based on the hypothesis that the flexible peptide adopts varying conformations and that its order parameter should be averaged over different conformers. The exchange is fast on the NMR time scale, but beyond the reach of MD simulations (1 μs–1 ms). To find the minimum number of conformations required to reproduce the experimental order parameters, all combinations of modeled structures were taken into account. As a result, five major dynorphin conformations were identified, revealing significant structural diversity in both N- and C-termini.

Analysis of the resulting complexes allowed us to conclude that the position of a phenol-like functional group in the orthosteric site was largely conserved, for antagonist-bound structures of KOP-JDTic, as well as DOP-DIPP and MOP-funaltrexamine. One of the structural models (available in O'Connor et al., 2015) revealed Tyr1 in a position near the phenol-piperidine fused-ring system of JDTic, resembling a previously proposed binding mode (Vardy et al., 2013). Another model suggested that the side-chain of Tyr1 is close to the allosteric sodium binding site. $D138^{3.32}$ makes polar contacts with Tyr1, Gly2 and Gly3 in both of these models. However, only the latter features polar contacts of both Tyr1 and Arg7 with $W287^{6.48}$ and $N322^{7.45}$.

6. Development of novel KOP ligands based on structural knowledge

Opioids possess powerful properties and are currently the most effective analgesics available, but development is still needed to reduce their undesired side-effects (Dogra & Yadav, 2015). KOP agonists produce analgesia but, for most, their use is limited by centrally-mediated adverse effects such as dysphoria. Nalfurafine (TRX-820), that does not induce dysphoria, has been registered in Japan since 2009 for the treatment of uremic pruritus (Kozono, Yoshitani, & Nakano, 2018). Several peripherally-restricted (to avoid undesired effects) KOP agonists are currently in clinical trials: asimadoline (EMD-61753) (Delvaux et al., 2004; Szarka et al., 2007) and difelikefalin (CR845) (Hesselink, 2017) for irritable bowel syndrome and chronic or post-operative pain (https://clinicaltrials.gov/). Another promising strategy is the development of KOP-biased ligands: by acting as agonists for the G protein-dependent signaling and not for G protein-independent pathways, they could induce analgesia without dysphoria (Dogra & Yadav, 2015). In addition, there is a strong potential for KOP antagonists as antidepressant, anxiolytic or anti-addiction drugs (Zheng et al., 2017). The knowledge obtained from the structures of inactive KOP (Wu et al., 2012), KOP-bound dynorphin (O'Connor et al., 2015) and G protein bound KOP (Che et al., 2018) offers new strategies for finding novel KOP ligands (Shang & Filizola, 2015).

Among the various ways of using structural knowledge in drug discovery, structure-based virtual screening has proven successful in designing GPCR ligands, such as for D_3 dopamine (Carlsson et al., 2011; Lane et al., 2013), β_2 adrenergic (Kolb et al., 2009; Weiss et al., 2013) and A_{2A} adenosine (Carlsson et al., 2010; Katritch et al., 2010) receptors. It has allowed the discovery of a G protein-biased MOP agonist that produces analgesia with diminished side-effects (Manglik et al., 2016). Studies using virtual screening from the inactive JDTic-bound KOP structure have discovered new KOP agonists (Negri et al., 2013) and new G protein-biased agonists scaffolds (White et al., 2013). More recently, ligand-guided receptor optimization was applied to the inactive JDTic-bound KOP structure to generate alternative orthosteric pocket models that served to increase the prediction power of the methods (Zheng et al., 2017). Virtual screening on the initial and alternative binding site models, followed by Tanimoto distances-based selection of novel chemotypes, revealed ligands in the micromolar range with a 32% hit rate. An initial optimization round generated 11 compounds with sub-micromolar affinities and functional assays defined two potent antagonists and one G protein-biased agonist. The accumulation of precise structural knowledge on the modulation of KOP signaling activity by orthosteric ligands and allosteric modulators will certainly play a major role in future drug discovery programs.

7. Conclusions and future perspectives

Future research into the dynorphin-KOP structure and dynamics must address one major question: what defines a ligand as a full, partial, unbiased or biased agonist, an antagonist or an inverse agonist (Wacker, Stevens, & Roth, 2017). As GPCR signaling involves multiple receptor conformations in dynamic exchange, extensive research aims to characterize their conformational landscape and their modulation by ligands and signaling partners (Weis & Kobilka, 2018). Crystallography and electron microscopy provide structures of lowest-energy populated states (Wacker et al., 2017). Spectroscopic methods such as NMR allow the characterization of dynamic properties (Weis & Kobilka, 2018), such as the weak allosteric coupling between the orthosteric site and the signaling interface, as described for MOP (Sounier et al., 2015). One may take advantage of yeast or bacterial expression systems to produce specifically labeled GPCRs and perform advanced relaxation-based analyses to assess receptor dynamics, as was done recently for BLT2 (Casiraghi et al., 2016) and A2A adenosine receptors (Clark et al., 2017; Eddy et al., 2018). We are currently applying the methodologies we developed to determine the conformation and dynamics of KOP-bound dynorphin (O'Connor et al., 2015) to the ternary dynorphin-KOP-Nb39 complex (where Nb39 is a nanobody which mimics G proteins and confers high affinity binding to agonists). We thus hope to explain the 10-fold gain in dynorphin binding affinity in the presence of G proteins. The conformational dynamics of G proteins and arrestins themselves can be modulated by the ligands (Hilger et al., 2018). The question of conformational landscape and allosteric coupling must therefore be extended to entire GPCR signaling complexes and posed in the context of real cellular environment where modulation by lipids, membrane domains and other receptors do occur. While using information derived from structural biology, one should always bear in mind that recombinant G protein-coupled receptors in vitro may not recapitulate all the properties of native receptors naturally expressed in tissues as was shown for instance in Broad et al. (2016).

References

Ahlbeck, K. (2011). Opioids: A two-faced Janus. *Current Medical Research and Opinion*, *27*(2), 439–448. https://doi.org/10.1185/03007995.2010.545379.

Aldrich, J. V., & McLaughlin, J. P. (2009). Peptide kappa opioid receptor ligands: Potential for drug development. *The AAPS Journal*, *11*(2), 312–322. https://doi.org/10.1208/s12248-009-9105-4.

Alexander, S. P. H., Christopoulos, A., Davenport, A. P., Kelly, E., Marrion, N. V., Peters, J. A., et al. (2017). The concise guide to pharmacology 2017/18: G protein-coupled receptors. *British Journal of Pharmacology*, *174*, S17–S129.

Al-Hasani, R., & Bruchas, M. R. (2011). Molecular mechanisms of opioid receptor-dependent signaling and behavior. *Anesthesiology*, *115*, 1363–1381.

Alonso, H., Bliznyuk, A. A., & Gready, J. E. (2006). Combining docking and molecular dynamic simulations in drug design. *Medicinal Research Reviews*, *26*, 531–568.

Attali, B., & Vogel, Z. (1989). Characterization of kappa opiate receptors in rat spinal cord-dorsal root ganglion cocultures and their regulation by chronic opiate treatment. *Brain Research*, *517*, 182–188.

Auge, S., Bersch, B., Tropis, M., & Milon, A. (2000). Characterization of substance P-membrane interaction by transferred nuclear Overhauser effect. *Biopolymers*, *54*(5), 297–306.

Axelrod, D., & Wang, M. D. (1994). Reduction-of-dimensionality kinetics at reaction-limited cell-surface receptors. *Biophysical Journal*, *66*(3), 588–600. https://doi.org/10.1016/S0006-3495(94)80834-3.

Bailey, S. J., & Husbands, S. M. (2018). Targeting opioid receptor signaling in depression: Do we need selective κ opioid receptor antagonists? *Neuronal Signaling*, *2*(2), NS20170145.

Ballesteros, J. A., & Weinstein, H. (1995). Integrated methods for the construction of three-dimensional models and computational probing of structure-function relations in G protein-coupled receptors. *Methods in Neurosciences*, *25*, 366–428.

Bals-Kubik, R., Ableitner, A., Herz, A., & Shippenberg, T. S. (1993). Neuroanatomical sites mediating the motivational effects of opioids as mapped by the conditioned place preference paradigm in rats. *The Journal of Pharmacology and Experimental Therapeutics*, *264*, 489–495.

Baneres, J. L., Martin, A., Hullot, P., Girard, J. P., Rossi, J. C., & Parello, J. (2003). Structure-based analysis of GPCR function: Conformational adaptation of both agonist and receptor upon leukotriene B-4 binding to recombinant BLT1. *Journal of Molecular Biology*, *329*, 801–814.

Baneres, J. L., Popot, J. L., & Mouillac, B. (2011). New advances in production and functional folding of G-protein-coupled receptors. *Trends in Biotechnology*, *29*, 314–322.

Barchfeld, C. C., & Medzihradsky, F. (1984). Receptor-mediated stimulation of brain GTPase by opiates in normal and dependent rats. *Biochemical and Biophysical Reasearch Communications*, *121*, 641–648.

Benyhe, S., Zador, F., & Otvos, F. (2015). Biochemistry of opioid (morphine) receptors: Binding, structure and molecular modelling. *Acta Biologica Szegediensis*, *59*, 17–37.

Bera, I., Marathe, M. V., Payghan, P. V., & Ghoshal, N. (2018). Identification of novel hits as highly prospective dual agonists for mu and kappa opioid receptors: An integrated in silico approach. *Journal of Biomolecular Structure and Dynamics*, *36*(2), 279–301.

Berman, Y., Mzhavia, N., Polonskaia, A., Furuta, M., Steiner, D. F., Pintar, J. E., et al. (2000). Defective prodynorphin processing in mice lacking prohormone convertase PC2. *Journal of Neurochemistry*, *75*, 1763–1770.

Bersch, B., Koehl, P., Nakatani, Y., Ourisson, G., & Milon, A. (1993). 1H nuclear magnetic resonance determination of the membrane-bound conformation of senktide, a highly selective neurokinin B agonist. *Journal of Biomolecular NMR*, *3*(4), 443–461.

Bjorneras, J., Kurnik, M., Oliveberg, M., Graslund, A., Maler, L., & Danielsson, J. (2014). Direct detection of neuropeptide dynorphin A binding to the second extracellular loop of the kappa opioid receptor using a soluble protein scaffold. *FEBS Journal*, *281*(3), 814–824. https://doi.org/10.1111/febs.12626.

Bokoch, M. P., Nygaard, R., Zou, Y. Z., Rasmussen, S. G. F., Pardo, L., Prosser, R. S., et al. (2010). Conformational changes in GPCR surface and core probed by [C-13]-methyl NMR spectroscopy. *Biophysical Journal*, *98*(3), 418a. https://doi.org/10.1016/j.bpj.2009.12.2257.

Bourinet, E., Soong, T. W., Stea, A., & Snutch, T. P. (1996). Determinants of the G protein-dependent opioid modulation of neuronal calcium channels. *Proceedings of the National Academy of Sciences of the United States of America*, *93*, 1486–1491.

Broad, J., Maurel, D., Kung, V. W. S., Hicks, G. A., Schemann, M., Barnes, M. R., et al. (2016). Human native kappa opioid receptor functions not predicted by recombinant receptors: Implications for drug design. *Scientific Reports*, *6*, 30797. https://doi.org/10.1038/srep30797.

Bruchas, M. R., & Chavkin, C. (2010). Kinase cascades and ligand-directed signaling at the kappa opioid receptor. *Psychopharmacology*, *210*, 137–147.

Bruchas, M. R., Land, B. B., & Chavkin, C. (2010). The dynorphin/kappa opioid system as a modulator of stress-induced and pro-addictive behaviors. *Brain Research*, *1314*, 44–55. https://doi.org/10.1016/j.brainres.2009.08.062.

Caffrey, M., & Cherezov, V. (2009). Crystallizing membrane proteins using lipidic mesophases. *Nature Protocols*, *4*(5), 706–731. https://doi.org/10.1038/nprot.2009.31.

Carlsson, J., Coleman, R. G., Setola, V., Irwin, J. J., Fan, H., Schlessinger, A., et al. (2011). Ligand discovery from a dopamine D3 receptor homology model and crystal structure. *Nature Chemical Biology*, 7(11), 769–778. https://doi.org/10.1038/nchembio.662.

Carlsson, J., Yoo, L., Gao, Z.-G., Irwin, J. J., Shoichet, B. K., & Jacobson, K. A. (2010). Structure-based discovery of A2A adenosine receptor ligands. *Journal of Medicinal Chemistry*, *53*(9), 3748–3755. https://doi.org/10.1021/jm100240h.

Casiraghi, M., Damian, M., Lescop, E., Point, E., Moncoq, K., Morellet, N., et al. (2016). Functional modulation of a G protein-coupled receptor conformational landscape in a lipid bilayer. *Journal of the American Chemical Society*, *138*(35), 11170–11175. https://doi.org/10.1021/jacs.6b04432.

Chavkin, C. (2013). Dynorphin–still an extraordinarily potent opioid peptide. *Molecular Pharmacology*, *83*(4), 729–736.

Chavkin, C., Bakhit, C., Weber, E., & Bloom, F. E. (1983). Relative contents and concomitant release of prodynorphin neoendorphin-derived peptides in rat hippocampus. *Proceedings of the National Academy of Sciences of the United States of America*, *80*(24), 7669–7673. https://doi.org/10.1073/pnas.80.24.7669.

Che, T., Majumdar, S., Zaidi, S. A., Ondachi, P., McCorvy, J. D., Wang, S., et al. (2018). Structure of the nanobody-stabilized active state of the kappa opioid receptor. *Cell*, *172*(1-2), 55. https://doi.org/10.1016/j.cell.2017.12.011.

Chen, Y., Mestek, A., Liu, J., Hurley, J. A., & Yu, L. (1993). Molecular-cloning and functional expression of a Mu-opioid receptor from rat-brain. *Molecular Pharmacology*, *44*(1), 8–12.

Cherezov, V. (2011). Lipidic cubic phase technologies for membrane protein structural studies. *Current Opinion in Structural Biology*, *21*(4), 559–566. https://doi.org/10.1016/j.sbi.2011.06.007.

Childers, S. R., & Snyder, S. H. (1978). Guanine nucleotides differentiate agonist and antagonist interactions with opiate receptors. *Life Sciences*, *23*, 759–762.

Choi, H., Murray, T. F., DeLander, G. E., Caldwell, V., & Aldrich, J. V. (1992). N-terminal alkylated derivatives of [D-Pro10]dynorphin A-(1-11) are highly selective for kappa-opioid receptors. *Journal of Medicinal Chemistry*, *35*(24), 4638–4639.

Chun, E., Thompson, A. A., Liu, W., Roth, C. B., Griffith, M. T., Katritch, V., et al. (2012). Fusion partner toolchest for the stabilization and crystallization of G protein-coupled receptors. *Structure*, *20*(6), 967–976. https://doi.org/10.1016/j.str.2012.04.010.

Civelli, O., Douglass, J., Goldstein, A., & Herbert, E. (1985). Sequence and expression of the rat prodynorphin gene. *Proceedings of the National Academy of Sciences of the United States of America*, *82*(12), 4291–4295.

Clark, L. D., Dikiy, I., Chapman, K., Rodstrom, K. E., Aramini, J., LeVine, M. V., et al. (2017). Ligand modulation of sidechain dynamics in a wild-type human GPCR. *eLife*, *6*, e28505. https://doi.org/10.7554/eLife.28505.

Claude, P. A., Wotta, D. R., Zhang, X. H., Prather, P. L., McGinn, T. M., Erickson, L. J., et al. (1996). Mutation of a conserved serine in TM4 of opioid receptors confers full agonistic properties to classical antagonists. *Proceedings of the National Academy of Sciences of the United States of America*, *93*(12), 5715–5719.

Corder, G., Castro, D. C., Bruchas, M. R., & Scherrer, G. (2018). Endogenous and exogenous opioids in pain. *Annual Review of Neuroscience*, *41*, 453–473. https://doi.org/10.1146/annurev-neuro-080317-061522.

Cox, B. M., Opheim, K. E., Teachemacher, H., & Goldstein, A. (1975). A peptide-like substance from pituitary that acts like morphine. 2. Purification and properties. *Life Sciences*, *16*, 1777–1782.

Czaplicki, J., & Milon, A. (1998). Simulations of transferred NOE in a ternary peptide-receptor-lipid complex. *Journal de Chimie Physique et de Physico-Chimie Biologique*, *95*, 196–207.

Czaplicki, J., & Milon, A. (2005). Quantitative analysis of transferred nuclear overhauser effects in complex spin systems by full relaxation matrix analysis. *Acta Physica Polonica A*, *108*(1), 25–32.

Dawaliby, R., Trubbia, C., Delporte, C., Masureel, M., Van Antwerpen, P., Kobilka, B. K., et al. (2016). Allosteric regulation of G protein-coupled receptor activity by phospholipids. *Nature Chemical Biology*, *12*(1), 35. https://doi.org/10.1038/Nchembio.1960.

Delvaux, M., Beck, A., Jacob, J., Bouzamondo, H., Weber, F. T., & Frexinos, J. (2004). Effect of asimadoline, a kappa opioid agonist, on pain induced by colonic distension in patients with irritable bowel syndrome. *Alimentary Pharmacology & Therapeutics*, *20*(2), 237–246. https://doi.org/10.1111/j.1365-2036.2004.01922.x.

Deupi, X., & Kobilka, B. K. (2010). Energy landscapes as a tool to integrate GPCR structure, dynamics, and function. *Physiology*, *25*(5), 293–303. https://doi.org/10.1152/physiol.00002.2010.

Dhawan, B. N., Cesselin, F., Raghubir, R., Reisine, T., Bradley, P. B., Portoghese, P. S., et al. (1996). International union of pharmacology. XII. Classification of opioid receptors. *Pharmacological Reviews*, *48*, 567–592.

Didenko, T., Liu, J. J., Horst, R., Stevens, R. C., & Wuthrich, K. (2013). Fluorine-19 NMR of integral membrane proteins illustrated with studies of GPCRs. *Current Opinion in Structural Biology*, *23*(5), 740–747. https://doi.org/10.1016/j.sbi.2013.07.011.

Dogra, S., & Yadav, P. N. (2015). Biased agonism at kappa opioid receptors: Implication in pain and mood disorders. *European Journal of Pharmacology*, *763*, 184–190. https://doi.org/10.1016/j.ejphar.2015.07.018.

Eddy, M. T., Lee, M. Y., Gao, Z. G., White, K. L., Didenko, T., Horst, R., et al. (2018). Allosteric coupling of drug binding and intracellular signaling in the A(2A) adenosine receptor. *Cell*, *172*(1-2), 68. https://doi.org/10.1016/j.cell.2017.12.004.

Erne, D., Sargent, D. F., & Schwyzer, R. (1985). Preferred conformation, orientation, and accumulation of dynorphin A-(1-13)-tridecapeptide on the surface of neutral lipid membranes. *Biochemistry*, *24*(16), 4261–4263.

Evans, C. J., Keith, D. E., Jr., Morrison, H., Magendzo, K., & Edwards, R. H. (1992). Cloning of a delta opioid receptor by functional expression. *Science*, *258*(5090), 1952–1955.

Fenalti, G., Giguere, P. M., Katritch, V., Huang, X. P., Thompson, A. A., Cherezov, V., et al. (2014). Molecular control of delta-opioid receptor signalling. *Nature*, *506*(7487), 191–196. https://doi.org/10.1038/nature12944.

Fenalti, G., Zatsepin, N. A., Betti, C., Giguere, P., Han, G. W., Ishchenko, A., et al. (2015). Structural basis for bifunctional peptide recognition at human delta-opioid receptor. *Nature Structural & Molecular Biology*, *22*(3), 265–268. https://doi.org/10.1038/nsmb.2965.

Feng, Y., He, X., Yang, Y., Chao, D., Lazarus, L. H., & Xia, Y. (2012). Current research on opioid receptor function. *Current Drug Targets*, *13*, 230–246.

Ferre, S., Casado, V., Devi, L. A., Filizola, M., Jockers, R., Lohse, M. J., et al. (2014). G protein-coupled receptor oligomerization revisited: Functional and pharmacological perspectives. *Pharmacological Reviews*, *66*(2), 413–434. https://doi.org/10.1124/pr.113.008052.

Fersht, A. R. (1999). *Structure and mechanism in protein science*. New York: Freeman.

Gentilucci, L., Tolomelli, A., De Marco, R., & Artali, R. (2012). Molecular docking of opiates and opioid peptides, a tool for the design of selective agonists and antagonists, and for the investigation of atypical ligand-receptor interactions. *Current Medicinal Chemistry*, *19*, 1587–1601.

Goldstein, A., Fischli, W., Lowney, L. I., Hunkapiller, M., & Hood, L. (1981). Porcine pituitary dynorphin: Complete amino acid sequence of the biologically active heptadecapeptide. *Proceedings of the National Academy of Sciences of the United States of America*, *78*(11), 7219–7223.

Goldstein, A., Tachibana, S., Lowney, L. I., Hunkapiller, M., & Hood, L. (1979). Dynorphin-(1-13), an extraordinarily potent opioid peptide. *Proceedings of the National Academy of Sciences of the United States of America*, *76*(12), 6666–6670.

Granier, S., & Kobilka, B. (2012). A new era of GPCR structural and chemical biology. *Nature Chemical Biology*, *8*(8), 670–673. https://doi.org/10.1038/nchembio.1025.

Granier, S., Manglik, A., Kruse, A. C., Kobilka, T. S., Thian, F. S., Weis, W. I., et al. (2012). Structure of the delta-opioid receptor bound to naltrindole. *Nature*, *485*(7398), 400–404. https://doi.org/10.1038/nature11111.

Henry, D. J., Grandy, D. K., Lester, H. A., Davidson, N., & Chavkin, C. (1995). K-opioid receptors couple to inwardly rectifying potassium channels when coexpressed by *Xenopus* oocytes. *Molecular Pharmacology*, *47*, 551–557.

Hesselink, J. M. K. (2017). CR845 (difelikefalin), a kappa receptors agonist in phase III by CARA therapeutics: A case of 'spin' in scientific writing? *Journal of Pharmacology & Clinical Research*, *2*(3), 555588. https://doi.org/10.19080/JPCR.2017.02.555588.

Hilger, D., Masureel, M., & Kobilka, B. K. (2018). Structure and dynamics of GPCR signaling complexes. *Nature Structural & Molecular Biology*, *25*(1), 4–12. https://doi.org/10.1038/s41594-017-0011-7.

Huang, W. J., Manglik, A., Venkatakrishnan, A. J., Laeremans, T., Feinberg, E. N., Sanborn, A. L., et al. (2015). Structural insights into mu-opioid receptor activation. *Nature*, *524*(7565), 315. https://doi.org/10.1038/nature14886.

Iadanza, M., Höltje, M., Ronsisvalle, G., & Höltje, H.-D. (2002). κ-Opioid receptor model in a phospholipid bilayer: Molecular dynamics simulation. *Journal of Medicinal Chemistry*, *45*(22), 4838–4846.

Ji, R.-R., Zhang, Q., Law, P. Y., Low, H. H., Elde, R., & Hökfelt, T. (1995). Expression of mu-, delta-, and kappa-opioid receptor-like immunoreactivities in rat dorsal root ganglia after carrageenan-induced inflammation. *The Journal of Neuroscience*, *15*, 8155–8166.

Johnson, S. (2017). *Design and synthesis of functionally selective kappa opioid receptor ligands*. Lawrence: University of Kansas. Retrieved from https://kuscholarworks.ku.edu/handle/1808/26154.

Jordan, B. A., & Devi, L. A. (1999). G protein coupled receptor heterodimerization modulates receptor function. *Nature*, *399*(6737), 697–700.

Kakidani, H., Furutani, Y., Takahashi, H., Noda, M., Morimoto, Y., Hirose, T., et al. (1982). Cloning and sequence analysis of cDNA for porcine beta-neo-endorphin/dynorphin precursor. *Nature*, *298*, 245–249.

Kallick, D. A. (1993). Conformation of dynorphin a(1-17) bound to dodecylphosphocholine micelles. *Journal of the American Chemical Society*, *115*(20), 9317–9318. https://doi.org/10.1021/ja00073a069.

Kane, B. E., Svensson, B., & Ferguson, D. M. (2006). Molecular recognition of opioid receptor ligands. *The AAPS Journal*, *8*(1), E126–E137.

Kang, Y. Y., Zhou, X. E., Gao, X., He, Y. Z., Liu, W., Ishchenko, A., et al. (2015). Crystal structure of rhodopsin bound to arrestin by femtosecond X-ray laser. *Nature*, *523*(7562), 561. https://doi.org/10.1038/nature14656.

Kaserer, T., Lantero, A., Schmidhammer, H., Spetea, M., & Schuster, D. (2016). μ Opioid receptor: Novel antagonists and structural modeling. *Scientific Reports*, *6*, 21548.

Katritch, V., Fenalti, G., Abola, E. E., Roth, B. L., Cherezov, V., & Stevens, R. C. (2014). Allosteric sodium in class A GPCR signaling. *Trends in Biochemical Sciences*, *39*(5), 233–244. https://doi.org/10.1016/j.tibs.2014.03.002.

Katritch, V., Jaakola, V.-P., Lane, J. R., Lin, J., Ijzerman, A. P., Yeager, M., et al. (2010). Structure-based discovery of novel chemotypes for adenosine A2A receptor antagonists. *Journal of Medicinal Chemistry*, *53*(4), 1799–1809. https://doi.org/10.1021/jm901647p.

Kieffer, B. L., Befort, K., Gaveriaux-Ruff, C., & Hirth, C. G. (1992). The delta-opioid receptor: Isolation of a cDNA by expression cloning and pharmacological characterization. *Proceedings of the National Academy of Sciences of the United States of America*, *89*(24), 12048–12052.

Knoll, A. T., & Carlezon, W. A. (2010). Dynorphin, stress, and depression. *Brain Research*, *1314*, 56–73. https://doi.org/10.1016/j.brainres.2009.09.074.

Kobilka, B., & Schertler, G. F. X. (2008). New G-protein-coupled receptor crystal structures: Insights and limitations. *Trends in Pharmacological Sciences*, *29*(2), 79–83. https://doi.org/10.1016/j.tips.2007.11.009.

Koehl, A., Hu, H., Maeda, S., Zhang, Y., Qu, Q., Paggi, J. M., et al. (2018). Structure of the μ-opioid receptor-G_i protein complex. *Nature*, *558*, 547–552. https://doi.org/10.1038/s41586-018-0219-7.

Kolb, P., Rosenbaum, D. M., Irwin, J. J., Fung, J. J., Kobilka, B. K., & Shoichet, B. K. (2009). Structure-based discovery of beta2-adrenergic receptor ligands. *Proceedings of the National Academy of Sciences of the United States of America*, *106*(16), 6843–6848. https://doi.org/10.1073/pnas.0812657106.

Kolinski, M., & Filipek, S. (2010). Study of a structurally similar kappa opioid receptor agonist and antagonist pair by molecular dynamics simulations. *Journal of Molecular Modeling*, *16*(10), 1567–1576.

Kong, H., Raynor, K., Yano, H., Takeda, J., Bell, G. I., & Reisine, T. (1994). Agonists and antagonists bind to different domains of the cloned kappa opioid receptor. *Proceedings of the National Academy of Sciences of the United States of America*, *91*(17), 8042–8046.

Kozono, H., Yoshitani, H., & Nakano, R. (2018). Post-marketing surveillance study of the safety and efficacy of nalfurafine hydrochloride (Remitch® capsules 2.5 μg) in 3,762 hemodialysis patients with intractable pruritus. *International Journal of Nephrology and Renovascular Disease*, *11*, 9–24. https://doi.org/10.2147/IJNRD.S145720.

Kreek, M. J., Levran, O., Reed, B., Schlussman, S. D., Zhou, Y., & Butelman, E. R. (2012). Opiate addiction and cocaine addiction: Underlying molecular neurobiology and genetics. *Journal of Clinical Investigation*, *122*(10), 3387–3393. https://doi.org/10.1172/jci60390.

Krishnan, V., & Nestler, E. J. (2010). Linking molecules to mood: New insight into the biology of depression. *American Journal of Psychiatry*, *167*(11), 1305–1320. https://doi.org/10.1176/appi.ajp.2009.10030434.

Krivov, G. G., Shapovalov, M. V., & Dunbrack, R. L. (2009). Improved prediction of protein side-chain conformations with SCWRL4. *Proteins*, 77(4), 778–795.

Lagane, B., Gaibelet, G., Meilhoc, E., Masson, J. M., Cezanne, L., & Lopez, A. (2000). Role of sterols in modulating the human mu-opioid receptor function in *Saccharomyces cerevisiae*. *Journal of Biological Chemistry*, *275*(43), 33197–33200. https://doi.org/10.1074/jbc.C000576200.

Lahti, R. A., Mickelson, M. M., McCall, J. M., & Von Voigtlander, P. F. (1985). U-69593 a highly selective ligand for the opioid kappa receptor. *European Journal of Pharmacology*, *109*, 281–284.

Lahti, R. A., Von Voigtlander, P. F., & Barsuhn, C. (1982). Properties of a selective kappa agonist, U-50,488H. *Life Sciences*, *31*, 2257.

Lancaster, C. R. D., Mishra, P. K., Hughes, D. W., Stpierre, S. A., Bothnerby, A. A., & Epand, R. M. (1991). Mimicking the membrane-mediated conformation of dynorphin A-(1-13)-peptide—Circular-dichroism and nuclear-magnetic-resonance studies in methanolic solution. *Biochemistry, 30*(19), 4715–4726. https://doi.org/10.1021/bi00233a012.

Lane, J. R., Chubukov, P., Liu, W., Canals, M., Cherezov, V., Abagyan, R., et al. (2013). Structure-based ligand discovery targeting orthosteric and allosteric pockets of dopamine receptors. *Molecular Pharmacology, 84*(6), 794–807. https://doi.org/10.1124/mol.113.088054.

Lapalu, S., Moisand, C., Mazarguil, H., Cambois, G., Mollereau, C., & Meunier, J. C. (1997). Comparison of the structure-activity relationships of nociceptin and dynorphin A using chimeric peptides. *FEBS Letters, 417*(3), 333–336.

Larson, D. L., Jones, R. M., Hjorth, S. A., Schwartz, T. W., & Portoghese, P. S. (2000). Binding of norbinaltorphimine (norBNI) congeners to wild-type and mutant mu and kappa opioid receptors: Molecular recognition loci for the pharmacophore and address components of kappa antagonists. *Journal of Medicinal Chemistry, 43*(8), 1573–1576.

Lavecchia, A., Greco, G., Novellino, E., Vittorio, F., & Ronsisvalle, G. (2000). Modeling of κ-opioid receptor/agonists interactions using pharmacophore-based and docking simulations. *Journal of Medicinal Chemistry, 43*(11), 2124–2134.

Law, P. Y., Wong, Y. H., & Loh, H. H. (2000). Molecular mechanisms and regulation of opioid receptor signaling. *Annual Review of Pharmacology and Toxicology, 40*, 389–430.

Li, S., Zhu, J., Chen, C. G., Chen, Y. W., Deriel, J. K., Ashby, B., et al. (1993). Molecular cloning and expression of a rat kappa opioid receptor. *The Biochemical Journal, 195*, 629–633.

Lind, J., Graslund, A., & Maler, L. (2006). Membrane interactions of dynorphins. *Biochemistry, 45*(51), 15931–15940. https://doi.org/10.1021/bi061199g.

Lutz, P.-E., & Kieffer, B. L. (2013). Opioid receptors: Distinct roles in mood disorders. *Trends in Neurosciences, 36*(3), 195–206. https://doi.org/10.1016/j.tins.2012.11.002.

Lv, X., Liu, J., Shi, Q., Tan, Q., Wu, D., Skinner, J. J., et al. (2016). In vitro expression and analysis of the 826 human G protein-coupled receptors. *Protein & Cell*, 7(5), 325–337. https://doi.org/10.1007/s13238-016-0263-8.

Manglik, A., Kruse, A. C., Kobilka, T. S., Thian, F. S., Mathiesen, J. M., Sunahara, R. K., et al. (2012). Crystal structure of the mu-opioid receptor bound to a morphinan antagonist. *Nature, 485*(7398), 321–326. https://doi.org/10.1038/nature10954.

Manglik, A., Lin, H., Aryal, D. K., McCorvy, J. D., Dengler, D., Corder, G., et al. (2016). Structure-based discovery of opioid analgesics with reduced side effects. *Nature, 537*(7619), 185. https://doi.org/10.1038/nature19112.

Mansour, A., Hoversten, M. T., Taylor, L. P., Watson, S. J., & Akil, H. (1995). The cloned mu, delta and kappa receptors and their endogenous ligands: Evidence for two opioid peptide recognition cores. *Brain Research, 27*, 89–98.

Margolis, E. B., Hjelmstad, G. O., Bonci, A., & Fields, H. L. (2003). Kappa-opioid agonists directly inhibit midbrain dopaminergic neurons. *The Journal of Neuroscience, 23*, 9981–9986.

Martin, W. R., Eades, C. G., Thompson, J. A., Huppler, R. E., & Gilbert, P. E. (1976). The effects of morphine- and nalorphine-like drugs in the nondependent and morphine-dependent chronic spinal dog. *The Journal of Pharmacology and Experimental Therapeutics, 197*(3), 517.

Martinez-Mayorga, K., Byler, K. G., Yongye, A. B., Giulianotti, M. A., Dooley, C. T., & Houghten, R. A. (2013). Ligand/kappa-opioid receptor interactions: Insights from the X-ray crystal structure. *European Journal of Medicinal Chemistry, 66*, 114–121.

Massou, S., Puech, V., Talmont, F., Demange, P., Lindley, N. D., Tropis, M., et al. (1999). Heterologous expression of a deuterated membrane-integrated receptor and partial deuteration in methylotrophic yeasts. *Journal of Biomolecular NMR, 14*(3), 231–239.

Meng, F., Xie, G. X., Thompson, R. C., Mansour, A., Goldstein, A., Watson, S. J., et al. (1993). Cloning and pharmacological characterization of a rat kappa opioid receptor. *Proceedings of the National Academy of Sciences of the United States of America*, *90*, 9954–9958.

Meral, D., Provasi, D., Prada-Gracia, D., Moller, J., Marino, K., Lohse, M. J., et al. (2018). Molecular details of dimerization kinetics reveal negligible populations of transient mu-opioid receptor homodimers at physiological concentrations. *Scientific Reports*, *8*, 7705. https://doi.org/10.1038/s41598-018-26070-8.

Merg, F., Filliol, D., Usynin, I., Bazov, I., Bark, N., Hurd, Y. L., et al. (2006). Big dynorphin as a putative endogenous ligand for the kappa-opioid receptor. *Journal of Neurochemistry*, *97*(1), 292–301. https://doi.org/10.1111/j.1471-4159.2006.03732.x.

Metzger, T. G., & Ferguson, D. M. (1995). On the role of extracellular loops of opioid receptors in conferring ligand selectivity. *FEBS Letters*, *375*(1–2), 1–4.

Metzger, T. G., Paterlini, M. G., Ferguson, D. M., & Portoghese, P. S. (2001). Investigation of the selectivity of oxymorphone- and naltrexone-derived ligands via site-directed mutagenesis of opioid receptors: Exploring the "address" recognition locus. *Journal of Medicinal Chemistry*, *44*(6), 857–862.

Meunier, J.-C., Mollereau, C., Toll, L., Suaudeau, C., Moisand, C., Alvinerie, P., et al. (1995). Isolation and structure of the endogenous agonist of opioid receptor-like ORL1 receptor. *Nature*, *377*, 532–535.

Milon, A., Miyazawa, T., & Higashijima, T. (1990). Transferred nuclear Overhauser effect analyses of membrane-bound enkephalin analogues by 1H nuclear magnetic resonance: Correlation between activities and membrane-bound conformations. *Biochemistry*, *29*(1), 65–75.

Minneman, K. P., & Iversen, L. L. (1976). Enkephalin and opiate narcotics increase cyclic GMP accumulation in slices of rat neostratium. *Nature*, *262*, 313–314.

Molineaux, C. J., & Cox, B. M. (1982). Subcellular localization of immunoreactive dynorphin and vasopressin in rat pituitary and hypothalamus. *Life Sciences*, *31*, 1765–1768.

Mollereau, C., Parmentier, M., Mailleux, P., Butour, J. L., Moisand, C., Chalon, P., et al. (1994). Orl1, a novel member of the opioid receptor family—Cloning, functional expression and localization. *FEBS Letters*, *341*(1), 33–38. https://doi.org/10.1016/0014-5793(94)80235-1.

Nagata, K., Randall, A., & Baldi, P. (2012). SIDEpro: A novel machine learning approach for the fast and accurate prediction of side-chain conformations. *Proteins*, *80*(1), 142–153.

Naito, A., & Nishimura, K. (2004). Conformational analysis of opioid peptides in the solid states and the membrane environments by NMR spectroscopy. *Current Topics in Medicinal Chemistry*, *4*(1), 135–145. https://doi.org/10.2174/1568026043451645.

Naqvi, T., Haq, W., & Mathur, K. B. (1998). Structure-activity relationship studies of dynorphin A and related peptides. *Peptides*, *19*(7), 1277–1292.

Negri, A., Rives, M.-L., Caspers, M. J., Prisinzano, T. E., Javitch, J. A., & Filizola, M. (2013). Discovery of a novel selective kappa-opioid receptor agonist using crystal structure-based virtual screening. *Journal of Chemical Information and Modeling*, *53*(3), 521–526. https://doi.org/10.1021/ci400019t.

Nestler, E. J., & Carlezon, W. A. (2006). The mesolimbic dopamine reward circuit in depression. *Biological Psychiatry*, *59*(12), 1151–1159. https://doi.org/10.1016/j.biopsych.2005.09.018.

Nilges, M., Clore, G. M., & Gronenborn, A. M. (1988). Determination of three-dimensional structures of proteins from interproton distance data by dynamical simulated annealing from a random array of atoms. Circumventing problems associated with folding. *FEBS Letters*, *239*(1), 129–136.

Nygaard, R., Zou, Y. Z., Dror, R. O., Mildorf, T. J., Arlow, D. H., Manglik, A., et al. (2013). The dynamic process of beta(2)-adrenergic receptor activation. *Cell*, *152*(3), 532–542. https://doi.org/10.1016/j.cell.2013.01.008.

O'Connor, C., White, K. L., Doncescu, N., Didenko, T., Roth, B. L., Czaplicki, G., et al. (2015). NMR structure and dynamics of the agonist dynorphin peptide bound to the human kappa opioid receptor. *Proceedings of the National Academy of Sciences of the United States of America*, *112*(38), 11852–11857. https://doi.org/10.1073/pnas.1510117112.

Owens, C. E., & Akil, H. (2002). Determinants of ligand selectivity at the kappa-receptor based on the structure of the orphanin FQ receptor. *Journal of Pharmacology and Experimental Therapeutics*, *300*(3), 992–999.

Palczewski, K., Kumasaka, T., Hori, T., Behnke, C. A., Motoshima, H., Fox, B. A., et al. (2000). Crystal structure of rhodopsin: A G protein-coupled receptor. *Science*, *289*(5480), 739–745. https://doi.org/10.1126/science.289.5480.739.

Pasternak, G. W., Snowman, A. M., & Snyder, S. H. (1975). Selective enhancement of H-3 opiate agonist binding by divalent-cations. *Molecular Pharmacology*, *11*(6), 735–744.

Paterlini, G. L., Portoghese, P. S., & Ferguson, D. M. (1997). Molecular simulation of dynorphin A-(1-10) binding to extracellular loop 2 of the kappa-opioid receptor. A model for receptor activation. *Journal of Medicinal Chemistry*, *40*(20), 3254–3262.

Patra, M. C., Kumar, K., Pasha, S., & Chopra, M. (2012). Comparative modeling of human kappa opioid receptor and docking analysis with the peptide YFa. *Journal of Molecular Graphics and Modelling*, *33*, 44–51.

Pattinson, K. T. S. (2008). Opioids and the control of respiration. *British Journal of Anaesthesia*, *100*(6), 747–758. https://doi.org/10.1093/bja/aen094.

Pickel, V. M., Chan, J., & Sesack, S. R. (1993). Cellular substrates for interactions between dynorphin terminals and dopamine dendrites in rat ventral tegmental area and substantia nigra. *Brain Research*, *602*, 275–289.

Pickel, V. M., Chan, J., Veznedaroglu, E., & Milner, T. A. (1995). Neuropeptide Y and dynorphin-immunoreactive large dense-core vesicles are strategically localized for presynaptic modulation in the hippocampal formation and substantia nigra. *Synapse*, *19*, 160–169.

Prather, P. L., McGinn, T. M., Claude, P. A., Liu-Chen, L. Y., Loh, H. H., & Law, P. Y. (1995). Properties of a kappa-opioid receptor expressed in CHO cells: Interaction with multiple G-proteins is not specific for any individual subunit and is similar to that of other opioid receptors. *Molecular Brain Research*, *29*, 336–346.

Pucadyil, T. J., & Chattopadhyay, A. (2006). Role of cholesterol in the function and organization of G-protein coupled receptors. *Progress in Lipid Research*, *45*(4), 295–333. https://doi.org/10.1016/j.plipres.2006.02.002.

Raman, M., Chen, W., & Cobb, M. H. (2007). Differential regulation and properties of MAPKs. *Oncogene*, *26*(22), 3100–3112. https://doi.org/10.1038/sj.onc.1210392.

Ramos-Colon, C. N., Lee, Y. S., Remesic, M., Hall, S. M., LaVigne, J., Davis, P., et al. (2016). Structure-activity relationships of [des-Arg(7)]dynorphin A analogues at the kappa opioid receptor. *Journal of Medicinal Chemistry*, *59*(22), 10291–10298. https://doi.org/10.1021/acs.jmedchem.6b01411.

Rasmussen, S. G. F., Choi, H. J., Fung, J. J., Pardon, E., Casarosa, P., Chae, P. S., et al. (2011). Structure of a nanobody-stabilized active state of the beta(2) adrenoceptor. *Nature*, *469*(7329), 175–180. https://doi.org/10.1038/nature09648.

Rasmussen, S. G. F., DeVree, B. T., Zou, Y. Z., Kruse, A. C., Chung, K. Y., Kobilka, T. S., et al. (2011). Crystal structure of the beta(2) adrenergic receptor-Gs protein complex. *Nature*, *477*(7366), 549–555. https://doi.org/10.1038/nature10361.

Raynor, K., Kong, H. Y., Chen, Y., Yasuda, K., Yu, L., Bell, G. I., et al. (1994). Pharmacological characterization of the cloned kappa-opioid, delta-opioid, and mu-opioid receptors. *Molecular Pharmacology*, *45*(2), 330–334.

Renugopalakrishnan, V., Rapaka, R. S., Huang, S. G., Moore, S., & Hutson, T. B. (1988). Dynorphin-A (1-13) peptide NH groups are solvent exposed—FT-IR and 500 MHz 1H NMR spectroscopic evidence. *Biochemical and Biophysical Research Communications*, *151*(3), 1220–1225. https://doi.org/10.1016/S0006-291x(88)80496-0.

Rhim, H., & Miller, R. J. (1994). Opioid receptors modulate diverse types of calcium channels in the nucleus tractus solitarius of the rat. *The Journal of Neuroscience*, *14*, 7608–7615.

Ribeiro, S. C., Kennedy, S. E., Smith, Y. R., Stohler, C. S., & Zubieta, J. K. (2005). Interface of physical and emotional stress regulation through the endogenous opioid system and μ-opioid receptors. *Progress in Neuro-Psychopharmacology and Biological Psychiatry*, *29*(8), 1264–1280. https://doi.org/10.1016/j.pnpbp.2005.08.011.

Rosenbaum, D. M., Cherezov, V., Hanson, M. A., Rasmussen, S. G. F., Thian, F. S., Kobilka, T. S., et al. (2007). GPCR engineering yields high-resolution structural insights into beta(2)-adrenergic receptor function. *Science*, *318*(5854), 1266–1273. https://doi.org/10.1126/science.1150609.

Rosenbaum, D. M., Rasmussen, S. G. F., & Kobilka, B. K. (2009). The structure and function of G-protein-coupled receptors. *Nature*, *459*(7245), 356–363. https://doi.org/10.1038/nature08144.

Rusin, L. I., Giovannucci, D. R., Stuenkel, E. L., & Moises, H. C. (1997). Kappa-opioid receptor activation modulates Ca21 currents and secretion in isolated neuroendocrine nerve terminals. *The Journal of Neuroscience*, *17*, 6565–6574.

Sadja, R., Alagem, N., & Reuveny, E. (2003). Gating of GIRK channels: Minireview details of an intricate, membrane-delimited signaling complex. *Neuron*, *39*, 9–12.

Salomon-Ferrer, R., Case, D. A., & Walker, R. C. (2013). An overview of the Amber biomolecular simulation package. *WIREs Computational Molecular Science*, *3*, 198–210.

Salomon-Ferrer, R., Götz, A. W., Poole, D., Le Grand, S., & Walker, R. C. (2013). Routine microsecond molecular dynamics simulations with AMBER on GPUs. 2. Explicit solvent particle mesh Ewald. *Journal of Chemical Theory and Computation*, *9*, 3878–3888.

Sankararamakrishnan, R., & Weinstein, H. (2000). Molecular dynamics simulations predict a tilted orientation for the helical region of dynorphin A(1-17) in dimyristoylphosphatidylcholine bilayers. *Biophysical Journal*, *79*(5), 2331–2344.

Sargent, D. F., & Schwyzer, R. (1986). Membrane lipid phase as catalyst for peptide-receptor interactions. *Proceedings of the National Academy of Sciences of the United States of America*, *83*(16), 5774–5778.

Schneider, S. P., Eckert, W. A., & Light, A. R. (1998). Opioid-activated postsynaptic, inward rectifying potassium currentsin whole cell recordings in substantia gelatinosa neurons. *Journal of Neurophysiology*, *80*, 2954–2962.

Schwyzer, R. (1977). ACTH: A short introductory review. *Annals of the New York Academy of Sciences*, *297*, 3–26.

Seki, T., Minami, M., Nakagawa, T., Ienaga, Y., Morisada, A., & Satoh, M. (1998). DAMGO recognizes four residues in the third extracellular loop to discriminate between mu- and kappa-opioid receptors. *European Journal of Pharmacology*, *350*(2–3), 301–310.

Shang, Y., & Filizola, M. (2015). Opioid receptors: Structural and mechanistic insights into pharmacology and signaling. *European Journal of Pharmacology*, *763*, 206–213. https://doi.org/10.1016/j.ejphar.2015.05.012.

Sharma, S. K., Jones, R. M., Metzger, T. G., Ferguson, D. M., & Portoghese, P. S. (2001). Transformation of a kappa-opioid receptor antagonist to a kappa-agonist by transfer of a guanidinium group from the 5'- to 6'-position of naltrindole. *Journal of Medicinal Chemistry*, *44*(13), 2073–2079.

Simonin, F., Gavériaux-Ruff, C., Befort, K., Mathes, H., Lannes, B., Micheletti, G., et al. (1995). Kappa-opioid receptor in humans: cDNA and genomic cloning, chromosomal assignment, functional expression, pharmacology, and expression pattern in the central nervous system. *Proceedings of the National Academy of Sciences of the United States of America, 92*, 7006–7010.

Smeets, C. J. L. M., Zmorzyńska, J., Melo, M. N., Stargardt, A., Dooley, C., Bakalkin, G., et al. (2016). Altered secondary structure of Dynorphin A associates with loss of opioid signalling and NMDA-mediated excitotoxicity in SCA23. *Human Molecular Genetics, 25*(13), 2728–2737.

Sounier, R., Mas, C., Steyaert, J., Laeremans, T., Manglik, A., Huang, W. J., et al. (2015). Propagation of conformational changes during mu-opioid receptor activation. *Nature, 524*(7565), 375. https://doi.org/10.1038/nature14680.

Spadaccini, R., Crescenzi, O., Picone, D., Tancredi, T., & Temussi, P. A. (1999). Solution structure of dynorphin A (1-17): A NMR study in a cryoprotective solvent mixture at 278 K. *Journal of Peptide Science, 5*(7), 306–312. https://doi.org/10.1002/(Sici)1099-1387(199907)5:7<306::Aid-Psc199>3.0.Co;2-B.

Spanagel, R., Herz, A., & Shippenberg, T. S. (1992). Opposing tonically active endogenous opioid systems modulate the mesolimbic dopaminergic pathway. *Proceedings of the National Academy of Sciences of the United States of America, 89*, 2046–2050.

Stevens, R. C., Cherezov, V., Katritch, V., Abagyan, R., Kuhn, P., Rosen, H., et al. (2013). The GPCR network: A large-scale collaboration to determine human GPCR structure and function. *Nature Reviews Drug Discovery, 12*(1), 25–34. https://doi.org/10.1038/nrd3859.

Szarka, L. A., Camilleri, M., Burton, D., Fox, J. C., McKinzie, S., Stanislav, T., et al. (2007). Efficacy of on-demand asimadoline, a peripheral kappa-opioid agonist, in females with irritable bowel syndrome. *Clinical Gastroenterology and Hepatology, 5*(11), 1268–1275. https://doi.org/10.1016/j.cgh.2007.07.011.

Talmont, F., Sidobre, S., Demange, P., Milon, A., & Emorine, L. J. (1996). Expression and pharmacological characterization of the human mu-opioid receptor in the methylotrophic yeast *Pichia pastoris*. *FEBS Letters, 394*(3), 268–272.

Tate, C. G. (2012). A crystal clear solution for determining G-protein-coupled receptor structures. *Trends in Biochemical Sciences, 37*(9), 343–352. https://doi.org/10.1016/j.tibs.2012.06.003.

Taussig, R., Iniguez-Lluhi, J. A., & Gilman, A. G. (1993). Inhibition of adenylyl cyclase by Gi alpha. *Science, 261*, 218–221.

Tavani, A., Petrillo, P., La Regina, A., & Sbacchi, M. (1990). Role of peripheral mu, delta, and kappa opioid receptors in opioid-induced inhibition of gastrointestinal transit in rats. *The Journal of Pharmacology and Experimental Therapeutics, 254*, 91–97.

Tessmer, M. R., Meyer, J.-P., Hruby, V. J., & Kallick, D. A. (1997). Structural model of a cyclic dynorphin A analog bound to dodecylphosphocholine micelles by NMR and restrained molecular dynamics. *Journal of Medicinal Chemistry, 40*(14), 2148–2155.

Thompson, A. A., Liu, W., Chun, E., Katritch, V., Wu, H. X., Vardy, E., et al. (2012). Structure of the nociceptin/orphanin FQ receptor in complex with a peptide mimetic. *Nature, 485*(7398), 395–U150. https://doi.org/10.1038/nature11085.

Toll, L., Berzetei-Gurske, I. P., Polgar, W. E., Brandt, S. R., Adapa, I. D., Rodriguez, L., et al. (1998). Standard binding and functional assays related to medications development division testing for potential cocaine and opiate narcotic treatment medications. *NIDA Research Monograph, 178*, 440–466.

Trott, O., & Olson, A. J. (2009). AutoDock Vina: Improving the speed and accuracy of docking with a new scoring function, efficient optimization, and multithreading. *Journal of Computational Chemistry, 31*(2), 455–461.

Vanderah, T. W. (2010). Delta and kappa opioid receptors as suitable drug targets for pain. *The Clinical Journal of Pain*, *26*, 10–15.

Vardy, E., Mosier, P. D., Frankowski, K. J., Wu, H. X., Katritch, V., Westkaemper, R. B., et al. (2013). Chemotype-selective modes of action of kappa-opioid receptor agonists. *Journal of Biological Chemistry*, *288*(48), 34470–34483. https://doi.org/10.1074/jbc.M113.515668.

Wacker, D., Stevens, R. C., & Roth, B. L. (2017). How ligands illuminate GPCR molecular pharmacology. *Cell*, *170*(3), 414–427. https://doi.org/10.1016/j.cell.2017.07.009.

Wagner, J. J., Evans, C. J., & Chavkin, C. (1991). Focal stimulation of the mossy fibers releases endogenous dynorphins that bind kappa-opioid receptors in guinea pig hippocampus. *Journal of Neurochemistry*, *57*, 333–343.

Wagner, J. J., Terman, G. W., & Chavkin, C. (1993). Endogenous dynorphins inhibit excitatory neurotransmission and block LTP induction in the hippocampus. *Nature*, *363*, 451–454.

Waldhoer, M., Bartlett, S. E., & Whistler, J. L. (2004). Opioid receptors. *Annual Review of Biochemistry*, *73*(1), 953–990. https://doi.org/10.1146/annurev.biochem.73.011303.073940.

Wan, X., Huang, X.-Q., Zhou, D.-H., Jiang, H.-L., Feng, Y.-P., Chen, K.-X., et al. (2000). Molecular modeling of kappa opioid receptor-Gi protein recognition. *Protein and Peptide Letters*, 7, 403–409.

Wang, J. B., Johnson, P. S., Wu, J. M., Wang, W. F., & Uhl, G. R. (1994). Human kappa opiate receptor second extracellular loop elevates dynorphin's affinity for human mu/kappa chimeras. *The Journal of Biological Chemistry*, *269*(42), 25966–25969.

Wang, D., Sun, X., & Sadee, W. (2007). Different effects of opioid antagonists on mu, delta and kappa opioid receptors with and without agonist pretreatment. *The Journal of Pharmacology and Experimental Therapeutics*, *321*, 544–552.

Weber, E., Evans, C. J., & Barchas, J. D. (1982). Prodominance of the amino-terminal octapeptide fragment of dynorphin in rat brain regions. *Nature*, *299*, 77–79.

Weis, W. I., & Kobilka, B. K. (2018). The molecular basis of G protein-coupled receptor activation. *Annual Review of Biochemistry*, *87*, 897–919. https://doi.org/10.1146/annurev-biochem060614-033910.

Weiss, D. R., Ahn, S., Sassano, M. F., Kleist, A., Zhu, X., Strachan, R., et al. (2013). Conformation guides molecular efficacy in docking screens of activated β-2 adrenergic G protein coupled receptor. *ACS Chemical Biology*, *8*(5), 1018–1026. https://doi.org/10.1021/cb400103f.

White, K. L., Scopton, A. P., Rives, M. L., Bikbulatov, R. V., Polepally, P. R., Brown, P. J., et al. (2013). Identification of novel functionally selective-opioid receptor scaffolds. *Molecular Pharmacology*, *85*(1), 83–90. https://doi.org/10.1124/mol.113.089649.

Whitnall, M. H., Gainer, H., Cox, B. M., & Molineaux, C. J. (1983). Dynorphin-A-(1-8) is contained within vasopressin neurosecretory vesicles in rat pituitary. *Science*, *222*, 1137–1139.

Wu, F., Song, G., Graaf, C. d., & Stevens, R. C. (2017). Structure and function of peptide-binding G protein-coupled receptors. *Journal of Molecular Biology*, *429*(17), 2726–2745.

Wu, H. X., Wacker, D., Mileni, M., Katritch, V., Han, G. W., Vardy, E., et al. (2012). Structure of the human kappa-opioid receptor in complex with JDTic. *Nature*, *485*(7398), 327–332. https://doi.org/10.1038/nature10939.

Xu, W., Yoon, S. I., Huang, P., Wang, Y. L., Chen, C. G., Chong, P. L. G., et al. (2006). Localization of the kappa opioid receptor in lipid rafts. *Journal of Pharmacology and Experimental Therapeutics*, *317*(3), 1295–1306. https://doi.org/10.1124/jpet.105.099507.

Yan, F., Mosier, P. D., Westkaemper, R. B., Stewart, J., Zjawiony, J. K., Vortherms, T. A., et al. (2005). Identification of the molecular mechanisms by which the diterpenoid salvinorin A binds to κ-opioid receptors. *Biochemistry*, *44*(24), 8643–8651. https://doi.org/10.1021/bi050490d.

Yasuda, K., Raynor, K., Kong, H., Breder, C. D., Takeda, J., Reisine, T., et al. (1993). Cloning and functional comparison of kappa and delta opioid receptors from mouse brain. *Proceedings of the National Academy of Sciences of the United States of America*, *90*, 6736–6740.

Yongye, A. B., & Martínez-Mayorga, K. (2012). Molecular aspects of opioid receptors and opioid receptor painkillers. In G. Racz (Ed.), *Pain Management—Current Issues and Opinions*. InTech. ISBN: 978-953307-813-7.

Zamir, N., Palkovits, M., Weber, E., Mezey, E., & Brownstein, M. J. (1984). A dynorphinergic pathway of Leu-enkphalin production in rat substantia nigra. *Nature*, *307*, 643–645.

Zamponi, G. W., & Snutch, T. P. (1998). Modulation of voltage-dependent calcium channels by G proteins. *Current Opinion in Neurobiology*, *8*, 351–356.

Zheng, Z., Huang, X.-P., Mangano, T. J., Zou, R., Chen, X., Zaidi, S. A., et al. (2017). Structure-based discovery of new antagonist and biased agonist chemotypes for the kappa opioid receptor. *Journal of Medicinal Chemistry*, *60*(7), 3070–3081. https://doi.org/10.1021/acs.jmedchem.7b00109.

Zhou, X. E., He, Y. Z., de Waal, P. W., Gao, X., Kang, Y. Y., Van Eps, N., et al. (2017). Identification of phosphorylation codes for arrestin recruitment by G protein-coupled receptors. *Cell*, *170*(3), 457. https://doi.org/10.1016/j.cell.2017.07.002.

Zhu, J., Chen, C., Xue, J. C., Kunapuli, S., De Riel, J. K., & Liu-Chen, L. Y. (1995). Cloning of a human kappa opioid receptor from the brain. *Life Sciences*, *56*, 201–207.

Zhu, J., Luo, L. Y., Li, J. G., Chen, C., & Liu-Chen, L. Y. (1997). Activation of the cloned human kappa opioid receptor by agonists enhances [35S]GTPgammaS binding to membranes: Determination of potencies and efficacies of ligands. *The Journal of Pharmacology and Experimental Therapeutics*, *282*, 676–684.

CHAPTER THREE

Molecular aspects of delta opioid receptors

Louis Gendron[a,f,*], Karim Nagi[b], Manel Zeghal[c,d], Patrick M. Giguère[c,d], Graciela Pineyro[e,f]

[a]Department of Pharmacology-Physiology, Université de Sherbrooke, Centre de Recherche du CHUS, Sherbrooke, QC, Canada
[b]College of Medicine, Member of QU Health, Qatar University, Doha, Qatar
[c]Department of Biochemistry, Microbiology and Immunology, University of Ottawa, Ottawa, ON, Canada
[d]Brain and Mind Research Institute, University of Ottawa, Ottawa, ON, Canada.
[e]CHU Sainte-Justine research center, Department of Pharmacology, Faculty of Medicine, University of Montreal, Montreal, QC, Canada
[f]Quebec Pain Research Network, Sherbrooke, QC, Canada
*Corresponding author: e-mail address: louis.gendron@usherbrooke.ca

Contents

Abstract

The delta opioid receptor (DOP) belongs to the Class A, rhodopsin-like family of G protein-coupled receptors. Although this receptor has a high level of similarity with the other opioid receptors, it displays unique aspects and functions. Indeed, as opposed to most membrane receptors, DOP is poorly addressed to the plasma membrane. In this chapter, we first review the molecular and cellular mechanisms regulating the

Vitamins and Hormones, Volume 111
ISSN 0083-6729
https://doi.org/10.1016/bs.vh.2019.06.001

expression and the cellular trafficking/sorting of DOP. We then summarize the structural insights of this receptor through the analysis of the existing crystal structures, with a particular focus on the role of the sodium binding site. Finally, we review the current signaling mechanisms mediating receptor function and desensitization.

1. The delta opioid receptor (DOP)

The opioid receptor subfamily is composed of three members belonging to the G protein-coupled receptors (GPCRs) family, namely, mu, delta, and kappa opioid receptors (MOP, DOP, and KOP, respectively). DOP is a 372 amino acid protein with seven transmembrane spanning domains (7TM) (Evans, Keith, Morrison, Magendzo, & Edwards, 1992; Kieffer, Befort, Gaveriaux-Ruff, & Hirth, 1992). The primary amino acid sequence of DOP is highly conserved (>90%) among most species, including mouse, rat, monkey and human. DOP is encoded by the *Oprd1* gene located to the distal region of chromosome 4 in mice, to chromosome 5 in rats and to the short arm of chromosome 1 in humans (Befort, Mattei, Roeckel, & Kieffer, 1994; Bzdega et al., 1993; Kaufman et al., 1994). The coding region of the DOP gene is composed of 3 exons (Simonin et al., 1994). Until recently, it was considered that only one protein was encoded by the *Oprd1* gene since no alternative splicing of the gene had been described. However, in mice and humans, conserved alternative transcriptional events were recently observed, leading to putative DOP splicing variants including a single TM and a 6TM domains isoforms (Piltonen et al., 2018). Whether or not these truncated proteins have a functional and/or a physiological relevance remains to be determined.

1.1 Regulation of *Oprd1* gene expression

The expression of the DOP gene is highly regulated (Gendron, Cahill, von Zastrow, Schiller, & Pineyro, 2016; Wei & Loh, 2011). In mice, the *Oprd1* gene spans over 32 kb. The promoter region of the gene contains a G+C rich region and a GC-box known to bind members of the Sp1 transcription factor family (Augustin et al., 1995; Liu, Shen, Augustin, Ko, & Loh, 1999; Smirnov, Im, & Loh, 2001). An E-box that binds upstream stimulatory factors (Liu et al., 1999), an Ets-1-binding site overlapping the E-box (Sun & Loh, 2001), and AP-1 and AP-2 transcription factors binding elements (Woltje, Kraus, & Hollt, 2000) have been described upstream of the start codon in various mouse cell lines. In the rat pheochromocytoma PC12 cells, an interaction of the promoter with the transcription factor NF-κB and its partner p300 was also observed (Chen, Law, & Loh, 2006, 2007; Chen, Monteith, Law, & Loh, 2010). In immune cells, the transcription of the mouse *Oprd1* gene is also controlled by Ikaros and Ikaros-2 (Sun & Loh, 2002, 2003). Finally, an interleukin-4 (IL-4) responsive element that binds signal transducer and activator of transcription 6 (STAT6) to control the expression of the DOP gene was also observed in cells of mouse and human origin (Borner, Woltje, Hollt, & Kraus, 2004). DNA methylation of the mouse DOP gene was also reported, suggesting that it is epigenetically regulated (Wang, Liu, Wei, Law, & Loh, 2005; Wang, Wei, & Loh, 2003).

1.2 Regulation of DOP expression

At the protein level, numerous post-translational modifications of DOP have been described. Firstly, palmitoylation of DOP on Cys333 plays an essential role in its membrane expression by promoting its export from the endoplasmic reticulum (ER) (Petaja-Repo et al., 2006). *N*-glycosylation of DOP on residues Asn18 and Asn33 represents an important step for proper folding of the receptor and its exit from the ER (Petaja-Repo, Hogue, Laperriere, Walker, & Bouvier, 2000). Although its role remains unknown, *O*-glycosylation of DOP has also been observed (Petaja-Repo et al., 2000). DOP further possesses a disulfide bond linking Cys121 to Cys198 which likely stabilizes extracellular loops 1 and 2 (Brandt, Golbraikh, Tager, & Lendeckel, 1999). Interestingly, two polymorphisms of DOP have been described (Simonin et al., 1994; Wei & Loh, 2011). Apart from the silent C for T exchange affecting codon 307 (Gly307Gly) (Mayer et al., 1997), a relatively frequent non-synonymous modification (G80T) at position 27 (Phe27Cys) was described (Gelernter & Kranzler, 2000). Such a transversion at position 27 significantly affects the maturation and the ligand-independent trafficking of DOP (Leskela et al., 2012; Leskela, Markkanen, Alahuhta, Tuusa, & Petaja-Repo, 2009; Sarajarvi et al., 2011). Thus, the Cys27 variant (Gelernter & Kranzler, 2000) displays greater precursor retention in the ER and enhanced turnover of mature surface receptors as compared to the Phe27 allele (Leskela et al., 2009).

2. Advances in the resolution of GPCR structures

GPCRs are large allosteric machines in that they dynamically bind extracellular and intracellular ligands or proteins to form complexes that propagate conformationally restricted cellular signals. This reciprocal allosteric system is the basis of the pluridimensionality and heterogeneity of GPCR signaling. GPCRs' structure and function have been probed mainly using biophysical and biochemical approaches. Over the last decade, breakthrough and refinement of protein chemistry, as well as innovative approaches have led to the structural determination of 50 different class A GPCRs (Munk et al., 2019). In spite of their characteristic 7TM architecture, GPCRs share little sequence homology (<20%). The most diverse family of GPCRs (class A, Rhodopsin-like) consists of 701 GPCRs that are divided into subgroups: the α-group, the β-group, the γ-group, and the δ-group. GPCRs can be further subdivided into clustered subfamilies based on the endogenous ligands they share (e.g., serotonin, adrenergic, opioid, dopaminergic, etc) (Katritch, Cherezov, & Stevens, 2013; Munk et al., 2019). The structures of several peptide-binding receptors of the γ-group have been solved, including structures of all three opioid receptors (DOP, MOP, KOP) and the closely related non-opioid NOP receptor. In fact, the sequence index for MOP, DOP, KOP and NOP within the TM domains is 76%, 73%, 74% and 67%, respectively. Opioid receptors—including NOP—are thus the most covered subfamily of receptors with a total of 13 structures, with 10 human receptors and 3 mouse receptors. As of 2019, MOP and KOP have been crystalized in both active and inactive states, while the DOP and NOP have only been reported in the inactive state (Table 1).

2.1 DOP structure insight

The first solved structure of the DOP was generated in 2012 by the group of Kobilka (Granier et al., 2012) using the mouse DOP fused with the T4 lysozyme (T4L) inserted within the third intracellular loop to increase stability and solubility of the purified receptor. The structure was solved bound to naltrindole, a DOP morphinan antagonist (Portoghese, Sultana, Nagase, & Takemori, 1988). The structures of all three opioid receptors show a very typical architecture common to other class A GPCRs.

More recently, two other structures of the human DOP were revealed. The first in complex with the antagonist naltrindole, and the other with the weak partial agonist DIPP-NH2 peptide. While DIPP-NH2 can be considered as antagonist at human DOP in most conditions, it has shown good partial agonist activity at the mouse DOP and weak partial agonist activity at a high expression level at the human DOP (Fenalti et al., 2015). This weak partial activity probably explains the outward movement of the helix-5

Table 1 Solved structures of opioid receptors (DOP, MOP, KOP) and related non-opioid receptor (NOP).

Receptor	Species	PBD code	Resolution	State	Associated signaling protein	Auxiliary protein: Fusion	Auxiliary protein: Antibodies	Ligand: Structure ligand	Ligand: Type	Methods	Na^+ in structure	Reference
MOP	M	6DDF	3.5	Act	Gi	–	–	DAMGO	Ago	Cryo-EM	N	Koehl et al. (2018)
MOP	M	6DDE	3.5	Act	Gi	–	Antibody	DAMGO	Ago	Cryo-EM	N	Koehl et al. (2018)
MOP	M	5C1M	2.1	Act	–	–	Antibody	BU72	Ago	Meso/LCP	N	Huang et al. (2015)
MOP	M	4DKL	2.8	Inact	–	T4-L		BF0	Antag	Meso/LCP	N	Manglik et al. (2012)
DOP	H	4N6H	1.8	Inact	–	BRIL	–	Naltrindole	Antag	Meso/LCP	Y	Fenalti et al. (2014)
DOP	H	4RWD	2.7	Inact	–	BRIL	–	DIPP-NH2	Antag	Meso/LCP	Y	Fenalti et al. (2015)
DOP	H	4RWA	3.3	Inact	–	BRIL	–	DIPP-NH2	Antag	Meso/LCP	N	Fenalti et al. (2015)
DOP	M	4EJ4	3.4	Inact	–	T4-L	–	Naltrindole	Antag	Meso/LCP	N	Granier et al. (2012)

Continued

Table 1 Solved structures of opioid receptors (DOP, MOP, KOP) and related non-opioid receptor (NOP).—cont'd

Receptor	Species	PBD code	Resolution	State	Associated signaling protein	Auxiliary protein		Ligand		Methods	Na^+ in structure	Reference
						Fusion	Antibodies	Structure ligand	Type			
KOP	H	6B73	3.1	Act	–	cyt b562	Nanobody	MP1104	Ago	Meso/LCP	N	Che et al. (2018)
KOP	H	4DJH	2.9	Inact	–	T4-L	–	JDTic	Antag	Meso/LCP	N	Wu et al. (2012)
NOP	H	5DHG	3	Inact	–	BRIL	–	N/A	N/A	Meso/LCP	N	Miller et al. (2015)
NOP	H	5DHH	3	Inact	–	BRIL	–	N/A	N/A	Meso/LCP	N	Miller et al. (2015)
NOP	H	4EA3	3	Inact	–	BRIL	–	C-24	Antag	Meso/LCP	N	Thompson et al. (2012)

Abbreviations: M: mouse; H: human; Act: active; Inact: inactive; Ago: agonist; Antag: antagonist; T4-L: T4-lysozyme; cyt b562: cytochrome *b*562.

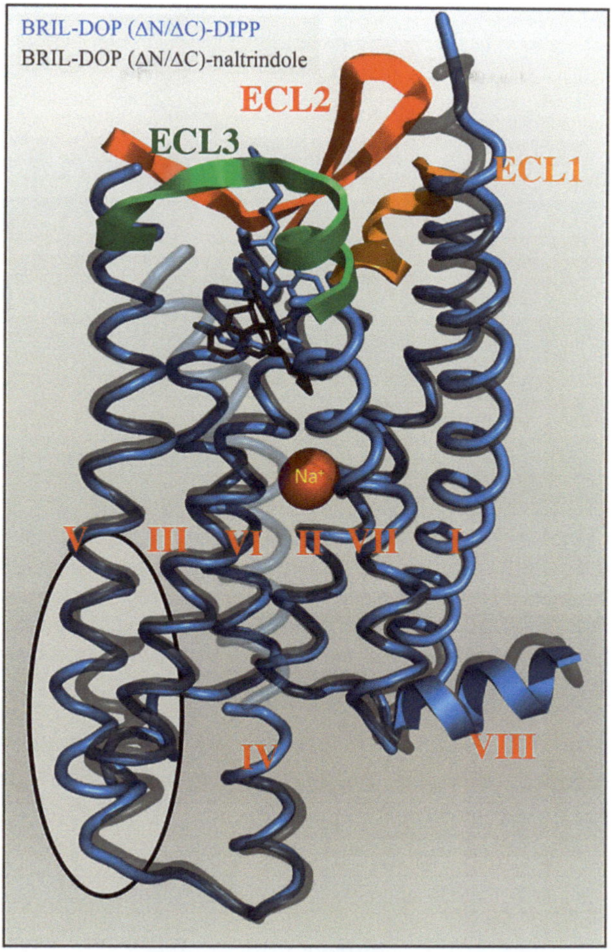

Fig. 1 Overall similarity between DOP-naltrindole and DOP-DIPP structures. BRIL-DOP (ΔN/ΔC) structure in complex with naltrindole (PDB: 4N6H) is shown in gray and DIPP (PDB: 4RWD) is shown in blue. Both structures were aligned to highlight the overall structure similarity and the most divergent change at the helix V caused by the weak partial agonist activity of the DIPP peptide, which is circled in black. (Note: N-terminal fused BRIL was removed.)

observed in this complex structure compared to the naltrindole structure (Fig. 1). Despite low sequence homology within the ECL2, all three receptors possess a conserved β-strand fold. The ECL3, a well-known molecular determinant of ligand selectivity, is not conserved at the sequential and structural levels. It also seems to be quite flexible, as poor resolution is obtained for the MOP and DOP, and no electronic density was revealed from the

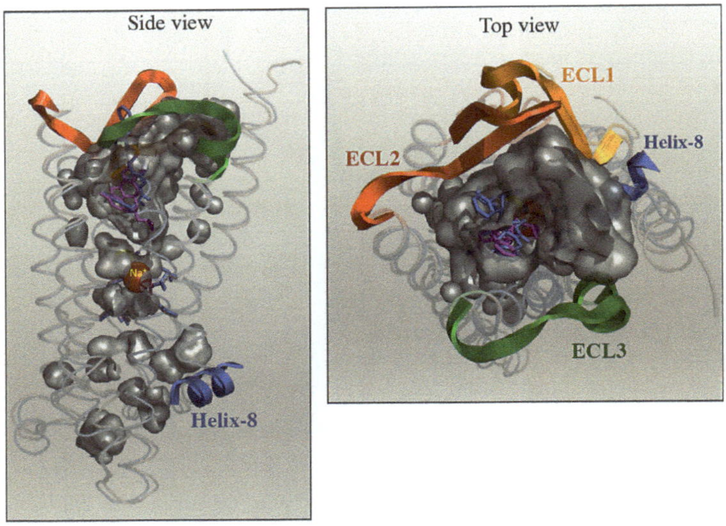

Fig. 2 Ligand-binding pocket of the DOP. BRIL-DOP (ΔN/ΔC) structure in complex with naltrindole (PDB: 4N6H) is shown aligned with the structure complex with DIPP (PDB: 4RWD). Surface cavity on the side view and top view of the receptor highlights the large binding pocket surface exposed by the three ECL domains: ECL1 (orange), ECL2 (red), ECL3 (green). The membrane-bound amphipathic helix-8 from the C-terminal tail is shown in blue. RMSD=0.922804.

KOP structure. For all opioid-related receptors, the arrangement of the extracellular loop, especially the ECL2, exposes a wide-open binding pocket, a characteristic of most peptide-binding class A GPCRs (Fig. 2).

2.1.1 The sodium cavity

A new major insight into the control of the DOP signaling behavior comes from the discovery of a sodium-dependent allosteric regulation site (Fenalti et al., 2014). This work highlights the presence of a cluster of sodium and water molecules housed in a cavity thought to only be present in the inactive conformation of the receptor. The crystallographic structure was used to specifically design point mutations that disrupt the sodium interaction. The critical involvement of the sodium acting as an allosteric modulator was then demonstrated. It was found that sodium stabilizes the inactive conformation of the receptor and the amino acids coordinating the sodium ions act as an efficacy switch, mostly at the arrestin pathway. This activity is believed to be related to the stabilization of the TM7 in an outward position in the presence of the sodium ion. Release of the sodium ion is believed to unleash TM7 in an inward movement, allowing β-arrestin recruitment at the receptor or stabilization of a high affinity state. Sodium in an identical position has also been found in inactive-state structure of the $A_{2A}AR$ (Liu et al., 2012), β1AR (Miller-Gallacher et al., 2014) and PAR1 (Zhang et al., 2012), receptors which belong to different branches of class A and thus supporting the common nature of Na^+ mechanisms in GPCRs. This cavity is built up from 16 of the 34 most conserved residues found in the majority of non-olfactory class A GPCRs and embraces important conserved motifs such at the CWxP and the NPxxY, highlighting its functional importance (Katritch et al., 2014; Nygaard, Frimurer, Holst, Rosenkilde, & Schwartz, 2009).

Sequence analysis indicate that this polar pocket is exceptionally well conserved in class A receptors (Fig. 3). The $D^{2.50}$ is conserved in 90% of class A receptors, $N^{1.50}$ (97%), $S^{3.39}$ (75%), $N^{7.45}$ (70% and 90% as polar side chain), $S^{7.46}$ (66% and 75% as any polar sidechain), and $N^{7.49}$ (75%). Among the 16 residues lining the cavity, only one at position 3.35 is less conserved with only 27% of non-olfactory GPCRs having an asparagine, including opioid receptors. This $N^{3.35}$ residue seems to have an important role within opioid receptors, especially in the DOP, as its mutation to alanine or valine generates a constitutively active receptor for the β-arrestin recruitment and seems to weakly couple to G protein (Fenalti et al., 2014). On the other end, mutation of the $D^{2.50}$, another sodium-cradling residue, has very

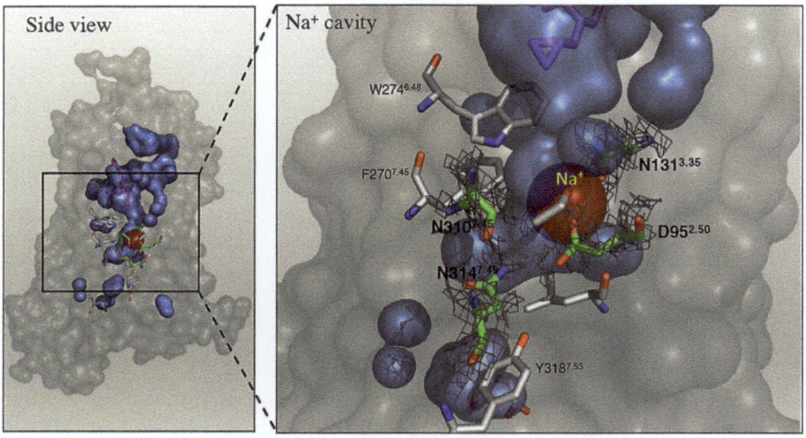

Fig. 3 Sodium cavity of the DOP. BRIL-DOP (ΔN/ΔC) structure in complex with naltrindole (PDB: 4N6H) is shown. Surface cavity of the allosteric Na^+-pocket and the orthosteric ligand binding is shown in blue, highlighting the close proximity of both cavities. Four of the five residues coordinating the Na^+ ions are shown in green, while the two residues of the CW274xP and N314PxxY318 motif are shown in gray.

divergent effects at different GPCRs, including opioid receptors. Mutation to alanine leads to a partially active DOP and converts morphinan antagonists such as naltrindole and naltriben to agonists (efficacy switch). Not as strong as for the DOP, a similar effect is also observed for the MOP. It also has to be noted that most mutations of sodium-coordinating residues within the DOP (D2.50, S3.39, N7.45, N7.49) also act as an efficacy switch converting antagonists to partial agonists (naltrindole/naltriben) (Fenalti et al., 2014; Sun et al., 2017).

Upon interaction of the agonist, changes in the orientation of critical residues coordinating the sodium ion and water molecules reduce affinity of said ion, leading to its release from the pocket. Based on the active structure of the MOP receptor (Huang et al., 2015), the $N^{3.35}$ residue undergoes an important rotation by about 7 Å comparable to rotameric toggle switch $NPxxY^{7.53}$, while minor conformational changes are observed for other residues crafting the pocket (Fig. 4).

It is still not clear what is happening with this ion once released. Movement toward the extracellular compartment is less probable, given the presence of the bound agonist that blocks this path. It is thus conceivable that the sodium ion moves toward the intracellular compartment (Sun et al., 2017).

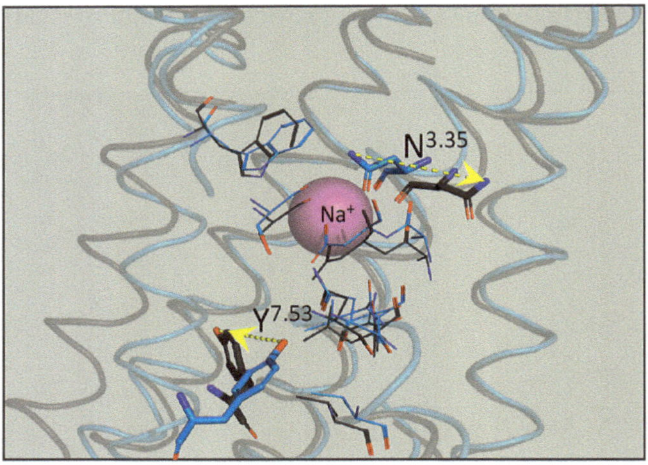

Fig. 4 Rotation of a critical residue coordinating Na^+ during receptor activation. Human DOP structure in complex with naltrindole (PDB: 4N6H) (blue) was aligned with active structure of the mouse MOP bound to the agonist BU72 (PDB: 5C1M) (gray). The Na^+ ion from the inactive DOP is shown (purple), as well as residues crafting the Na^+ cavity for both receptors are represented by sticks. Major rotameric movement (~7 Å) of the residue $N^{3.35}$—coordinating the Na^+ in the inactive state structure—is highlighted and compared to the toggle switch $NPxxY^{7.53}$ (yellow arrow).

While based on sequence homology and other crystal structures the sodium pocket might be present in many GPCRs, the shape of this allosteric cavity seems to have important structural variation (Katritch et al., 2014). The suggested functional importance of the allosteric pocket in receptor activation, and its high conservation in GPCRs, make it an attractive target for discovery of small molecules with unique pharmacological properties. This low degree of structural homology makes this site optimal for the design of new allosteric modulators that would bind non-favorable druggable receptors such as chemokine receptors, or receptors within the same group (binding the same ligand) such as aminergic GPCRs (Giguere, Kroeze, & Roth, 2014; Katritch et al., 2013, 2014). The proximity of this cavity to the orthosteric site opens the possibility to create bitopic ligands that can act on both sites simultaneously (Valant, Sexton, & Christopoulos, 2009). A recent structure published by Hori et al. revealed the action of a sodium-mimicking ligand stabilizing the inactive state of the leukotriene B4 receptor BLT1, supporting this principle. The benzamidine substitution of the ligand BIIL260 extends directly within the sodium cavity, interacting with known sodium-coordinating residues (Hori et al., 2018) (Fig. 5). The difficulty of generating selective opioid ligands could thus benefit from bitopic ligands.

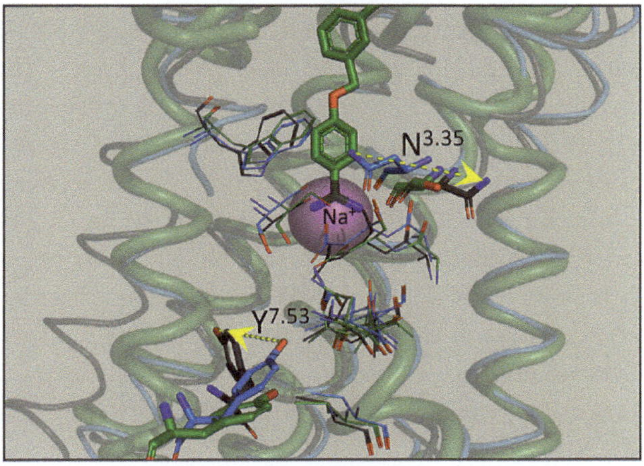

Fig. 5 Potential development of bitopic allosteric ligand. Human DOP structure in complex with naltrindole (PDB: 4N6H) (blue) was aligned with the inactive structure of human BLT1 bound to the ligand BIIL260 (PDB: 5X33) (green). The Na^+ ion from the inactive DOP is shown (purple) as well as some residues crafting the Na^+ cavity for both receptors are shown as lines. BLT1 receptor has a serine in place of $N^{3.35}$ in DOP. The benzamine moiety of BIIL260 in BLT1 structure is located in lieu of the Na^+ in the DOP structure, mimicking the Na^+ ion to stabilize the inactive conformation.

2.1.2 Biased agonism

An emerging concept in drug pharmacology has shown some promise for biased agonists, including for the opioid system. As of 2019, four G protein-biased MOP agonists have been described including TRV130 (DeWire et al., 2013), PZM21 (Manglik et al., 2016), mitragynine pseudoindoxyl (Váradi et al., 2016) and SR-17018 (Schmid et al., 2017). While those biased agonists were found to display good analgesia and reduced respiratory depression and constipation, they all retain opioid-like abuse potential (Negus & Freeman, 2018). Those ligands were mainly developed based on their selectivity to MOP receptor and their biased signaling toward G proteins. This highlights the fact that while biased agonism is a powerful and promising tool to manipulate receptor pharmacological output and reduces side effects, other approaches have to be developed. For instance, 3-Iodobenzoyl naltrexamine (IBNtxA) is a potent non-selective agonist interacting with all three opioid receptors (KOP, DOP, MOP) with potent analgesic response while conferring reduced side effects, including physical dependence and reinforcing behavior. Recently, the active state structure of the KOP bound to MP1104, a close analog to IBNTxA, was published (Che et al., 2018). IBNTxA is a high affinity ligand with balanced signaling at KOP, if compared to Salvinorin A (arrestin-biased if compared to Dynorphin-A), and G protein-biased at the DOP and MOP receptors when compared to DADLE and DAMGO, respectively. Interestingly, MP1104 is balanced at MOP, suggesting that the substituted hydroxyl group at position C14 found in IBNTxA mediated this effect. Molecular docking revealed that position 7.35 could be a key determinant for this discrepancy (Che et al., 2018). This position is occupied by a Trp in MOP, Tyr in KOP and Leu in DOP and NOP. This region at the tip of helix-7 is well-described for its role in ligand selectivity, as well as for functional selectivity at the top of helix-6 (position 6.58). This region seems to have an important role in controlling β-arrestin-recruitment efficacy by regulating helix-6 rotation and helix-7 inward movement (Cheng et al., 2016).

Recent advances in structural biology of opioid receptors are thus opening the door for the rational design of safer and better narcotic analgesics. Their development is slowed down due to the missing pharmacological signature needed to achieve this goal, but further characterization of opioid receptor biology and physiology, and advances in pain neuronal circuitry holds promise of a bright future.

2.2 Importance of highly conserved amino acids

Other than the previously described highly conserved amino acid residues found within each transmembrane domains of DOP, one should note the presence of a DRY motif (Asp145-Arg146-Tyr147), located at the intracellular end of the DOP TM3, the CWxP (Cys273-Trp274-Ala275-Pro276), located within TM6, and the $NP(x)_2Y(x)_6F$ (Asn314-Pro315-$(x)_2$-Tyr318-$(x)_6$-Phe325), at the end of TM7. The role of the DRY motif and the impact of a mutation within the motif is receptor-dependent (Rovati, Capra, & Neubig, 2007). Indeed, mutation of the aspartate residue alters the activity and the G protein coupling of MOP (Li et al., 2001). However, the impact of a mutation within the DRY motif of DOP has not been reported. Similarly, the role of the CWxP motif has not been specifically studied in DOP. In other class A GPCRs, the role of the highly conserved cysteine residue is unclear but is thought to participate in the rearrangement of the TM6 and TM7 interface following activation of the receptor (Olivella, Caltabiano, & Cordomi, 2013). Perhaps solving the active state structure of DOP will shed light on the role of this highly conserved residue. The proline residue comprised within this motif would participate in G protein activation by producing a movement of TM6 away from TM3 (Moreira, 2014). The tryptophan residue within this motif is by far the most documented. Referred to as the "tryptophan switch" or the "aromatic lock," its replacement normally impedes the activity of the receptor (Ahuja & Smith, 2009; Holst et al., 2010). Finally, the $NP(x)_2Y(x)_6F$ motif of DOP includes residue Asn314 which, as mentioned above, is part of the conserved polar network inside the receptor and therefore plays an important role in stabilizing the protein. As discussed below, this motif also participates in the agonist-induced receptor internalization and ERK signaling, with phosphorylation of Tyr318 being an essential step in mediating these effects (Kramer et al., 2000).

3. Trafficking of DOP to the plasma membrane

To reach the cell surface, GPCRs need to go through a process of maturation. At term, only properly folded proteins can reach the Golgi complex and undergo essential post-translational modifications (e.g., glycosylation). The Golgi apparatus provides DOP with the necessary chaperone proteins to escort the receptor to the plasma membrane. As opposed to most GPCRs, DOP is mainly targeted to neuronal plasma membranes via the regulated pathway, using specialized secretory vesicles. Indeed, using a wide variety of approaches, it was shown that the majority of DOP are localized to intracellular sites throughout the central and peripheral nervous systems (Arvidsson et al., 1995; Cahill et al., 2001; Cheng, Liu-Chen, & Pickel, 1997; Cheng et al., 1995; Commons, 2003; Commons, Beck, Rudoy, & Van Bockstaele, 2001; Elde et al., 1995; Gendron et al., 2006; Guan et al., 2005; Lucido, Morinville, Gendron, Stroh, & Beaudet, 2005; Pasquini et al., 1992; Petaja-Repo et al., 2000; Petaja-Repo et al., 2002; Petaja-Repo et al., 2006; Wang & Pickel, 2001; Zerari et al., 1994; Zhang, Bao, Arvidsson, Elde, & Hokfelt, 1998). The low density of DOP at the plasma membrane is consistent with the fact that DOP commonly has modest effects in modulating nociception and reward (Cahill, Holdridge, & Morinville, 2007; Pradhan, Befort, Nozaki, Gaveriaux-Ruff, & Kieffer, 2011). It is thus believed that the majority of the receptors are awaiting targeting to the plasma membrane in response to some kind of "physiological stressors."

Evidence that DOP travels through the regulated pathway first came from studies on sensory neurons. DOP was often found to be present in substance P-containing primary afferent neurons (Elde et al., 1995; Guan et al., 2005; He et al., 2011; Wang et al., 2010) and on large dense core vesicles (Arvidsson et al., 1995; Cheng et al., 1995; Zhang et al., 1998; Zhao et al., 2011). However, whether or not DOP is present in substance P-containing sensory neurons remains unclear. Indeed, in DOP-eGFP knock in mice, DOP was not observed in substance P-containing sensory neurons (Scherrer et al., 2009), although DOP immunolabeling was present in substance P neurons within the myenteric plexus of the small intestine (Poole et al., 2011). Adding to the dispute, *in situ* hybridization and single cell PCR studies report different results as of DOP expression in substance P neurons (Scherrer et al., 2009; Wang et al., 2010). Studies also report conflicting results over whether DOP expression exists in calcitonin gene-related peptide (CGRP) sensory neurons (Pettinger, Gigout, Linley, & Gamper, 2013; Scherrer et al., 2009). Whether or not DOP is expressed in peptidergic sensory neurons, other evidence support the fact that DOP are targeted to the plasma membrane via the regulated secretory pathway. More specifically, an increase in DOP at the cell surface was observed after trigeminal sensory neurons were stimulated with the inflammatory mediator bradykinin (Patwardhan et al., 2005), a finding replicated by others in dorsal root sensory neurons transfected with DOP fused with eGFP (Pettinger et al., 2013).

Various reports suggest that the maturation and the membrane trafficking of DOP as well as its enhanced function can be facilitated or augmented. For example, prolonged exposure to morphine or other mu opioid receptor agonists, physiological stressors, as well as learning-related plasticity all promoted DOP insertion into the plasma membrane and/or increased the DOP-mediated effects (reviewed in Gendron et al., 2016) (Table 2). Although the exact molecular and cellular mechanisms involved in the regulation of DOP trafficking remain to be defined, a significant number of partners and proteins interacting with DOP and affecting its trafficking and membrane targeting have been identified (Table 3). Novel approaches such as proximity labeling now allow spatiotemporal identification of novel GPCR partners (Lobingier et al., 2017). Although describing the physiological role and relevance of such protein partners needs further investigations, one could expect significant advance in this area within the next few years.

In addition to a wide variety of protein partners potentially participating in the maturation and trafficking of DOP, it was observed that ligands able to reach and bind intracellular DOP were promoting its transport to the plasma membrane. Such ligands, called pharmacological chaperones, were shown to promote the maturation of DOP within the ER and to increase its cell surface density (Petaja-Repo et al., 2002). Interestingly, the properties of pharmacological chaperones were found to be independent of their intrinsic signaling efficacy as both agonists (e.g., buprenorphine, TAN-67, SNC80) and antagonists (e.g., naltrexone, NTB, NTI) could promote receptor maturation. The exact mechanism through which a ligand can act as a chaperone remains unknown. One can speculate that binding of a ligand stabilizes DOP and facilitates folding, therefore improving its maturation and export.

Another important aspect of GPCRs is their capability in assembling as homo- and/or heteromers. The association between two GPCRs have been shown to change both their pharmacology and signaling properties (Fujita, Gomes, & Devi, 2014, 2015; Gomes et al., 2015). As an example, the GABAB receptor is an heteromer composed of GABAB(1) and GABAB(2) subunits. Their assembly is necessary for surface trafficking, GABA binding, and G-protein coupling (Jones et al., 1998; Kaupmann et al., 1998; Kuner et al., 1999; White et al., 1998). DOP exists as monomers, homomers, as well as heteromers (extensively reviewed in Gendron et al., 2016). Association of DOP with MOP is by far one of the most studied heteromer. DOP-MOP heteromers were shown to display a distinct pharmacological profile (George et al., 2000). Most interestingly, all the criteria

Table 2 Stimuli known to increase cell surface DOP and/or its function.

Stimulus	Target	References
2-MeSADP	P2Y1	Bao et al. (2003)
Bradykinin	B1 and/or B2	Cayla et al. (2012), Patwardhan et al. (2005), Pettinger et al. (2013), and Rowan et al. (2009)
Chronic ethanol exposure	n/d	van Rijn, Brissett, and Whistler (2012)
Chronic inflammation	Various	Cahill, Morinville, Hoffert, O'Donnell, and Beaudet (2003), Cao, Hong, Dray, and Perkins (2001), Fraser, Gaudreau, Clarke, Menard, and Perkins (2000), Gendron et al. (2007), Gendron et al. (2006), and Hurley and Hammond (2000) Hylden, Thomas, Iadarola, Nahin, and Dubner (1991), Lucido et al. (2005), Pradhan, Smith, McGuire, Evans, and Walwyn (2013), and Qiu, Sora, Ren, Uhl, and Dubner (2000)
Deltorphin	DOP	Bao et al. (2003) and Mittal et al. (2013)
Forced swim stress	n/d	Commons (2003)
Hypoxic preconditioning	n/d	Gao et al. (2012)
Morphine	MOP	Bie, Zhu, and Pan (2009), Cahill et al. (2001), Chieng and Christie (2009), Gendron et al. (2006), Hack, Bagley, Chieng, and Christie (2005), Ma, Zhang, Kalyuzhny, and Pan (2006), Morinville et al. (2004), Walwyn et al. (2005), and Zhang and Pan (2012)
NGF	TrkA	Bie et al. (2010)
Predictive learning	n/d	Bertran-Gonzalez, Laurent, Chieng, Christie, and Balleine (2013) and Laurent, Wong, and Balleine (2015)
Repeated foot shock	n/d	Margolis, Mitchell, Hjelmstad, and Fields (2011)
Substance P	TRPV1	Bao et al. (2003), Gendron et al. (2006), and Guan et al. (2005)
Trypsin	PAR-2	Patwardhan et al. (2006)
Water deprivation	n/d	Stein, Hiller, and Simon (1992)

Table 3 Proteins and chaperones interacting with DOP.

Protein name	Role	References
Cofilin	Activation of cofilin increases the targeting of DOP at the membrane	Mittal et al. (2013)
COPI complex	COPI interacts with DOP and decreases its membrane expression	Shiwarski, Crilly, Dates, and Puthenveedu (2019) and St-Louis et al. (2017)
Na^+/H^+ exchange regulatory factor-1 (NHERF-1)	NHERF-1 is a scaffold protein interacting with DOP to increase its cell surface expression	Bie et al. (2010)
Cyclin-dependent kinase 5 (Cdk5)	Cdk5 phosphorylates DOP on Thr161 and increases its membrane expression as well as DOP-mediated analgesia	Beaudry et al. (2015), Chen et al. (2012), and Xie et al. (2009)
SERCA2b	Participates in DOP folding and maturation	Tuusa, Leskela, and Petaja-Repo (2010)
Calnexin	Binds to newly synthesized, incompletely folded DOP	Helenius and Aebi (2004), Leskela et al. (2009), Leskela, Markkanen, Pietila, Tuusa, and Petaja-Repo (2007), and Markkanen and Petaja-Repo (2008)
Receptor transport protein 4 (RTP4)	Regulates MOP-DOP membrane expression	Decaillot, Rozenfeld, Gupta, and Devi (2008)
GPCR-associated sorting protein (GASP)	Retains DOP in the endosomes	He et al. (2013), Marley and von Zastrow (2010), and Whistler et al. (2002)
Chaperonin containing TCP-1 subunit eta (CCT7)	Unknown	Genier et al. (2016)

established to confirm the existence of heteromers, *in vitro* (Ferre et al., 2009; Pin et al., 2007) and *in vivo* (Gomes et al., 2015), have been fulfilled for this pair of receptors. The DOP-MOP complex should thus be considered a distinct target with specific properties, roles, and functions. It is indeed already considered a novel therapeutic target for various conditions and pathologies (Fujita et al., 2015).

4. Molecular mechanisms of DOP signaling

DOP canonically couples to Gαi/o proteins to modulate multiple enzymatic and channel effectors. An overview of these canonical pathways and the *in vivo* responses they induce upon DOP activation is provided below (Fig. 6).

4.1 Enzymatic effectors

4.1.1 Adenylate cyclases

Consequences of cyclase modulation by DOP depend on the type of adenylyl cyclase (ACs) considered. In striatal medium spiny neurons where DOP are co-expressed with AC5 (Mons et al., 1995; Scherrer et al., 2006), stimulation of the receptor inhibits cAMP production and causes locomotor activation (Kim et al., 2006; Olianas, Dedoni, Olianas, & Onali, 2012). In contrast, in medial prefrontal cortex (Olianas et al., 2012) and hippocampal neurons (Yao et al., 2003), stimulation of the receptor induces Gβγ-dependent activation of AC1/AC2/AC4 (Olianas, Dedoni, & Onali, 2013; Ujcikova et al., 2014). One can speculate that resulting increases in cAMP within these limbic structures may contribute to antidepressant actions of DOP agonists by enhancing PKA-dependent plasticity and survival mechanisms (Olianas et al., 2013). In support of this notion is the observation that knockout of hippocampal DOP reduces performance in learning and memory tasks (Pellissier, Pujol, Becker, & Le Merrer, 2018). In addition to these acute effects, persistent DOP activation leads to potentiation of cAMP production, a phenomenon known as cellular tolerance. Mechanism underlying this adaptive response are diverse (see Gendron et al., 2016) and their exact contribution to analgesic tolerance by DOP agonists remains to be elucidated.

4.1.2 Kinase cascades

Like the majority of GPCRs, DOP activates evolutionary-conserved MAPKs that control cellular responses to environmentally stressful stimuli. Thus, DOP activation of the ERK cascade during exposure to hypoxic/metabolic insult upregulates the antiapoptotic factor Bcl2, reduces caspase 3 activity, and increases neuronal survival *in vitro* (Zheng et al., 2012) and *in vivo* (Su, Wang, Zheng, Zhao, & Wang, 2007). DOP may similarly engage the PI3K-Akt pathway to upregulate Bcl2, inactivate BAD, reduce apoptosis, and decrease volume of cerebral infarction (Lv et al., 2017) or

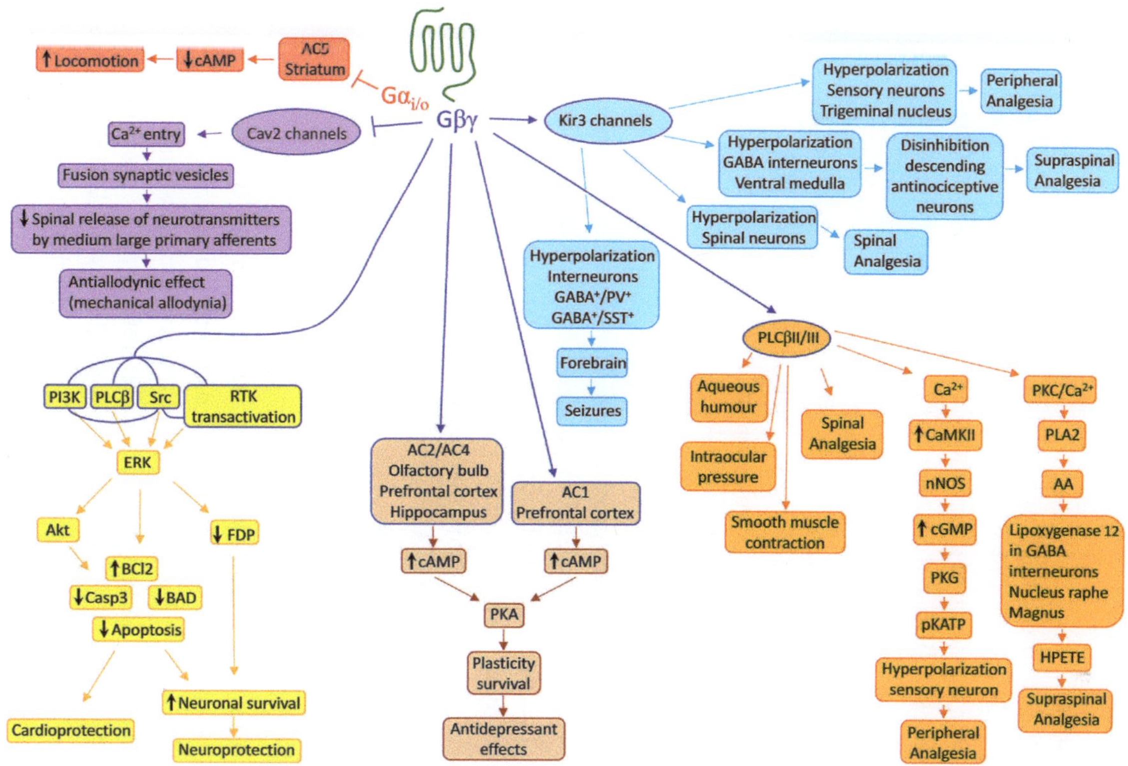

Fig. 6 Molecular mechanisms of DOP signaling. Abbreviations—AC: adenylyl cyclase, cAMP: cyclic adenosine monophosphate, Cav: voltage-dependent calcium channels, PI3K: phosphoinositide 3-kinase, PLCβ: phospholipase Cβ, Src: proto-oncogene tyrosine-protein kinase, RTK: receptor tyrosine kinase, ERK: extracellular signal-regulated kinase, Akt: protein kinase B, FADD, Fas-associated death domain, Bcl2: *B-cell lymphoma 2, Casp3:* Caspase-3, BAD: Bcl-2 associated agonist of cell death, PKA: cAMP-dependent protein kinase A, Kir3: G protein-coupled inward rectifier potassium channels, GABA: gamma-aminobutyric acid, PV: parvalbumin, SST: somatostatin, CaMKII: Ca^{2+}/calmodulin-dependent protein kinase II, nNOS: neuronal nitric oxide synthase, cGMP: cyclic guanosine monophosphate, PKG: cGMP-dependent *protein* kinase G, pKATP: phosphorylated ATP-sensitive potassium channel, PKC: *protein kinase C, PLA2: phospholipase A2, AA*: arachidonic acid, HPETE: hydroperoxyeicosatetraenoic acid.

support cardioprotection (Philipp et al., 2006; Suo, Sun, & Yang, 2014). Molecular mechanisms linking DOP to MAPK activation to cardioprotection/neuroprotection are cell- and species-specific, and were recently reviewed (Gendron et al., 2016).

4.1.3 Phospholipases

Phospholipase C (PLC) hydrolyzes phosphatidylinositol 4,5-bisphosphate to diacylglycerol and inositol triphosphate, leading to Ca^{2+} mobilization from intracellular stores and activation of conventional PKCs. DOP engages Gβγ-sensitive PLCβII and PLCβIII, and through this pathway, they can control aqueous humor dynamics and intraocular pressure (Dortch-Carnes & Potter, 2003), visceral smooth muscle contraction (Murthy & Makhlouf, 1996), as well as peripheral (Gong et al., 2015), spinal (Overland et al., 2009) and supraspinal nociception (Ohsawa, Nagase, & Kamei, 1998). In primary afferents, DOP-dependent stimulation of PLCβ and concomitant Ca^{+2} mobilization stimulates CaMKII and downstream activation of K_{ATP} channels (Fig. 6), leading to hyperpolarization of primary afferents and reduced transmission of painful stimuli to the spinal cord (Gong et al., 2015). In interneurons of the nucleus raphe magnus (NRM), stimulation of PLCβ by DOP agonists induces PKC-dependent activation of *phospholipase A2* (PLA2). The resulting arachidonic acid (AA) is metabolized by 12-lipoxygenase to hydroperoxyeicosatetraenoic acids (HPETEs) which inhibit GABA release, and contribute to supraspinal analgesia by DOP agonists (Zhang & Pan, 2012). Finally, DOP promotes their own internalization via the activation of *phospholipase D2* (Yang et al., 2010) and subsequent phosphorylation of Rab5 effectors (Yang et al., 2010).

4.2 Channel effectors

4.2.1 G protein-coupled inward rectifier potassium channels (GIRK)

GIRK (or Kir3) channels are activated via direct interaction with Gβγ dimers released from heterotrimeric Gi/o proteins. Constitutive association of DOP receptors with channel subunits and heterotrimeric G proteins (Nagi & Pineyro, 2014, 2015) not only allows for rapid and specific activation of the channels, but it also supports the transfer of conformational information codified by the interaction of the ligand with the receptor to the rest of the signaling complex (Richard-Lalonde et al., 2013).

The contribution of Kir3 channels to the analgesic effect of opioids has been demonstrated both in preclinical models and in humans (Nagi & Pineyro, 2014). In the latter, variations in the gene coding for Kir3.2

subunits (*KCNJ6*) influences opioid dose requirements for the management of both acute postoperative pain (Bruehl et al., 2013) and chronic pain. Animal models have proven useful in evaluating different ways in which DOP activation of Kir3 channels may induce analgesia. Thus, injection of the Kir3 channel blocker tertiapin-Q, into the masseter muscle of capsaicin-treated rats prevented analgesia produced by peripheral injection of a DOP agonist, directly linking Kir3 activation in peripheral nociceptors to DOP analgesia (Chung et al., 2014). Analgesic effects induced by intrathecal administration of deltorphin II were also blocked by tertiapin-Q and absent in mice lacking either Kir3.1 or Kir3.2 subunits (Marker, Lujan, Loh, & Wickman, 2005), pointing to a possible spinal component of Kir3 analgesia by DOP agonists. DOP agonists may also induce Kir3 channel currents in GABAergic neurons of the ventral medulla contributing to opioid analgesia by disinhibiting descending antinociceptive neurons in this nucleus (Marinelli et al., 2005; Pedersen, Vaughan, & Christie, 2011).

DOP also modulates Kir3 channel activity in parvalbumin-positive interneurons of the hippocampus (Svoboda, Adams, & Lupica, 1999), and in doing so they enhance pyramidal/granule cells excitability (Drake, Chavkin, & Milner, 2007). Because of this facilitation, activation of DOP on GABAergic neurons may produce seizures that limit the clinical use of DOP agonists as analgesics. The role of forebrain interneurons in the pro-convulsive effects of DOP agonists has been recently certified by the observation that seizures were blunted in mice in which the receptors had been specifically silenced in this particular neuronal population (Chung et al., 2015).

4.2.2 Voltage-dependent Cav2 channels

Voltage-gated Cav2 channels are also DOP effectors. Following activation of the receptor, Gβγ subunits released from Gi/o proteins bind to the intracellular loop that links TM1 and TM2 of Cav2α1 channel subunits. The association shifts channel activation to more depolarized potentials and slows down calcium current activation (Tejedor-Real et al., 1998). Reduction in Ca^{2+} entry decreases fusion of synaptic vesicles with the plasma membrane and inhibits neurotransmitter release from synaptic terminals (Tejedor-Real et al., 1998).

Electrophysiological studies show that activation of endogenously expressed DOP inhibits voltage-gated calcium currents in acutely dissociated (Bardoni et al., 2014) as well as primary DRG neuron cultures (Mittal et al., 2013). Consistent with DOP inhibition of neurotransmitter release by axon terminals, receptor agonists blocked postsynaptic currents evoked in dorsal horn lamina II–III neurons following stimulation of low threshold mechanoreceptors (Bardoni et al., 2014). Use of specific toxins further indicated that DOP may inhibit both Cav2.1 (P/Q-type) and Cav2.2 (N-type) channels on presynaptic terminals of DRG neurons (Acosta & Lopez, 1999; Wu, Chen, & Pan, 2008). Furthermore, consistent with their predominant expression in large diameter neurons, DOP agonists produced greater inhibition of Ca^{2+} channel currents in medium-large than small-size neurons (Acosta & Lopez, 1999; Bardoni et al., 2014). Responsiveness of Cav2 channels to DOP agonists is however typically enhanced by the presence of inflammatory pain (Pradhan et al., 2013) or exposure to the inflammatory mediator bradykinin (Brackley, Gomez, Akopian, Henry, & Jeske, 2016). How responsiveness is enhanced remains to be fully elucidated, but different studies suggest greater DOP expression at the membrane (Gendron, Mittal, Beaudry, & Walwyn, 2015) and disruption of the constitutive association between DOP and GRK2 that exists in peripheral sensory neurons as possible mechanisms (Brackley et al., 2016; Brackley, Sarrami, Gomez, Guerrero, & Jeske, 2017).

5. Molecular mechanisms underlying regulation of DOP signaling

5.1 Kinase-driven desensitization of DOP signaling

Phosphorylation triggers both ligand-dependent and independent desensitization of DOP. In both cases, phosphorylation quickly attains its maximum within 5–10 min of activation of the regulatory kinase (Law et al., 2000), and essentially involves Ser/Thr residues within the C-terminal domain (Kouhen et al., 2000; Law et al., 2000). Out of seven plausible phosphorylation sites in the C-terminal region, Thr358 and Ser363 are actual GRK2 substrates (Guo et al., 2000; Zhang, Wang, Chen, Chen, & Ma, 2008). On the other hand, Thr361, Glu355 and Asp364 function as GRK2/GRK3/GRK5 interaction sites (Guo et al., 2000; Li et al., 2003). In addition, G protein activation and release of $G\beta\gamma$ is required for recruitment of GRK2/3 to agonist-occupied membrane DOP (Li et al., 2003).

The rate of phosphorylation of endogenously expressed DOP parallels the rate of desensitization of DOP responses including AC and Cav2 channel inhibition (Cai, Zhang, Wu, & Pei, 1996; Morikawa et al., 1998). The parallel time course of phosphorylation and desensitization, together with the observations that truncation of the receptor or Ala substitution of all C-terminal Ser/Thr residues reduce the rate and maximal desensitization attained upon agonist exposure (Qiu, Loh, & Law, 2007), support the long-held notion that phosphorylation is essential for rapid loss of receptor responsiveness to agonist. Moreover, since these same mutations also interfere with βarr recruitment to the receptor (Qiu et al., 2007; Zhang et al., 2008), DOP desensitization can be associated to both C-terminal phosphorylation and βarr association to this portion of the receptor (Qiu et al., 2007). Among the different phosphorylation sites targeted by GRKs, Ser363 is essential for rapid desensitization by efficacious agonists both *in vitro* (Kouhen et al., 2000; Law et al., 2000) and *in vivo* (Pradhan et al., 2009). The functional importance of this residue for *in vivo* regulation of DOP responses is exemplified by the fact that only ligands which induce rapid phosphorylation of Ser363 induce analgesic tolerance (Pradhan et al., 2009). Together with GRKs, full agonists also activate the tyrosine kinase Src, which phosphorylates DOP at Tyr318 (NPxxY domain). Tyr phosphorylation contributes to functional desensitization of the receptor (Audet, Paquin-Gobeil, Landry-Paquet, Schiller, & Pineyro, 2005), at least in part by interfering with resensitization and recycling to the membrane (Archer-Lahlou et al., 2009).

Additionally, GRK2 may regulate DOP in a ligand-independent manner. Indeed, in peripheral sensory neurons of the trigeminal ganglia, GRK2 constitutively associates to inactive membrane DOP (Brackley et al., 2016). This interaction is independent of GRK2 kinase activity, but requires phosphorylation of the kinase at Ser685 by AKAP79/150-bound PKA, which ensures that GRK2 phosphorylation occurs in proximity of membrane DOP (Brackley et al., 2017). By directly associating to ligand-free surface DOP, phosphorylated GRK2 is thought to prevent adequate activation of Gβγ signaling and contribute to the analgesic incompetence of DOP agonists in the absence of inflammation (Brackley et al., 2016). Interestingly, inflammatory stimuli like bradykinin, which sequester GRK2 into a complex with Raf kinase inhibitory protein (RKIP), restore DOP signaling and its analgesic competence (Brackley et al., 2016).

DOP may also undergo ligand-independent desensitization by PKCs, whose phosphorylation target Ser344 does not overlap GRK modulation sites (Pei, Kieffer, Lefkowitz, & Freedman, 1995). A diversity of stimuli such as signaling by D2 receptors, α2AR or ligand-gated ion channels such as AMPA or NMDA can desensitize DOP through this mechanism (Gendron et al., 2016).

5.2 RGS-mediated desensitization of DOP signaling

Regulators of G protein signaling (RGS) bind to Gα subunits and enhance their GTPase activity by accelerating the switch from the active GTP-bound to the inactive GDP-bound state of the G protein. RGS4 directly interacts with the first 26 amino acids of the DOP C-terminal tail (Georgoussi et al., 2006) and inhibits cAMP and ERK signaling by the receptor (Leontiadis, Papakonstantinou, & Georgoussi, 2009; Wang, Liu-Chen, & Traynor, 2009). RGS4 is also the main regulator of DOP-mediated *in vivo* responses such as antidepressant effects and analgesia that are typically enhanced in mice that lack RGS4 or express RGS4-resistant forms of the Gαo subunit (Dripps et al., 2018, 2017). On the other hand, seizures induced by DOP activation were not modified in these same transgenic animals (Dripps et al., 2018, 2017), pointing to specificity in RGS4 regulation of DOP signaling.

5.3 Trafficking and relocalization of DOP signaling from the membrane to intracellular compartments

5.3.1 *Internalization and postendocytic trafficking*

Upon recruitment to the receptor, βarr undergoes conformational changes that link DOP to the endocytic machinery, drive the receptor into clathrin-coated pits and promote its subsequent sequestration into endosomal vesicles (Henry, White, Marsh, von Zastrow, & Hislop, 2011; Rosciglione, Theriault, Boily, Paquette, & Lavoie, 2014). Phosphorylated DOP undergoes internalization by both βarr1 and βarr2, while only βarr2 is capable of promoting slow internalization of non-phosphorylated DOP (Charfi et al., 2014; Qiu et al., 2007; Zhang et al., 2008). Considerable evidence has associated DOP internalization with lysosomal delivery and long-lasting signaling desensitization (Scherrer et al., 2006; Tsao & von Zastrow, 2000). The association between internalization and degradation supports the classification of DOP ligands as internalizing and non-internalizing as a means of predicting their potential to induce acute analgesic tolerance (Pradhan et al., 2009). However, DOP are not only "hard wired" for lysosomal delivery. Actually, as much as 60–70% of internalized DOP can recycle to the membrane and continue to signal upon agonist washout (Archer-Lahlou et al., 2009; Tsao & von Zastrow, 2000). DOP recycling is typically favored by short periods of agonist exposure, and depends on the type of agonist bound to the receptor (Audet et al., 2012). For example, if deltorphin II is the ligand promoting internalization, it can be degraded by endosomal endothelin-converting enzyme-2 (ECE2) and ligand-free DOP can recycle back to the membrane (Gupta, Gomes, Wardman, & Devi, 2014). Not all peptidic ligands are ECE2 substrates but the acidic endosomal environment may still promote ligand dissociation (Mellman, Fuchs, & Helenius, 1986) and leave the receptor available for relocalization to the membrane. On the other hand, if internalized DOP are occupied by a synthetic ligand like SNC80, which is not degraded by endosomal convertases and remains tightly bound, the receptor is excluded from recycling and sorted for degradation (Gupta et al., 2014; Pradhan et al., 2009). The stability of DOP-β arr2 interaction is also a factor determining ligand-specific recycling of DOP (Audet et al., 2012). In particular, ligands that induce transient association between DOP, βarr2 and Gβγ support recycling, and those that lock the receptor in a stable DOP-βarr2-Gβγ complex are excluded from this path (Audet et al., 2012). Among highly internalizing ligands, those that favor DOP recycling do not produce acute analgesic tolerance (Audet et al., 2012; Beaudry et al., 2009), while those that fail to support membrane recovery of the receptor induce rapid and sustained tolerance to their antinociceptive actions (Audet et al., 2012; Beaudry et al., 2009; Pradhan et al., 2009). Hence, internalization is informative as a predictor of tolerance, only if considered on relation to postendocytic routing of the internalized receptors.

The postendocytic itinerary followed by DOP depends on their interaction with distinct sorting machineries. One of such machineries is the GPCR-associated binding protein-1 (GASP-1) (Whistler et al., 2002), which binds the carboxy-tail of DOP and prevents recycling from early endosomes while simultaneously linking the receptor to the ESCRT machinery via its interaction with dysbindin (Marley & von Zastrow, 2010). By packaging the receptor into intraluminal vesicles (ILVs) of multivesicular bodies (MVBs), ESCRT targets DOP for lysosomal degradation (Henry et al., 2011; Rosciglione et al., 2014). As DOP progresses from early endosomes to MVBs, they may alternatively remain on the endosomal limiting membrane (Henry et al., 2011). Receptors on the limiting membrane immediately encounter the Rab9/TIP47 complex, while those in ILVs may continue to the lysosome or may regain the limiting membrane by the intervention of the ESCRT accessory protein ALIX, and become once again available for interaction with Rab9/TIP47 complex (Charfi, Abdallah, Gendron, & Pineyro, 2018; Rosciglione et al., 2014). This sorting complex shuttles cargo from late endosomes/MVBs to the TGN

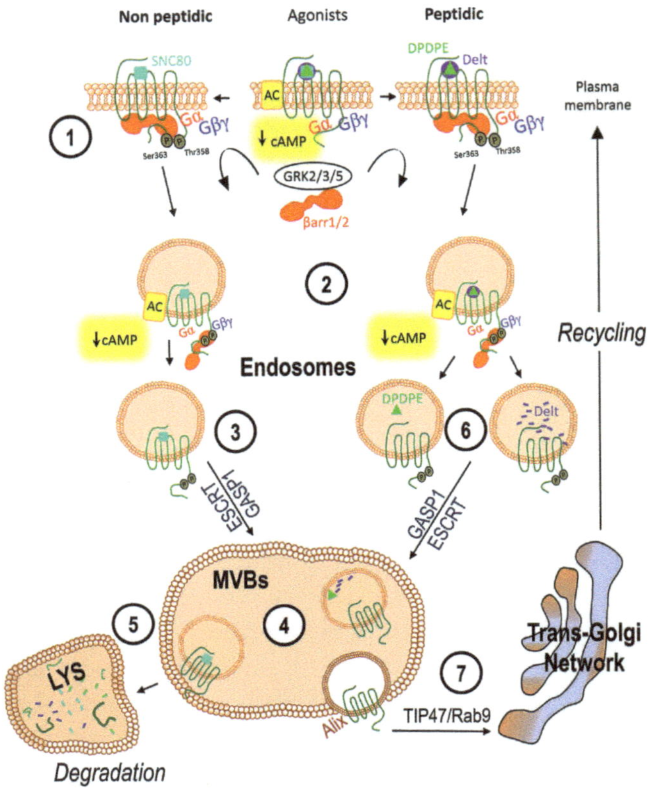

Fig. 7 Trafficking and relocalization of DOP signaling from the membrane to intracellular compartments (proposed series of events; details in text). (1) Ligand-bound DOP recruits βarrs upon phosphorylation by GRKs, and G-protein signaling at the membrane stops when the finger loop of βarr dislodges Gα from the receptor (core conformation). (2) The "tail conformation" of βarr supports receptor internalization and formation of megacomplexes that support signaling by ligand-bound DOP as they progress through the endosomal compartment. (3) Some non-peptidic ligands like SNC-80 remain bound to the receptor as GASP1 and ESCRT machinery sort them to intraluminal vesicles (ILVs) of (4) multivesicular (MVBs) bodies, and from there they enter (5) lysosomes (LYS) for final degradation. (6) A great majority of peptidic ligands dissociate from the receptor at the acidic pH of endosomal compartment, and some of these ligands can be degraded as the receptor enters ILVs. (7) Ligand-free DOP within ILVs regain the limiting membrane by the action of the cytosolic protein ALIX and are sorted to the *trans*-Golgi network (TGN), from where they can regain the membrane for a new signaling cycle.

(Diaz & Pfeffer, 1998; Lombardi et al., 1993), and redirects DOP out of the degradation path and into the *trans*-Golgi network from where they can regain the membrane (Charfi et al., 2018) (Fig. 7). In neurons, this final step is dependent on ROCK-II activity and the organization of actin filaments that direct cargo from the TGN to the membrane. *In vivo,* inhibition of TGN to membrane transport precipitates acute analgesic tolerance following treatment with DPDPE, an agonist that normally supports recycling and displays no acute tolerance (Charfi et al., 2018). Thus, even DOP that have progressed toward degradation can still recycle to the membrane and help maintain analgesic response.

5.3.2 Intracellular signaling

According to the traditional view, GPCRs effectively signal to G proteins only when they are at the plasma membrane. However, cumulative evidence indicates that numerous GPCRs modulate cAMP production after sequestration to endosomal compartments (Lobingier & von Zastrow, 2019). DOP are no exception to this recently described signaling modality, which was evidenced for DOP ligands in immortalized cell lines and in cultured medium spiny neurons (Stoeber et al., 2018). In particular, the use of a biosensor that recognizes DOP in its active state revealed that upon agonist exposure, this conformation first accumulates at the plasma membrane followed by a subsequent wave of active DOP that appears in the endosomal compartment (Stoeber et al., 2018). Moreover, washout of the agonist bound to membrane DOP, or its displacement by an antagonist that does not cross the plasma membrane, does not modify the endosomal wave of receptor activation nor the associated modulation of intracellular cAMP levels (Stoeber et al., 2018). The two discrete signaling surges that successively appear at the membrane and endosomal compartments have been associated with "core" and "tail" modalities of βarr interaction with ligand bound GPCRs (Cahill et al., 2017; Shukla et al., 2014; Thomsen et al., 2016). Indeed, recruitment to surface receptors in the "core" conformation allows βarr to displace Gα from its signaling position and terminate membrane signaling (Shukla et al., 2014). In the tail conformation, βarr sequesters the receptor to the endosome (Cahill et al., 2017), and forms a βarr-receptor-Gβγ megacomplex that leaves Gα subunits free to interact and signal via the receptor's core (Thomsen et al., 2016). DOP ligands may form transient or stable DOP-βarr2-Gβγ complexes (Audet et al., 2012). As long as these complexes remain present in the endosome, they provide the conditions for intracellular inhibition of cAMP production by this agonist (Stoeber et al., 2018; Tudashki, Robertson, Schiller, & Pineyro, 2014).

6. Conclusions

DOP was the first opioid receptor to be cloned and the first in subfamily for which a crystal structure was solved. Despite the fact that there are currently no approved drugs for clinical use, this receptor represents a promising target for the development of novel therapies for pain, as well as for a number of psychiatric disorders such as, among others, anxiety and depression. Recent advances in the field have significantly increased our knowledge of the molecular insights controlling opioid receptor activation and coupling to its effectors. Once a crystalline structure of DOP bound to an agonist (this is to say in its active state) will be solved, important details/changes triggering the activation of the receptor should be revealed. In particular, one could expect that such knowledge will be helpful for the design of—most needed—novel drugs biased toward the desired effects associated with the activation of the receptor.

Acknowledgments

The authors acknowledge the support of the Canadian Institutes of Health Research (CIHR grants # MOP-136871 and PJT-162103 awarded to L.G., MOP-142219 awarded to P.M.G., and MOP-324876 and MOP-311997 awarded to G.P.) and the Natural Sciences and Engineering Research Council of Canada (NSERC grant #RGPIN-2018-06100 awarded to L.G.). L.G. is the recipient of a Senior Salary support from the Fonds de la Recherche Québec—Santé and P.M.G. holds a tier 2 Canada Research Chair (CRC2).

References

Acosta, C. G., & Lopez, H. S. (1999). delta opioid receptor modulation of several voltage-dependent Ca(2+) currents in rat sensory neurons. *The Journal of Neuroscience*, *19*(19), 8337–8348.

Ahuja, S., & Smith, S. O. (2009). Multiple switches in G protein-coupled receptor activation. *Trends in Pharmacological Sciences*, *30*(9), 494–502. https://doi.org/10.1016/j.tips.2009.06.003.

Archer-Lahlou, E., Audet, N., Amraei, M. G., Huard, K., Paquin-Gobeil, M., & Pineyro, G. (2009). Src promotes delta opioid receptor (DOR) desensitization by interfering with receptor recycling. *Journal of Cellular and Molecular Medicine*, *13*(1), 147–163. https://doi.org/10.1111/j.1582-4934.2008.00308.x.

Arvidsson, U., Dado, R. J., Riedl, M., Lee, J. H., Law, P. Y., Loh, H. H., et al. (1995). delta-Opioid receptor immunoreactivity: Distribution in brainstem and spinal cord, and relationship to biogenic amines and enkephalin. *The Journal of Neuroscience*, *15*(2), 1215–1235.

Audet, N., Charfi, I., Mnie-Filali, O., Amraei, M., Chabot-Dore, A. J., Millecamps, M., et al. (2012). Differential association of receptor-Gbetagamma complexes with beta-arrestin2 determines recycling bias and potential for tolerance of delta opioid receptor agonists. *The Journal of Neuroscience*, *32*(14), 4827–4840. https://doi.org/10.1523/JNEUROSCI.3734-11.2012.

Audet, N., Paquin-Gobeil, M., Landry-Paquet, O., Schiller, P. W., & Pineyro, G. (2005). Internalization and Src activity regulate the time course of ERK activation by delta opioid receptor ligands. *The Journal of Biological Chemistry*, *280*(9), 7808–7816. https://doi.org/10.1074/jbc.M411695200.

Augustin, L. B., Felsheim, R. F., Min, B. H., Fuchs, S. M., Fuchs, J. A., & Loh, H. H. (1995). Genomic structure of the mouse delta opioid receptor gene. *Biochemical and Biophysical Research Communications*, *207*(1), 111–119. https://doi.org/10.1006/bbrc.1995.1160.

Bao, L., Jin, S. X., Zhang, C., Wang, L. H., Xu, Z. Z., Zhang, F. X., et al. (2003). Activation of delta opioid receptors induces receptor insertion and neuropeptide secretion. *Neuron*, *37*(1), 121–133.

Bardoni, R., Tawfik, V. L., Wang, D., Francois, A., Solorzano, C., Shuster, S. A., et al. (2014). Delta opioid receptors presynaptically regulate cutaneous mechanosensory neuron input to the spinal cord dorsal horn. *Neuron*, *81*(6), 1312–1327. https://doi.org/10.1016/j.neuron.2014.01.044.

Beaudry, H., Mercier-Blais, A. A., Delaygue, C., Lavoie, C., Parent, J. L., Neugebauer, W., et al. (2015). Regulation of mu and delta opioid receptor functions: Involvement of cyclin-dependent kinase 5. *British Journal of Pharmacology*, *172*(10), 2573–2587. https://doi.org/10.1111/bph.13088.

Beaudry, H., Proteau-Gagne, A., Li, S., Dory, Y., Chavkin, C., & Gendron, L. (2009). Differential noxious and motor tolerance of chronic delta opioid receptor agonists in rodents. *Neuroscience*, *161*(2), 381–391.

Befort, K., Mattei, M. G., Roeckel, N., & Kieffer, B. (1994). Chromosomal localization of the delta opioid receptor gene to human 1p34.3-p36.1 and mouse 4D bands by in situ hybridization. *Genomics*, *20*(1), 143–145.

Bertran-Gonzalez, J., Laurent, V., Chieng, B. C., Christie, M. J., & Balleine, B. W. (2013). Learning-related translocation of delta-opioid receptors on ventral striatal cholinergic interneurons mediates choice between goal-directed actions. *The Journal of Neuroscience*, *33*(41), 16060–16071.

Bie, B., Zhang, Z., Cai, Y. Q., Zhu, W., Zhang, Y., Dai, J., et al. (2010). Nerve growth factor-regulated emergence of functional delta-opioid receptors. *The Journal of Neuroscience*, *30*(16), 5617–5628. https://doi.org/10.1523/JNEUROSCI.5296-09.2010.

Bie, B., Zhu, W., & Pan, Z. Z. (2009). Rewarding morphine-induced synaptic function of delta-opioid receptors on central glutamate synapses. *The Journal of Pharmacology and Experimental Therapeutics*, *329*(1), 290–296. https://doi.org/10.1124/jpet.108.148908.

Borner, C., Woltje, M., Hollt, V., & Kraus, J. (2004). STAT6 transcription factor binding sites with mismatches within the canonical 5'-TTC...GAA-3′ motif involved in regulation of delta- and mu-opioid receptors. *Journal of Neurochemistry*, *91*(6), 1493–1500. https://doi.org/10.1111/j.1471-4159.2004.02846.x.

Brackley, A. D., Gomez, R., Akopian, A. N., Henry, M. A., & Jeske, N. A. (2016). GRK2 constitutively governs peripheral delta opioid receptor activity. *Cell Reports*, *16*(10), 2686–2698. https://doi.org/10.1016/j.celrep.2016.07.084.

Brackley, A. D., Sarrami, S., Gomez, R., Guerrero, K. A., & Jeske, N. A. (2017). Identification of a signaling cascade that maintains constitutive delta-opioid receptor incompetence in peripheral sensory neurons. *The Journal of Biological Chemistry*, *292*(21), 8762–8772. https://doi.org/10.1074/jbc.M117.776799.

Brandt, W., Golbraikh, A., Tager, M., & Lendeckel, U. (1999). A molecular mechanism for the cleavage of a disulfide bond as the primary function of agonist binding to G-protein-coupled receptors based on theoretical calculations supported by experiments. *European Journal of Biochemistry*, *261*(1), 89–97.

Bruehl, S., Denton, J. S., Lonergan, D., Koran, M. E., Chont, M., Sobey, C., et al. (2013). Associations between KCNJ6 (GIRK2) gene polymorphisms and pain-related phenotypes. *Pain*, *154*(12), 2853–2859. https://doi.org/10.1016/j.pain.2013.08.026.

Bzdega, T., Chin, H., Kim, H., Jung, H. H., Kozak, C. A., & Klee, W. A. (1993). Regional expression and chromosomal localization of the delta opiate receptor gene. *Proceedings of the National Academy of Sciences of the United States of America*, *90*(20), 9305–9309.

Cahill, T. J., 3rd, Thomsen, A. R., Tarrasch, J. T., Plouffe, B., Nguyen, A. H., Yang, F., et al. (2017). Distinct conformations of GPCR-beta-arrestin complexes mediate desensitization, signaling, and endocytosis. *Proceedings of the National Academy of Sciences of the United States of America*, *114*(10), 2562–2567. https://doi.org/10.1073/pnas.1701529114.

Cahill, C. M., Holdridge, S. V., & Morinville, A. (2007). Trafficking of delta-opioid receptors and other G-protein-coupled receptors: Implications for pain and analgesia. *Trends in Pharmacological Sciences*, *28*(1), 23–31.

Cahill, C. M., Morinville, A., Hoffert, C., O'Donnell, D., & Beaudet, A. (2003). Up-regulation and trafficking of delta opioid receptor in a model of chronic inflammation: Implications for pain control. *Pain*, *101*(1–2), 199–208.

Cahill, C. M., Morinville, A., Lee, M. C., Vincent, J. P., Collier, B., & Beaudet, A. (2001). Prolonged morphine treatment targets delta opioid receptors to neuronal plasma membranes and enhances delta-mediated antinociception. *The Journal of Neuroscience*, *21*(19), 7598–7607.

Cai, Y., Zhang, Y., Wu, Y., & Pei, G. (1996). delta Opioid receptor in neuronal cells undergoes acute and homologous desensitization. *Biochemical and Biophysical Research Communications*, *219*(2), 342–347. https://doi.org/10.1006/bbrc.1996.0235.

Cao, C. Q., Hong, Y., Dray, A., & Perkins, M. (2001). Spinal delta-opioid receptors mediate suppression of systemic SNC80 on excitability of the flexor reflex in normal and inflamed rat. *European Journal of Pharmacology*, *418*(1–2), 79–87.

Cayla, C., Labuz, D., Machelska, H., Bader, M., Schafer, M., & Stein, C. (2012). Impaired nociception and peripheral opioid antinociception in mice lacking both kinin B1 and B2 receptors. *Anesthesiology*, *116*(2), 448–457.

Charfi, I., Abdallah, K., Gendron, L., & Pineyro, G. (2018). Delta opioid receptors recycle to the membrane after sorting to the degradation path. *Cellular and Molecular Life Sciences*, *75*(12), 2257–2271. https://doi.org/10.1007/s00018-017-2732-5.

Charfi, I., Nagi, K., Mnie-Filali, O., Thibault, D., Balboni, G., Schiller, P. W., et al. (2014). Ligand- and cell-dependent determinants of internalization and cAMP modulation by delta opioid receptor (DOR) agonists. *Cellular and Molecular Life Sciences*, *71*(8), 1529–1546. https://doi.org/10.1007/s00018-013-1461-7.

Che, T., Majumdar, S., Zaidi, S. A., Ondachi, P., McCorvy, J. D., Wang, S., et al. (2018). Structure of the nanobody-stabilized active state of the kappa opioid receptor. *Cell*, *172*(1–2), 55–67. e15. https://doi.org/10.1016/j.cell.2017.12.011.

Chen, Y. L., Law, P. Y., & Loh, H. H. (2006). Sustained activation of phosphatidylinositol 3-kinase/Akt/nuclear factor kappaB signaling mediates G protein-coupled delta-opioid receptor gene expression. *The Journal of Biological Chemistry*, *281*(6), 3067–3074. https://doi.org/10.1074/jbc.M506721200.

Chen, Y. L., Law, P. Y., & Loh, H. H. (2007). Action of NF-kappaB on the delta opioid receptor gene promoter. *Biochemical and Biophysical Research Communications*, *352*(3), 818–822. https://doi.org/10.1016/j.bbrc.2006.11.103.

Chen, Y. L., Monteith, N., Law, P. Y., & Loh, H. H. (2010). Dynamic association of p300 with the promoter of the G protein-coupled rat delta opioid receptor gene during NGF-induced neuronal differentiation. *Biochemical and Biophysical Research Communications*, *396*(2), 294–298. https://doi.org/10.1016/j.bbrc.2010.04.083.

Chen, H. J., Xie, W. Y., Hu, F., Zhang, Y., Wang, J., & Wang, Y. (2012). Disruption of delta-opioid receptor phosphorylation at threonine 161 attenuates morphine tolerance in rats with CFA-induced inflammatory hypersensitivity. *Neuroscience Bulletin*, *28*(2), 182–192. https://doi.org/10.1007/s12264-012-1216-8.

Cheng, P. Y., Liu-Chen, L. Y., & Pickel, V. M. (1997). Dual ultrastructural immunocytochemical labeling of mu and delta opioid receptors in the superficial layers of the rat cervical spinal cord. *Brain Research*, *778*(2), 367–380.

Cheng, J., Sun, X., Li, W., Liu, G., Tu, Y., & Tang, Y. (2016). Molecular switches of the kappa opioid receptor triggered by 6'-GNTI and 5'-GNTI. *Scientific Reports*, *6*, 18913. https://doi.org/10.1038/srep18913.

Cheng, P. Y., Svingos, A. L., Wang, H., Clarke, C. L., Jenab, S., Beczkowska, I. W., et al. (1995). Ultrastructural immunolabeling shows prominent presynaptic vesicular localization of delta-opioid receptor within both enkephalin- and nonenkephalin-containing axon terminals in the superficial layers of the rat cervical spinal cord. *The Journal of Neuroscience*, *15*(9), 5976–5988.

Chieng, B., & Christie, M. J. (2009). Chronic morphine treatment induces functional delta-opioid receptors in amygdala neurons that project to periaqueductal grey. *Neuropharmacology*, *57*(4), 430–437.

Chung, P. C., Boehrer, A., Stephan, A., Matifas, A., Scherrer, G., Darcq, E., et al. (2015). Delta opioid receptors expressed in forebrain GABAergic neurons are responsible for SNC80-induced seizures. *Behavioural Brain Research*, *278*, 429–434. https://doi.org/10.1016/j.bbr.2014.10.029.

Chung, M. K., Cho, Y. S., Bae, Y. C., Lee, J., Zhang, X., & Ro, J. Y. (2014). Peripheral G protein-coupled inwardly rectifying potassium channels are involved in delta-opioid receptor-mediated anti-hyperalgesia in rat masseter muscle. *European Journal of Pain*, *18*(1), 29–38. https://doi.org/10.1002/j.1532-2149.2013.00343.x.

Commons, K. G. (2003). Translocation of presynaptic delta opioid receptors in the ventrolateral periaqueductal gray after swim stress. *The Journal of Comparative Neurology*, *464*(2), 197–207.

Commons, K. G., Beck, S. G., Rudoy, C., & Van Bockstaele, E. J. (2001). Anatomical evidence for presynaptic modulation by the delta opioid receptor in the ventrolateral periaqueductal gray of the rat. *The Journal of Comparative Neurology*, *430*(2), 200–208.

Decaillot, F. M., Rozenfeld, R., Gupta, A., & Devi, L. A. (2008). Cell surface targeting of mu-delta opioid receptor heterodimers by RTP4. *Proceedings of the National Academy of Sciences of the United States of America*, *105*(41), 16045–16050. https://doi.org/10.1073/pnas.0804106105.

DeWire, S. M., Yamashita, D. S., Rominger, D. H., Liu, G., Cowan, C. L., Graczyk, T. M., et al. (2013). A G protein-biased ligand at the mu-opioid receptor is potently analgesic with reduced gastrointestinal and respiratory dysfunction compared with morphine. *The Journal of Pharmacology and Experimental Therapeutics*, *344*(3), 708–717. https://doi.org/10.1124/jpet.112.201616.

Diaz, E., & Pfeffer, S. R. (1998). TIP47: A cargo selection device for mannose 6-phosphate receptor trafficking. *Cell*, *93*(3), 433–443.

Dortch-Carnes, J., & Potter, D. E. (2003). Delta-opioid agonist-stimulated inositol phosphate formation in isolated, rabbit iris-ciliary bodies: Role of G(i/o) proteins and Gbetagamma-subunits. *Experimental Eye Research*, *77*(6), 647–652.

Drake, C. T., Chavkin, C., & Milner, T. A. (2007). Opioid systems in the dentate gyrus. *Progress in Brain Research*, *163*, 245–263. https://doi.org/10.1016/S0079-6123(07)63015-5.

Dripps, I. J., Boyer, B. T., Neubig, R. R., Rice, K. C., Traynor, J. R., & Jutkiewicz, E. M. (2018). Role of signalling molecules in behaviours mediated by the delta opioid receptor agonist SNC80. *British Journal of Pharmacology*, *175*(6), 891–901. https://doi.org/10.1111/bph.14131.

Dripps, I. J., Wang, Q., Neubig, R. R., Rice, K. C., Traynor, J. R., & Jutkiewicz, E. M. (2017). The role of regulator of G protein signaling 4 in delta-opioid receptor-mediated behaviors. *Psychopharmacology*, *234*(1), 29–39. https://doi.org/10.1007/s00213-016-4432-5.

Elde, R., Arvidsson, U., Riedl, M., Vulchanova, L., Lee, J. H., Dado, R., et al. (1995). Distribution of neuropeptide receptors. New views of peptidergic neurotransmission made possible by antibodies to opioid receptors. *Annals of the New York Academy of Sciences*, *757*, 390–404.

Evans, C. J., Keith, D. E., Jr., Morrison, H., Magendzo, K., & Edwards, R. H. (1992). Cloning of a delta opioid receptor by functional expression. *Science*, *258*(5090), 1952–1955.

Fenalti, G., Giguere, P. M., Katritch, V., Huang, X. P., Thompson, A. A., Cherezov, V., et al. (2014). Molecular control of delta-opioid receptor signalling. *Nature*, *506*(7487), 191–196. https://doi.org/10.1038/nature12944.

Fenalti, G., Zatsepin, N. A., Betti, C., Giguere, P., Han, G. W., Ishchenko, A., et al. (2015). Structural basis for bifunctional peptide recognition at human delta-opioid receptor. *Nature Structural & Molecular Biology*, *22*(3), 265–268. https://doi.org/10.1038/nsmb.2965.

Ferre, S., Baler, R., Bouvier, M., Caron, M. G., Devi, L. A., Durroux, T., et al. (2009). Building a new conceptual framework for receptor heteromers. *Nature Chemical Biology*, *5*(3), 131–134. https://doi.org/10.1038/nchembio0309-131.

Fraser, G. L., Gaudreau, G. A., Clarke, P. B., Menard, D. P., & Perkins, M. N. (2000). Antihyperalgesic effects of delta opioid agonists in a rat model of chronic inflammation. *British Journal of Pharmacology*, *129*(8), 1668–1672.

Fujita, W., Gomes, I., & Devi, L. A. (2014). Revolution in GPCR signalling: Opioid receptor heteromers as novel therapeutic targets: IUPHAR review 10. *British Journal of Pharmacology*, *171*(18), 4155–4176. https://doi.org/10.1111/bph.12798.

Fujita, W., Gomes, I., & Devi, L. A. (2015). Heteromers of mu-delta opioid receptors: New pharmacology and novel therapeutic possibilities. *British Journal of Pharmacology*, *172*(2), 375–387. https://doi.org/10.1111/bph.12663.

Gao, C. J., Niu, L., Ren, P. C., Wang, W., Zhu, C., Li, Y. Q., et al. (2012). Hypoxic preconditioning attenuates global cerebral ischemic injury following asphyxial cardiac arrest through regulation of delta opioid receptor system. *Neuroscience*, *202*, 352–362. https://doi.org/10.1016/j.neuroscience.2011.11.060.

Gelernter, J., & Kranzler, H. R. (2000). Variant detection at the delta opioid receptor (OPRD1) locus and population genetics of a novel variant affecting protein sequence. *Human Genetics*, *107*(1), 86–88.

Gendron, L., Cahill, C. M., von Zastrow, M., Schiller, P. W., & Pineyro, G. (2016). Molecular pharmacology of delta-Opioid receptors. *Pharmacological Reviews*, *68*(3), 631–700. https://doi.org/10.1124/pr.114.008979.

Gendron, L., Esdaile, M. J., Mennicken, F., Pan, H., O'Donnell, D., Vincent, J.-P., et al. (2007). Morphine priming in rats with chronic inflammation reveals a dichotomy between antihyperalgesic and antinociceptive properties of deltorphin. *Neuroscience*, *144*, 263–274. https://doi.org/10.1016/j.neuroscience.2006.08.077.

Gendron, L., Lucido, A. L., Mennicken, F., O'Donnell, D., Vincent, J. P., Stroh, T., et al. (2006). Morphine and pain-related stimuli enhance cell surface availability of somatic delta-opioid receptors in rat dorsal root ganglia. *The Journal of Neuroscience*, *26*(3), 953–962.

Gendron, L., Mittal, N., Beaudry, H., & Walwyn, W. (2015). Recent advances on the delta opioid receptor: From trafficking to function. *British Journal of Pharmacology*, *172*(2), 403–419. https://doi.org/10.1111/bph.12706.

Genier, S., Degrandmaison, J., Moreau, P., Labrecque, P., Hebert, T. E., & Parent, J. L. (2016). Regulation of GPCR expression through an interaction with CCT7, a subunit of the CCT/TRiC complex. *Molecular Biology of the Cell*, *27*(24), 3800–3812. https://doi.org/10.1091/mbc.E16-04-0224.

George, S. R., Fan, T., Xie, Z., Tse, R., Tam, V., Varghese, G., et al. (2000). Oligomerization of mu- and delta-opioid receptors. Generation of novel functional properties. *The Journal of Biological Chemistry*, *275*(34), 26128–26135.

Georgoussi, Z., Leontiadis, L., Mazarakou, G., Merkouris, M., Hyde, K., & Hamm, H. (2006). Selective interactions between G protein subunits and RGS4 with the C-terminal domains of the mu- and delta-opioid receptors regulate opioid receptor signaling. *Cellular Signalling, 18*(6), 771–782. https://doi.org/10.1016/j.cellsig.2005.07.003.

Giguere, P. M., Kroeze, W. K., & Roth, B. L. (2014). Tuning up the right signal: Chemical and genetic approaches to study GPCR functions. *Current Opinion in Cell Biology, 27*, 51–55. https://doi.org/10.1016/j.ceb.2013.11.006.

Gomes, I., Ayoub, M. A., Fujita, W., Jaeger, W. C., Pfleger, K. D., & Devi, L. A. (2015). G protein-coupled receptor heteromers. *Annual Review of Pharmacology and Toxicology, 56*, 103–425. https://doi.org/10.1146/annurev-pharmtox-011613-135952.

Gong, L., Gao, F., Li, J., Li, J., Yu, X., Ma, X., et al. (2015). Oxytocin-induced membrane hyperpolarization in pain-sensitive dorsal root ganglia neurons mediated by Ca(2+)/nNOS/NO/KATP pathway. *Neuroscience, 289*, 417–428. https://doi.org/10.1016/j.neuroscience.2014.12.058.

Granier, S., Manglik, A., Kruse, A. C., Kobilka, T. S., Thian, F. S., Weis, W. I., et al. (2012). Structure of the delta-opioid receptor bound to naltrindole. *Nature, 485*(7398), 400–404. https://doi.org/10.1038/nature11111.

Guan, J. S., Xu, Z. Z., Gao, H., He, S. Q., Ma, G. Q., Sun, T., et al. (2005). Interaction with vesicle luminal protachykinin regulates surface expression of delta-opioid receptors and opioid analgesia. *Cell, 122*(4), 619–631.

Guo, J., Wu, Y., Zhang, W., Zhao, J., Devi, L. A., Pei, G., et al. (2000). Identification of G protein-coupled receptor kinase 2 phosphorylation sites responsible for agonist-stimulated delta-opioid receptor phosphorylation. *Molecular Pharmacology, 58*(5), 1050–1056.

Gupta, A., Gomes, I., Wardman, J., & Devi, L. A. (2014). Opioid receptor function is regulated by post-endocytic peptide processing. *The Journal of Biological Chemistry, 289*(28), 19613–19626. https://doi.org/10.1074/jbc.M113.537704.

Hack, S. P., Bagley, E. E., Chieng, B. C., & Christie, M. J. (2005). Induction of delta-opioid receptor function in the midbrain after chronic morphine treatment. *The Journal of Neuroscience, 25*(12), 3192–3198.

He, C., Wei, Y., Sun, K., Li, B., Dong, X., Zou, Z., et al. (2013). Beclin 2 functions in autophagy, degradation of G protein-coupled receptors, and metabolism. *Cell, 154*(5), 1085–1099. https://doi.org/10.1016/j.cell.2013.07.035.

He, S. Q., Zhang, Z. N., Guan, J. S., Liu, H. R., Zhao, B., Wang, H. B., et al. (2011). Facilitation of mu-opioid receptor activity by preventing delta-opioid receptor-mediated codegradation. *Neuron, 69*(1), 120–131. https://doi.org/10.1016/j.neuron.2010.12.001.

Helenius, A., & Aebi, M. (2004). Roles of N-linked glycans in the endoplasmic reticulum. *Annual Review of Biochemistry, 73*, 1019–1049. https://doi.org/10.1146/annurev.biochem.73.011303.073752.

Henry, A. G., White, I. J., Marsh, M., von Zastrow, M., & Hislop, J. N. (2011). The role of ubiquitination in lysosomal trafficking of delta-opioid receptors. *Traffic, 12*(2), 170–184. https://doi.org/10.1111/j.1600-0854.2010.01145.x.

Holst, B., Nygaard, R., Valentin-Hansen, L., Bach, A., Engelstoft, M. S., Petersen, P. S., et al. (2010). A conserved aromatic lock for the tryptophan rotameric switch in TM-VI of seven-transmembrane receptors. *The Journal of Biological Chemistry, 285*(6), 3973–3985. https://doi.org/10.1074/jbc.M109.064725.

Hori, T., Okuno, T., Hirata, K., Yamashita, K., Kawano, Y., Yamamoto, M., et al. (2018). Na(+)-mimicking ligands stabilize the inactive state of leukotriene B4 receptor BLT1. *Nature Chemical Biology, 14*(3), 262–269. https://doi.org/10.1038/nchembio.2547.

Huang, W., Manglik, A., Venkatakrishnan, A. J., Laeremans, T., Feinberg, E. N., Sanborn, A. L., et al. (2015). Structural insights into micro-opioid receptor activation. *Nature, 524*(7565), 315–321. https://doi.org/10.1038/nature14886.

Hurley, R. W., & Hammond, D. L. (2000). The analgesic effects of supraspinal mu and delta opioid receptor agonists are potentiated during persistent inflammation. *The Journal of Neuroscience*, *20*(3), 1249–1259.

Hylden, J. L., Thomas, D. A., Iadarola, M. J., Nahin, R. L., & Dubner, R. (1991). Spinal opioid analgesic effects are enhanced in a model of unilateral inflammation/hyperalgesia: Possible involvement of noradrenergic mechanisms. *European Journal of Pharmacology*, *194*(2–3), 135–143.

Jones, K. A., Borowsky, B., Tamm, J. A., Craig, D. A., Durkin, M. M., Dai, M., et al. (1998). GABA(B) receptors function as a heteromeric assembly of the subunits GABA(B)R1 and GABA(B)R2. *Nature*, *396*(6712), 674–679. https://doi.org/10.1038/25348.

Katritch, V., Cherezov, V., & Stevens, R. C. (2013). Structure-function of the G protein-coupled receptor superfamily. *Annual Review of Pharmacology and Toxicology*, *53*, 531–556. https://doi.org/10.1146/annurev-pharmtox-032112-135923.

Katritch, V., Fenalti, G., Abola, E. E., Roth, B. L., Cherezov, V., & Stevens, R. C. (2014). Allosteric sodium in class A GPCR signaling. *Trends in Biochemical Sciences*, *39*(5), 233–244. https://doi.org/10.1016/j.tibs.2014.03.002.

Kaufman, D. L., Xia, Y. R., Keith, D. E., Jr., Newman, D., Evans, C. J., & Lusis, A. J. (1994). Localization of the delta-opioid receptor gene to mouse chromosome 4 by linkage analysis. *Genomics*, *19*(2), 405–406. https://doi.org/10.1006/geno.1994.1087.

Kaupmann, K., Malitschek, B., Schuler, V., Heid, J., Froestl, W., Beck, P., et al. (1998). GABA(B)-receptor subtypes assemble into functional heteromeric complexes. *Nature*, *396*(6712), 683–687. https://doi.org/10.1038/25360.

Kieffer, B. L., Befort, K., Gaveriaux-Ruff, C., & Hirth, C. G. (1992). The delta-opioid receptor: Isolation of a cDNA by expression cloning and pharmacological characterization. *Proceedings of the National Academy of Sciences of the United States of America*, *89*(24), 12048–12052.

Kim, K. S., Lee, K. W., Lee, K. W., Im, J. Y., Yoo, J. Y., Kim, S. W., et al. (2006). Adenylyl cyclase type 5 (AC5) is an essential mediator of morphine action. *Proceedings of the National Academy of Sciences of the United States of America*, *103*(10), 3908–3913. https://doi.org/10.1073/pnas.0508812103.

Koehl, A., Hu, H., Maeda, S., Zhang, Y., Qu, Q., Paggi, J. M., et al. (2018). Structure of the micro-opioid receptor-Gi protein complex. *Nature*, *558*(7711), 547–552. https://doi.org/10.1038/s41586-018-0219-7.

Kouhen, O. M., Wang, G., Solberg, J., Erickson, L. J., Law, P. Y., & Loh, H. H. (2000). Hierarchical phosphorylation of delta-opioid receptor regulates agonist-induced receptor desensitization and internalization. *The Journal of Biological Chemistry*, *275*(47), 36659–36664. https://doi.org/10.1074/jbc.M006788200.

Kramer, H. K., Andria, M. L., Kushner, S. A., Esposito, D. H., Hiller, J. M., & Simon, E. J. (2000). Mutation of tyrosine 318 (Y318F) in the delta-opioid receptor attenuates tyrosine phosphorylation, agonist-dependent receptor internalization, and mitogen-activated protein kinase activation. *Brain Research. Molecular Brain Research*, *79*(1–2), 55–66.

Kuner, R., Kohr, G., Grunewald, S., Eisenhardt, G., Bach, A., & Kornau, H. C. (1999). Role of heteromer formation in GABAB receptor function. *Science*, *283*(5398), 74–77.

Laurent, V., Wong, F. L., & Balleine, B. W. (2015). delta-Opioid receptors in the accumbens shell mediate the influence of both excitatory and inhibitory predictions on choice. *British Journal of Pharmacology*, *172*(2), 562–570. https://doi.org/10.1111/bph.12731.

Law, P. Y., Kouhen, O. M., Solberg, J., Wang, W., Erickson, L. J., & Loh, H. H. (2000). Deltorphin II-induced rapid desensitization of delta-opioid receptor requires both phosphorylation and internalization of the receptor. *The Journal of Biological Chemistry*, *275*(41), 32057–32065. https://doi.org/10.1074/jbc.M002395200.

Leontiadis, L. J., Papakonstantinou, M. P., & Georgoussi, Z. (2009). Regulator of G protein signaling 4 confers selectivity to specific G proteins to modulate mu- and delta-opioid receptor signaling. *Cellular Signalling, 21*(7), 1218–1228. https://doi.org/10.1016/j.cellsig.2009.03.013.

Leskela, T. T., Lackman, J. J., Vierimaa, M. M., Kobayashi, H., Bouvier, M., & Petaja-Repo, U. E. (2012). Cys-27 variant of human delta-opioid receptor modulates maturation and cell surface delivery of Phe-27 variant via heteromerization. *The Journal of Biological Chemistry, 287*(7), 5008–5020. https://doi.org/10.1074/jbc.M111.305656.

Leskela, T. T., Markkanen, P. M., Alahuhta, I. A., Tuusa, J. T., & Petaja-Repo, U. E. (2009). Phe27Cys polymorphism alters the maturation and subcellular localization of the human delta opioid receptor. *Traffic, 10*(1), 116–129. https://doi.org/10.1111/j.1600-0854.2008.00846.x.

Leskela, T. T., Markkanen, P. M., Pietila, E. M., Tuusa, J. T., & Petaja-Repo, U. E. (2007). Opioid receptor pharmacological chaperones act by binding and stabilizing newly synthesized receptors in the endoplasmic reticulum. *The Journal of Biological Chemistry, 282*(32), 23171–23183.

Li, J., Huang, P., Chen, C., de Riel, J. K., Weinstein, H., & Liu-Chen, L. Y. (2001). Constitutive activation of the mu opioid receptor by mutation of D3.49(164), but not D3.32(147): D3.49(164) is critical for stabilization of the inactive form of the receptor and for its expression. *Biochemistry, 40*(40), 12039–12050.

Li, J., Xiang, B., Su, W., Zhang, X., Huang, Y., & Ma, L. (2003). Agonist-induced formation of opioid receptor-G protein-coupled receptor kinase (GRK)-G beta gamma complex on membrane is required for GRK2 function in vivo. *The Journal of Biological Chemistry, 278*(32), 30219–30226. https://doi.org/10.1074/jbc.M302385200.

Liu, W., Chun, E., Thompson, A. A., Chubukov, P., Xu, F., Katritch, V., et al. (2012). Structural basis for allosteric regulation of GPCRs by sodium ions. *Science, 337*(6091), 232–236. https://doi.org/10.1126/science.1219218.

Liu, H. C., Shen, J. T., Augustin, L. B., Ko, J. L., & Loh, H. H. (1999). Transcriptional regulation of mouse delta-opioid receptor gene. *The Journal of Biological Chemistry, 274*(33), 23617–23626.

Lobingier, B. T., Huttenhain, R., Eichel, K., Miller, K. B., Ting, A. Y., von Zastrow, M., et al. (2017). An approach to spatiotemporally resolve protein interaction networks in living cells. *Cell, 169*(2), 350–360. e312. https://doi.org/10.1016/j.cell.2017.03.022.

Lobingier, B. T., & von Zastrow, M. (2019). When trafficking and signaling mix: How subcellular location shapes G protein-coupled receptor activation of heterotrimeric G proteins. *Traffic, 20*(2), 130–136. https://doi.org/10.1111/tra.12634.

Lombardi, D., Soldati, T., Riederer, M. A., Goda, Y., Zerial, M., & Pfeffer, S. R. (1993). Rab9 functions in transport between late endosomes and the trans Golgi network. *The EMBO Journal, 12*(2), 677–682.

Lucido, A. L., Morinville, A., Gendron, L., Stroh, T., & Beaudet, A. (2005). Prolonged morphine treatment selectively increases membrane recruitment of delta-opioid receptors in mouse basal ganglia. *Journal of Molecular Neuroscience, 25*(3), 207–214.

Lv, M. R., Li, B., Wang, M. G., Meng, F. G., Yu, J. J., Guo, F., et al. (2017). Activation of the PI3K-Akt pathway promotes neuroprotection of the delta-opioid receptor agonist against cerebral ischemia-reperfusion injury in rat models. *Biomedicine & Pharmacotherapy, Biomedecine & Pharmacotherapie, 93*, 230–237. https://doi.org/10.1016/j.biopha.2017.05.121.

Ma, J., Zhang, Y., Kalyuzhny, A. E., & Pan, Z. Z. (2006). Emergence of Functional {delta}-opioid receptors induced by chronic morphine. *Molecular Pharmacology, 69*, 1137–1145.

Manglik, A., Kruse, A. C., Kobilka, T. S., Thian, F. S., Mathiesen, J. M., Sunahara, R. K., et al. (2012). Crystal structure of the micro-opioid receptor bound to a morphinan antagonist. *Nature, 485*(7398), 321–326. https://doi.org/10.1038/nature10954.

Manglik, A., Lin, H., Aryal, D. K., McCorvy, J. D., Dengler, D., Corder, G., et al. (2016). Structure-based discovery of opioid analgesics with reduced side effects. *Nature*, *537*(7619), 185–190. https://doi.org/10.1038/nature19112.

Margolis, E. B., Mitchell, J. M., Hjelmstad, G. O., & Fields, H. L. (2011). A novel opioid receptor-mediated enhancement of GABAA receptor function induced by stress in ventral tegmental area neurons. *The Journal of Physiology*, *589*(Pt. 17), 4229–4242. https://doi.org/10.1113/jphysiol.2011.209023.

Marinelli, S., Connor, M., Schnell, S. A., Christie, M. J., Wessendorf, M. W., & Vaughan, C. W. (2005). Delta-opioid receptor-mediated actions on rostral ventromedial medulla neurons. *Neuroscience*, *132*(2), 239–244. https://doi.org/10.1016/j.neuroscience.2005.01.015.

Marker, C. L., Lujan, R., Loh, H. H., & Wickman, K. (2005). Spinal G-protein-gated potassium channels contribute in a dose-dependent manner to the analgesic effect of mu- and delta- but not kappa-opioids. *The Journal of Neuroscience*, *25*(14), 3551–3559. https://doi.org/10.1523/JNEUROSCI.4899-04.2005.

Markkanen, P. M., & Petaja-Repo, U. E. (2008). N-glycan-mediated quality control in the endoplasmic reticulum is required for the expression of correctly folded delta-opioid receptors at the cell surface. *The Journal of Biological Chemistry*, *283*(43), 29086–29098. https://doi.org/10.1074/jbc.M801880200.

Marley, A., & von Zastrow, M. (2010). Dysbindin promotes the post-endocytic sorting of G protein-coupled receptors to lysosomes. *PLoS One*, *5*(2). e9325. https://doi.org/10.1371/journal.pone.0009325.

Mayer, P., Rochlitz, H., Rauch, E., Rommelspacher, H., Hasse, H. E., Schmidt, S., et al. (1997). Association between a delta opioid receptor gene polymorphism and heroin dependence in man. *Neuroreport*, *8*(11), 2547–2550.

Mellman, I., Fuchs, R., & Helenius, A. (1986). Acidification of the endocytic and exocytic pathways. *Annual Review of Biochemistry*, *55*, 663–700. https://doi.org/10.1146/annurev.bi.55.070186.003311.

Miller, R. L., Thompson, A. A., Trapella, C., Guerrini, R., Malfacini, D., Patel, N., et al. (2015). The importance of ligand-receptor conformational pairs in stabilization: Spotlight on the N/OFQ G protein-coupled receptor. *Structure*, *23*(12), 2291–2299. https://doi.org/10.1016/j.str.2015.07.024.

Miller-Gallacher, J. L., Nehme, R., Warne, T., Edwards, P. C., Schertler, G. F., Leslie, A. G., et al. (2014). The 2.1 A resolution structure of cyanopindolol-bound beta1-adrenoceptor identifies an intramembrane Na+ ion that stabilises the ligand-free receptor. *PLoS One*, *9*(3). e92727https://doi.org/10.1371/journal.pone.0092727.

Mittal, N., Roberts, K., Pal, K., Bentolila, L. A., Fultz, E., Minasyan, A., et al. (2013). Select G-protein-coupled receptors modulate agonist-induced signaling via a ROCK, LIMK, and beta-arrestin 1 pathway. *Cell Reports*, *5*(4), 1010–1021. https://doi.org/10.1016/j.celrep.2013.10.015.

Mons, N., Harry, A., Dubourg, P., Premont, R. T., Iyengar, R., & Cooper, D. M. (1995). Immunohistochemical localization of adenylyl cyclase in rat brain indicates a highly selective concentration at synapses. *Proceedings of the National Academy of Sciences of the United States of America*, *92*(18), 8473–8477.

Moreira, I. S. (2014). Structural features of the G-protein/GPCR interactions. *Biochimica et Biophysica Acta*, *1840*(1), 16–33. https://doi.org/10.1016/j.bbagen.2013.08.027.

Morikawa, H., Fukuda, K., Mima, H., Shoda, T., Kato, S., & Mori, K. (1998). Desensitization and resensitization of delta-opioid receptor-mediated Ca2+ channel inhibition in NG108–15 cells. *British Journal of Pharmacology*, *123*(6), 1111–1118. https://doi.org/10.1038/sj.bjp.0701733.

Morinville, A., Cahill, C. M., Aibak, H., Rymar, V. V., Pradhan, A., Hoffert, C., et al. (2004). Morphine-induced changes in delta opioid receptor trafficking are linked to somatosensory processing in the rat spinal cord. *The Journal of Neuroscience, 24*(24), 5549–5559.

Munk, C., Mutt, E., Isberg, V., Nikolajsen, L. F., Bibbe, J. M., Flock, T., et al. (2019). An online resource for GPCR structure determination and analysis. *Nature Methods, 16*(2), 151–162. https://doi.org/10.1038/s41592-018-0302-x.

Murthy, K. S., & Makhlouf, G. M. (1996). Opioid mu, delta, and kappa receptor-induced activation of phospholipase C-beta 3 and inhibition of adenylyl cyclase is mediated by Gi2 and G(o) in smooth muscle. *Molecular Pharmacology, 50*(4), 870–877.

Nagi, K., & Pineyro, G. (2014). Kir3 channel signaling complexes: Focus on opioid receptor signaling. *Frontiers in Cellular Neuroscience, 8*, 186. https://doi.org/10.3389/fncel.2014.00186.

Nagi, K., & Pineyro, G. (2015). Practical guide for calculating and representing biased signaling by GPCR ligands: A stepwise approach. *Methods, 92*, 78–86. https://doi.org/10.1016/j.ymeth.2015.09.010.

Negus, S. S., & Freeman, K. B. (2018). Abuse potential of biased Mu opioid receptor agonists. *Trends in Pharmacological Sciences, 39*(11), 916–919. https://doi.org/10.1016/j.tips.2018.08.007.

Nygaard, R., Frimurer, T. M., Holst, B., Rosenkilde, M. M., & Schwartz, T. W. (2009). Ligand binding and micro-switches in 7TM receptor structures. *Trends in Pharmacological Sciences, 30*(5), 249–259. https://doi.org/10.1016/j.tips.2009.02.006.

Ohsawa, M., Nagase, H., & Kamei, J. (1998). Role of intracellular calcium in modification of mu and delta opioid receptor-mediated antinociception by diabetes in mice. *The Journal of Pharmacology and Experimental Therapeutics, 286*(2), 780–787.

Olianas, M. C., Dedoni, S., Olianas, A., & Onali, P. (2012). delta-Opioid receptors stimulate the metabolic sensor AMP-activated protein kinase through coincident signaling with G(q/11)-coupled receptors. *Molecular Pharmacology, 81*(2), 154–165. https://doi.org/10.1124/mol.111.075267.

Olianas, M. C., Dedoni, S., & Onali, P. (2013). Coincidence signaling of dopamine D1-like and M1 muscarinic receptors in the regulation of cyclic AMP formation and CREB phosphorylation in mouse prefrontal cortex. *Neurosignals, 21*(1–2), 61–74. https://doi.org/10.1159/000335208.

Olivella, M., Caltabiano, G., & Cordomi, A. (2013). The role of cysteine 6.47 in class A GPCRs. *BMC Structural Biology, 13*, 3. https://doi.org/10.1186/1472-6807-13-3.

Overland, A. C., Kitto, K. F., Chabot-Dore, A. J., Rothwell, P. E., Fairbanks, C. A., Stone, L. S., et al. (2009). Protein kinase C mediates the synergistic interaction between agonists acting at alpha2-adrenergic and delta-opioid receptors in spinal cord. *The Journal of Neuroscience, 29*(42), 13264–13273.

Pasquini, F., Bochet, P., Garbay-Jaureguiberry, C., Roques, B. P., Rossier, J., & Beaudet, A. (1992). Electron microscopic localization of photoaffinity-labelled delta opioid receptors in the neostriatum of the rat. *The Journal of Comparative Neurology, 326*(2), 229–244. https://doi.org/10.1002/cne.903260206.

Patwardhan, A. M., Berg, K. A., Akopain, A. N., Jeske, N. A., Gamper, N., Clarke, W. P., et al. (2005). Bradykinin-induced functional competence and trafficking of the delta-opioid receptor in trigeminal nociceptors. *The Journal of Neuroscience, 25*(39), 8825–8832. https://doi.org/10.1523/JNEUROSCI.0160-05.2005.

Patwardhan, A. M., Diogenes, A., Berg, K. A., Fehrenbacher, J. C., Clarke, W. P., Akopian, A. N., et al. (2006). PAR-2 agonists activate trigeminal nociceptors and induce functional competence in the delta opioid receptor. *Pain, 125*(1–2), 114–124.

Pedersen, N. P., Vaughan, C. W., & Christie, M. J. (2011). Opioid receptor modulation of GABAergic and serotonergic spinally projecting neurons of the rostral ventromedial medulla in mice. *Journal of Neurophysiology*, *106*(2), 731–740. https://doi.org/10.1152/jn.01062.2010.

Pei, G., Kieffer, B. L., Lefkowitz, R. J., & Freedman, N. J. (1995). Agonist-dependent phosphorylation of the mouse delta-opioid receptor: Involvement of G protein-coupled receptor kinases but not protein kinase C. *Molecular Pharmacology*, *48*(2), 173–177.

Pellissier, L. P., Pujol, C. N., Becker, J. A. J., & Le Merrer, J. (2018). Delta opioid receptors: Learning and motivation. *Handbook of Experimental Pharmacology*, *247*, 227–260. https://doi.org/10.1007/164_2016_89.

Petaja-Repo, U. E., Hogue, M., Bhalla, S., Laperriere, A., Morello, J. P., & Bouvier, M. (2002). Ligands act as pharmacological chaperones and increase the efficiency of delta opioid receptor maturation. *The EMBO Journal*, *21*(7), 1628–1637. https://doi.org/10.1093/emboj/21.7.1628.

Petaja-Repo, U. E., Hogue, M., Laperriere, A., Walker, P., & Bouvier, M. (2000). Export from the endoplasmic reticulum represents the limiting step in the maturation and cell surface expression of the human delta opioid receptor. *The Journal of Biological Chemistry*, *275*(18), 13727–13736.

Petaja-Repo, U. E., Hogue, M., Leskela, T. T., Markkanen, P. M., Tuusa, J. T., & Bouvier, M. (2006). Distinct subcellular localization for constitutive and agonist-modulated palmitoylation of the human delta opioid receptor. *The Journal of Biological Chemistry*, *281*(23), 15780–15789.

Pettinger, L., Gigout, S., Linley, J. E., & Gamper, N. (2013). Bradykinin controls pool size of sensory neurons expressing functional delta-opioid receptors. *The Journal of Neuroscience*, *33*(26), 10762–10771.

Philipp, S., Critz, S. D., Cui, L., Solodushko, V., Cohen, M. V., & Downey, J. M. (2006). Localizing extracellular signal-regulated kinase (ERK) in pharmacological preconditioning's trigger pathway. *Basic Research in Cardiology*, *101*(2), 159–167. https://doi.org/10.1007/s00395-005-0566-z.

Piltonen, M., Parisien, M., Gregoire, S., Chabot-Dore, A. J., Jafarnejad, S. M., Berube, P., et al. (2018). Alternative splicing of the delta-opioid receptor gene suggests existence of new functional isoforms. *Molecular Neurobiology*, *56*, 2855–2869. https://doi.org/10.1007/s12035-018-1253-z.

Pin, J. P., Neubig, R., Bouvier, M., Devi, L., Filizola, M., Javitch, J. A., et al. (2007). International union of basic and clinical pharmacology. LXVII. Recommendations for the recognition and nomenclature of G protein-coupled receptor heteromultimers. *Pharmacological Reviews*, *59*(1), 5–13. https://doi.org/10.1124/pr.59.1.5.

Poole, D. P., Pelayo, J. C., Scherrer, G., Evans, C. J., Kieffer, B. L., & Bunnett, N. W. (2011). Localization and regulation of fluorescently labeled delta opioid receptor, expressed in enteric neurons of mice. *Gastroenterology*, *141*(3), 982–991. e981–988. https://doi.org/10.1053/j.gastro.2011.05.042.

Portoghese, P. S., Sultana, M., Nagase, H., & Takemori, A. E. (1988). Application of the message-address concept in the design of highly potent and selective non-peptide delta opioid receptor antagonists. *Journal of Medicinal Chemistry*, *31*(2), 281–282.

Pradhan, A. A., Becker, J. A., Scherrer, G., Tryoen-Toth, P., Filliol, D., Matifas, A., et al. (2009). In vivo delta opioid receptor internalization controls behavioral effects of agonists. *PLoS One*, *4*(5). e5425https://doi.org/10.1371/journal.pone.0005425.

Pradhan, A. A., Befort, K., Nozaki, C., Gaveriaux-Ruff, C., & Kieffer, B. L. (2011). The delta opioid receptor: An evolving target for the treatment of brain disorders. *Trends in Pharmacological Sciences*, *32*(10), 581–590. https://doi.org/10.1016/j.tips.2011.06.008.

Pradhan, A., Smith, M., McGuire, B., Evans, C., & Walwyn, W. (2013). Chronic inflammatory injury results in increased coupling of delta opioid receptors to voltage-gated Ca2+ channels. *Molecular Pain, 9*, 8.

Qiu, Y., Loh, H. H., & Law, P. Y. (2007). Phosphorylation of the delta-opioid receptor regulates its beta-arrestins selectivity and subsequent receptor internalization and adenylyl cyclase desensitization. *The Journal of Biological Chemistry, 282*(31), 22315–22323. https://doi.org/10.1074/jbc.M611258200.

Qiu, C., Sora, I., Ren, K., Uhl, G., & Dubner, R. (2000). Enhanced delta-opioid receptor-mediated antinociception in mu-opioid receptor-deficient mice. *European Journal of Pharmacology, 387*(2), 163–169.

Richard-Lalonde, M., Nagi, K., Audet, N., Sleno, R., Amraei, M., Hogue, M., et al. (2013). Conformational dynamics of Kir3.1/Kir3.2 channel activation via delta-opioid receptors. *Molecular Pharmacology, 83*(2), 416–428. https://doi.org/10.1124/mol.112.081950.

Rosciglione, S., Theriault, C., Boily, M. O., Paquette, M., & Lavoie, C. (2014). Galphas regulates the post-endocytic sorting of G protein-coupled receptors. *Nature Communications, 5*, 4556. https://doi.org/10.1038/ncomms5556.

Rovati, G. E., Capra, V., & Neubig, R. R. (2007). The highly conserved DRY motif of class A G protein-coupled receptors: Beyond the ground state. *Molecular Pharmacology, 71*(4), 959–964. https://doi.org/10.1124/mol.106.029470.

Rowan, M. P., Ruparel, N. B., Patwardhan, A. M., Berg, K. A., Clarke, W. P., & Hargreaves, K. M. (2009). Peripheral delta opioid receptors require priming for functional competence in vivo. *European Journal of Pharmacology, 602*(2–3), 283–287.

Sarajarvi, T., Tuusa, J. T., Haapasalo, A., Lackman, J. J., Sormunen, R., Helisalmi, S., et al. (2011). Cysteine 27 variant of the delta-opioid receptor affects amyloid precursor protein processing through altered endocytic trafficking. *Molecular and Cellular Biology, 31*(11), 2326–2340. https://doi.org/10.1128/MCB.05015-11.

Scherrer, G., Imamachi, N., Cao, Y. Q., Contet, C., Mennicken, F., O'Donnell, D., et al. (2009). Dissociation of the opioid receptor mechanisms that control mechanical and heat pain. *Cell, 137*(6), 1148–1159.

Scherrer, G., Tryoen-Toth, P., Filliol, D., Matifas, A., Laustriat, D., Cao, Y. Q., et al. (2006). Knockin mice expressing fluorescent delta-opioid receptors uncover G protein-coupled receptor dynamics in vivo. *Proceedings of the National Academy of Sciences of the United States of America, 103*(25), 9691–9696. https://doi.org/10.1073/pnas.0603359103.

Schmid, C. L., Kennedy, N. M., Ross, N. C., Lovell, K. M., Yue, Z., Morgenweck, J., et al. (2017). Bias factor and therapeutic window correlate to predict safer opioid analgesics. *Cell, 171*(5), 1165–1175. e1113. https://doi.org/10.1016/j.cell.2017.10.035.

Shiwarski, D. J., Crilly, S. E., Dates, A., & Puthenveedu, M. A. (2019). Dual RXR motifs regulate nerve growth factor-mediated intracellular retention of the delta opioid receptor. *Molecular Biology of the Cell, 30*(5), 680–690. https://doi.org/10.1091/mbc.E18-05-0292.

Shukla, A. K., Westfield, G. H., Xiao, K., Reis, R. I., Huang, L. Y., Tripathi-Shukla, P., et al. (2014). Visualization of arrestin recruitment by a G-protein-coupled receptor. *Nature, 512*(7513), 218–222. https://doi.org/10.1038/nature13430.

Simonin, F., Befort, K., Gaveriaux-Ruff, C., Matthes, H., Nappey, V., Lannes, B., et al. (1994). The human delta-opioid receptor: Genomic organization, cDNA cloning, functional expression, and distribution in human brain. *Molecular Pharmacology, 46*(6), 1015–1021.

Smirnov, D., Im, H. J., & Loh, H. H. (2001). delta-Opioid receptor gene: Effect of Sp1 factor on transcriptional regulation in vivo. *Molecular Pharmacology, 60*(2), 331–340.

Stein, E. A., Hiller, J. M., & Simon, E. J. (1992). Effects of stress on opioid receptor binding in the rat central nervous system. *Neuroscience, 51*(3), 683–690.

St-Louis, E., Degrandmaison, J., Grastilleur, S., Genier, S., Blais, V., Lavoie, C., et al. (2017). Involvement of the coatomer protein complex I in the intracellular traffic of the delta opioid receptor. *Molecular and Cellular Neurosciences*, *79*, 53–63. https://doi.org/10.1016/j.mcn.2016.12.005.

Stoeber, M., Jullie, D., Lobingier, B. T., Laeremans, T., Steyaert, J., Schiller, P. W., et al. (2018). A genetically encoded biosensor reveals location bias of opioid drug action. *Neuron*, *98*(5), 963–976. e965. https://doi.org/10.1016/j.neuron.2018.04.021.

Su, D. S., Wang, Z. H., Zheng, Y. J., Zhao, Y. H., & Wang, X. R. (2007). Dose-dependent neuroprotection of delta opioid peptide [D-Ala2, D-Leu5] enkephalin in neuronal death and retarded behavior induced by forebrain ischemia in rats. *Neuroscience Letters*, *423*(2), 113–117. https://doi.org/10.1016/j.neulet.2007.06.044.

Sun, X., Laroche, G., Wang, X., Agren, H., Bowman, G. R., Giguere, P. M., et al. (2017). Propagation of the allosteric modulation induced by sodium in the delta-Opioid receptor. *Chemistry*, *23*(19), 4615–4624. https://doi.org/10.1002/chem.201605575.

Sun, P., & Loh, H. H. (2001). Transcriptional regulation of mouse delta-opioid receptor gene: Role of Ets-1 in the transcriptional activation of mouse delta-opioid receptor gene. *The Journal of Biological Chemistry*, *276*(48), 45462–45469. https://doi.org/10.1074/jbc.M104793200.

Sun, P., & Loh, H. H. (2002). Transcriptional regulation of mouse delta-opioid receptor gene. Role of Ikaros in the stimulated transcription of mouse delta-opioid receptor gene in activated T cells. *The Journal of Biological Chemistry*, *277*(15), 12854–12860. https://doi.org/10.1074/jbc.M112268200.

Sun, P., & Loh, H. H. (2003). Transcriptional regulation of mouse delta-opioid receptor gene. Ikaros-2 and upstream stimulatory factor synergize in trans-activating mouse delta-opioid receptor gene in T cells. *The Journal of Biological Chemistry*, *278*(4), 2304–2308. https://doi.org/10.1074/jbc.M208162200.

Suo, C., Sun, L., & Yang, S. (2014). Alpinetin activates the delta receptor instead of the kappa and mu receptor pathways to protect against rat myocardial cell apoptosis. *Experimental and Therapeutic Medicine*, 7(1), 109–116. https://doi.org/10.3892/etm.2013.1359.

Svoboda, K. R., Adams, C. E., & Lupica, C. R. (1999). Opioid receptor subtype expression defines morphologically distinct classes of hippocampal interneurons. *The Journal of Neuroscience*, *19*(1), 85–95.

Tejedor-Real, P., Mico, J. A., Smadja, C., Maldonado, R., Roques, B. P., & Gilbert-Rahola,-J. (1998). Involvement of delta-opioid receptors in the effects induced by endogenous enkephalins on learned helplessness model. *European Journal of Pharmacology*, *354*(1), 1–7.

Thompson, A. A., Liu, W., Chun, E., Katritch, V., Wu, H., Vardy, E., et al. (2012). Structure of the nociceptin/orphanin FQ receptor in complex with a peptide mimetic. *Nature*, *485*(7398), 395–399. https://doi.org/10.1038/nature11085.

Thomsen, A. R. B., Plouffe, B., Cahill, T. J., 3rd, Shukla, A. K., Tarrasch, J. T., Dosey, A. M., et al. (2016). GPCR-G protein-beta-arrestin super-complex mediates sustained G protein signaling. *Cell*, *166*(4), 907–919. https://doi.org/10.1016/j.cell.2016.07.004.

Tsao, P. I., & von Zastrow, M. (2000). Type-specific sorting of G protein-coupled receptors after endocytosis. *The Journal of Biological Chemistry*, *275*(15), 11130–11140.

Tudashki, H. B., Robertson, D. N., Schiller, P. W., & Pineyro, G. (2014). Endocytic profiles of delta-opioid receptor ligands determine the duration of rapid but not sustained cAMP responses. *Molecular Pharmacology*, *85*(1), 148–161. https://doi.org/10.1124/mol.113.089003.

Tuusa, J. T., Leskela, T. T., & Petaja-Repo, U. E. (2010). Human delta opioid receptor biogenesis is regulated via interactions with SERCA2b and calnexin. *The FEBS Journal*, *277*(13), 2815–2829. https://doi.org/10.1111/j.1742-4658.2010.07699.x.

Ujcikova, H., Brejchova, J., Vosahlikova, M., Kagan, D., Dlouha, K., Sykora, J., et al. (2014). Opioid-receptor (OR) signaling cascades in rat cerebral cortex and model cell lines: The role of plasma membrane structure. *Physiological Research*, *63*(Suppl. 1), S165–S176.

Valant, C., Sexton, P. M., & Christopoulos, A. (2009). Orthosteric/allosteric bitopic ligands: Going hybrid at GPCRs. *Molecular Interventions*, *9*(3), 125–135. https://doi.org/10.1124/mi.9.3.6.

van Rijn, R. M., Brissett, D. I., & Whistler, J. L. (2012). Emergence of functional spinal delta opioid receptors after chronic ethanol exposure. *Biological Psychiatry*, *71*(3), 232–238.

Váradi, A., Marrone, G. F., Palmer, T. C., Narayan, A., Szabó, M. R., Le Rouzic, V., et al. (2016). Mitragynine/corynantheidine pseudoindoxyls as opioid analgesics with Mu agonism and delta antagonism, which do not recruit β-arrestin-2. *Journal of Medicinal Chemistry*, *59*(18), 8381–8397. https://doi.org/10.1021/acs.jmedchem.6b00748.

Walwyn, W., Maidment, N. T., Sanders, M., Evans, C. J., Kieffer, B. L., & Hales, T. G. (2005). Induction of delta opioid receptor function by up-regulation of membrane receptors in mouse primary afferent neurons. *Molecular Pharmacology*, *68*(6), 1688–1698. https://doi.org/10.1124/mol.105.014829.

Wang, G., Liu, T., Wei, L. N., Law, P. Y., & Loh, H. H. (2005). DNA methylation-related chromatin modification in the regulation of mouse delta-opioid receptor gene. *Molecular Pharmacology*, *67*(6), 2032–2039. https://doi.org/10.1124/mol.105.011056.

Wang, Q., Liu-Chen, L. Y., & Traynor, J. R. (2009). Differential modulation of mu- and delta-opioid receptor agonists by endogenous RGS4 protein in SH-SY5Y cells. *The Journal of Biological Chemistry*, *284*(27), 18357–18367. https://doi.org/10.1074/jbc.M109.015453.

Wang, H., & Pickel, V. M. (2001). Preferential cytoplasmic localization of delta-opioid receptors in rat striatal patches: Comparison with plasmalemmal mu-opioid receptors. *The Journal of Neuroscience*, *21*(9), 3242–3250.

Wang, G., Wei, L. N., & Loh, H. H. (2003). Transcriptional regulation of mouse delta-opioid receptor gene by CpG methylation: Involvement of Sp3 and a methyl-CpG-binding protein, MBD2, in transcriptional repression of mouse delta-opioid receptor gene in Neuro2A cells. *The Journal of Biological Chemistry*, *278*(42), 40550–40556. https://doi.org/10.1074/jbc.M302879200.

Wang, H. B., Zhao, B., Zhong, Y. Q., Li, K. C., Li, Z. Y., Wang, Q., et al. (2010). Coexpression of delta- and mu-opioid receptors in nociceptive sensory neurons. *Proceedings of the National Academy of Sciences of the United States of America*, *107*(29), 13117–13122. https://doi.org/10.1073/pnas.1008382107.

Wei, L. N., & Loh, H. H. (2011). Transcriptional and epigenetic regulation of opioid receptor genes: Present and future. *Annual Review of Pharmacology and Toxicology*, *51*, 75–97. https://doi.org/10.1146/annurev-pharmtox-010510-100605.

Whistler, J. L., Enquist, J., Marley, A., Fong, J., Gladher, F., Tsuruda, P., et al. (2002). Modulation of postendocytic sorting of G protein-coupled receptors. *Science*, *297*(5581), 615–620. https://doi.org/10.1126/science.1073308.

White, J. H., Wise, A., Main, M. J., Green, A., Fraser, N. J., Disney, G. H., et al. (1998). Heterodimerization is required for the formation of a functional GABA(B) receptor. *Nature*, *396*(6712), 679–682. https://doi.org/10.1038/25354.

Woltje, M., Kraus, J., & Hollt, V. (2000). Regulation of mouse delta-opioid receptor gene transcription: Involvement of the transcription factors AP-1 and AP-2. *Journal of Neurochemistry*, *74*(4), 1355–1362.

Wu, Z. Z., Chen, S. R., & Pan, H. L. (2008). Distinct inhibition of voltage-activated Ca2+ channels by delta-opioid agonists in dorsal root ganglion neurons devoid of functional T-type Ca2+ currents. *Neuroscience*, *153*(4), 1256–1267. https://doi.org/10.1016/j.neuroscience.2008.03.031.

Wu, H., Wacker, D., Mileni, M., Katritch, V., Han, G. W., Vardy, E., et al. (2012). Structure of the human kappa-opioid receptor in complex with JDTic. *Nature*, *485*(7398), 327–332. https://doi.org/10.1038/nature10939.

Xie, W. Y., He, Y., Yang, Y. R., Li, Y. F., Kang, K., Xing, B. M., et al. (2009). Disruption of Cdk5-associated phosphorylation of residue threonine-161 of the delta-opioid receptor: Impaired receptor function and attenuated morphine antinociceptive tolerance. *The Journal of Neuroscience*, *29*(11), 3551–3564. https://doi.org/10.1523/JNEUROSCI.0415-09.2009.

Yang, L., Seifert, A., Wu, D., Wang, X., Rankovic, V., Schroder, H., et al. (2010). Role of phospholipase D2/phosphatidic acid signal transduction in micro- and delta-opioid receptor endocytosis. *Molecular Pharmacology*, *78*(1), 105–113. https://doi.org/10.1124/mol.109.063107.

Yao, L., Fan, P., Jiang, Z., Mailliard, W. S., Gordon, A. S., & Diamond, I. (2003). Addicting drugs utilize a synergistic molecular mechanism in common requiring adenosine and Gi-beta gamma dimers. *Proceedings of the National Academy of Sciences of the United States of America*, *100*(24), 14379–14384. https://doi.org/10.1073/pnas.2336093100.

Zerari, F., Zouaoui, D., Gastard, M., Apartis, E., Fischer, J., Herbrecht, F., et al. (1994). Ultrastructural study of delta-opioid receptors in the dorsal horn of the rat spinal cord using monoclonal anti-idiotypic antibodies. *Journal of Chemical Neuroanatomy*, 7(3), 159–170.

Zhang, X., Bao, L., Arvidsson, U., Elde, R., & Hokfelt, T. (1998). Localization and regulation of the delta-opioid receptor in dorsal root ganglia and spinal cord of the rat and monkey: Evidence for association with the membrane of large dense-core vesicles. *Neuroscience*, *82*(4), 1225–1242.

Zhang, Z., & Pan, Z. Z. (2012). Signaling cascades for delta-opioid receptor-mediated inhibition of GABA synaptic transmission and behavioral antinociception. *Molecular Pharmacology*, *81*(3), 375–383. https://doi.org/10.1124/mol.111.076307.

Zhang, C., Srinivasan, Y., Arlow, D. H., Fung, J. J., Palmer, D., Zheng, Y., et al. (2012). High-resolution crystal structure of human protease-activated receptor 1. *Nature*, *492*(7429), 387–392. https://doi.org/10.1038/nature11701.

Zhang, X., Wang, F., Chen, X., Chen, Y., & Ma, L. (2008). Post-endocytic fates of delta-opioid receptor are regulated by GRK2-mediated receptor phosphorylation and distinct beta-arrestin isoforms. *Journal of Neurochemistry*, *106*(2), 781–792. https://doi.org/10.1111/j.1471-4159.2008.05431.x.

Zhao, B., Wang, H. B., Lu, Y. J., Hu, J. W., Bao, L., & Zhang, X. (2011). Transport of receptors, receptor signaling complexes and ion channels via neuropeptide-secretory vesicles. *Cell Research*, *21*(5), 741–753. https://doi.org/10.1038/cr.2011.29.

Zheng, Y. J., Wang, X. R., Chen, H. Z., Wu, X. J., Zhao, Y. H., & Su, D. S. (2012). Protective effects of the delta opioid peptide [D-Ala2, D-Leu5]enkephalin in an ex vivo model of ischemia/reperfusion in brain slices. *CNS Neuroscience & Therapeutics*, *18*(9), 762–766. https://doi.org/10.1111/j.1755-5949.2012.00360.x.

CHAPTER FOUR

The role of heat shock protein 90 in regulating pain, opioid signaling, and opioid antinociception

John M. Streicher*
Department of Pharmacology, College of Medicine, University of Arizona, Tucson, AZ, United States
*Corresponding author: e-mail address: JStreicher@email.arizona.edu

Contents

Abstract

Heat shock protein 90 (Hsp90) is one of the central signal transduction regulators of the cell. Via client interactions with hundreds of proteins, including receptors, receptor regulatory kinases, and downstream signaling regulators, Hsp90 has a crucial and wide-ranging impact on signaling in response to numerous drugs with impacts on resultant physiology and behavior. Despite this importance, however, Hsp90 has barely been studied in the context of pain and the opioid receptor system, leaving open the possibility that Hsp90 could be manipulated to improve pain therapeutic outcomes, a current area of massive medical need. In this review, we will highlight the known roles of Hsp90 in directly regulating the initiation and maintenance of the pain state. We will also explore how Hsp90 regulates signaling and antinociceptive responses to opioid analgesic drugs, with a special emphasis on ERK MAPK signaling. Understanding this new and growing area will improve our understanding of how Hsp90 regulates signaling and physiology, and also may provide new ways to treat pain, and perhaps reduce the severe impact of the ongoing opioid addiction and overdose crisis.

Vitamins and Hormones, Volume 111
ISSN 0083-6729
https://doi.org/10.1016/bs.vh.2019.05.010

1. Introduction

Heat shock protein 90 (Hsp90) is a crucial and central aspect of the protein regulatory machinery of the cell. Hsp90 was first described as a chaperone protein which acts late in the protein folding process to fully mature nascent proteins into their active conformations (Li & Buchner, 2013). Hsp90 is ubiquitously expressed, albeit with differences among four gene isoforms (Kim et al., 2016; Mishra, Ghosh, Stothert, Dickey, & Blagg, 2017; Shervington et al., 2008), and normally comprises 2–3% of the total protein content of the cell, potentially rising to 5–6% of total content under heat stress (Li & Buchner, 2013). Hsp90 has been most intensively studied in the field of cancer, where it was found that Hsp90 can be upregulated to promote proliferation and enhance protection in cancer cells. For this reason, Hsp90 inhibitors are a strong area for clinical development, with numerous 2nd and 3rd generation inhibitor compounds in clinical trials for the treatment of cancer (Kang et al., 2009; Sidera & Patsavoudi, 2014; Walsby, Lazenby, Pepper, Knapper, & Burnett, 2013).

However, Hsp90 has roles that extend beyond chaperone and protein folding activity. Hsp90 is a multifunctional and central regulator of acute signal transduction. Hsp90 directly regulates numerous receptor regulatory kinases, such as the G Protein Receptor Kinases (GRK), as well as the heterotrimeric G proteins, both downstream of G Protein Coupled Receptors (GPCRS), the largest class of receptors in the genome (Busconi, Guan, & Denker, 2000; Chen et al., 2013; Salim & Eikenburg, 2007; Vaiskunaite, Kozasa, & Voyno-Yasenetskaya, 2001; Waheed & Jones, 2002; Wu et al., 2012). Through these interactions, Hsp90 can broadly regulate hundreds of different receptor systems. Hsp90 also strongly regulates the formation of selective signaling complexes, which brings together different signaling proteins that are required for normal signal transduction. Notably, the normal activity of Raf1 kinase requires Hsp90 to complex this protein with the small GTPase Ras (Mitra, Ghosh, Gayen, Roy, & Mandal, 2016). Hsp90 can also regulate protein localization within different compartments of the cell, which can be crucial for normal signaling function and imparts specific signaling within the different compartments. For example, Hsp90 is responsible for selectively translocating $G\alpha_{12}$ to lipid rafts, which is required for calcium signaling and transcriptional activation (Montgomery et al., 2014; Sabath et al., 2008). These additional features establish Hsp90 as a central and crucial signaling regulator of the cell.

However, despite this importance, Hsp90 has been understudied in the context of neuroscience, and barely studied at all in the fields of pain and the treatment of pain, especially opioid signaling. In this review, we will explore the limited field of study showing how this crucial signaling regulator impacts the initiation and maintenance of pain, and how Hsp90 regulates opioid signaling and antinociception. This emerging data suggest new approaches to manipulate Hsp90 to enhance the treatment of pain, which is an area of massive unmet medical need.

2. Hsp90 in pain

Pain, especially chronic pain, is a massive health problem in the United States and worldwide. In the US alone, nearly 100 million patients suffer from some sort of chronic pain, with an economic cost of more than $600 billion yearly (Gaskin & Richard, 2012). Moderate to severe chronic pain can seriously impair patient quality of life (Breivik et al., 2009). Opioid drugs like morphine are the gold standard for the treatment of moderate to severe chronic pain, especially post-surgical and cancer pain (Kroenke, Krebs, & Bair, 2009). However, opioid drugs are limited for their therapeutic efficacy by the development of serious side effects, particularly with dose-escalation. These side effects include tolerance, dependence and withdrawal, nausea, constipation, and especially reward and addiction (Kroenke et al., 2009; National Collaborating Centre for Cancer, 2012; Raehal & Bohn, 2011; Raehal, Walker, & Bohn, 2005). In addition, opioid drugs may not be effective for certain chronic pain types, especially migraine and musculoskeletal pain (Berthelot, Darrieutort-Lafitte, Le Goff, & Maugars, 2015; Charles & Pradhan, 2016). At the same time, alternative analgesic drugs like NSAIDS can lack efficacy in strong pain, and drugs like tricyclic antidepressants and gabapentinoids only work for a minority of the patient population (Kroenke et al., 2009). These limitations for the treatment of pain combined with vast medical need, and a burgeoning opioid abuse crisis, emphasize the great need to understand the mechanisms of pain to improve our ability to design novel and efficacious treatments without the limitations of existing drugs.

Hsp90 has been shown by several different research groups to regulate the initiation and onset of different types of pain. In general, these studies show that Hsp90 inhibitors are beneficial in reducing the onset or severity of the pain, suggesting the use of Hsp90 inhibitors to directly manage pain. Linda Watkins and colleagues have published the most studies on this topic. In early studies, they found that Hsp90 acted to promote Toll-Like Receptor 4 (TLR4) signaling, which is a critical pro-inflammatory receptor that initiates or enhances inflammatory and neuropathic pain. The use of an Hsp90 activator enhanced TLR4-mediated pain, and an Hsp90 inhibitor significantly reduced the hyperalgesia and allodynia produced by TLR4 activation (Hutchinson et al., 2009; Lewis et al., 2010). These results suggested that Hsp90 inhibitors could be used to ameliorate the initiation or extent of inflammatory-related pain, such as neuropathic pain. This hypothesis was supported by a recent study from the Watkins group which showed that Hsp90 inhibition could reverse the long-term allodynia produced by morphine in a chronic constriction injury (CCI) model of neuropathic pain (Grace et al., 2018). Similar to these inflammatory findings, another group found that Hsp90 inhibition reduced viral titers and inflammation in a Chikungunya virus infection and pain model (Rathore et al., 2014).

Another group that has been active in this field is the collaborative team of the Hsp90 medicinal chemist Brian Blagg from the University of Notre Dame and the diabetic pain scientist Rick Dobrowsky from the University of Kansas. Their work focused on the novel C-terminal targeting novobiocin-scaffold derived Hsp90 inhibitor KU-32 and derivatives (Anyika, McMullen, Forsberg, Dobrowsky, & Blagg, 2016; Burlison, Neckers, Smith, Maxwell, & Blagg, 2006). So far this family of inhibitors has seemed to avoid the toxicity found in first generation geldanamycin-scaffold derivatives, suggesting their potential for non-cancer-related chronic therapy (Jhaveri et al., 2014). This team found that KU-32 and related compounds blocked or ameliorated the allodynia, hyperalgesia, and sensory hypoalgesia (numbness) caused by diabetic peripheral neuropathy, induced by streptozotocin treatment (Urban et al., 2010, 2012). Intriguingly, the mechanism of this effect is not through modulation of inflammatory signaling as above; instead these Hsp90 inhibitors are broadly neuroprotective, and block apoptosis, improve mitochondrial energetics, and preserve neurons in response to a variety of insults (Ansar et al., 2007; Lu, Ansar, Michaelis, & Blagg, 2009; Ma, Pan, Anyika, Blagg, & Dobrowsky, 2015). These inhibitors still block cancer cell proliferation, showing that Hsp90 inhibitors can be selectively toxic to cancer cells while preserving normal cells (Samadi et al., 2011; Zhao, Anyika, Girgis, & Blagg, 2014). These inhibitors also block neuronal toxicity in response to Aβ protein, suggesting their neuroprotective potential in neurodegenerative diseases like Alzheimer's Disease (Ansar et al., 2007). This hypothesis is partially supported by a study from another group showing that a geldanamycin-scaffold Hsp90 inhibitor is protective in a stroke model (Kim, Kim, & Yenari, 2015). These collective results show that new generations of Hsp90 inhibitors can be used to manage chronic pain, and a KU-32 derivative is currently in human clinical trials for the treatment of diabetic peripheral neuropathy (B. Blagg, personal communication).

A few other studies have also reported a role for Hsp90 in regulating pain. Gene variants of the mitochondrial Hsp90 isoform TRAP1 were associated with chronic pain in functional disorders, although a mechanism was not identified (Boles et al., 2015). Another group found that the transcription factor ATF3 promoted Hsp90 expression in the dorsal root ganglion of a rat model of monoarthritis. Treatment with an Hsp90 inhibitor reduced the mechanical allodynia in these rats, and further decreased glial inflammatory activation and neuronal sensitization (Nascimento et al., 2017).

Although the literature on Hsp90 in pain is still limited, it is also consistent. In several different pain models, Hsp90 inhibitors decreased the initiation or extent of the pain state. This occurs through at least two different mechanisms by reducing inflammation and a direct neuroprotective effect. These results are promising, and when combined with new generations of Hsp90 inhibitors lacking toxic side effects, suggest new and effective avenues for treating pain. Given the great medical need for new pain treatments, additional study is greatly needed to define the role of Hsp90 in more diverse pain models, and define the limits of therapy, such as potential tolerance or other side effects.

3. Hsp90 in opioid signaling and antinociception

While a consistent picture has so far emerged on the role of Hsp90 in regulating pain, less is known about how Hsp90 regulates opioid receptor signaling and antinociception. As explained above, opioid drugs are the gold standard for the treatment of pain, but are limited by low efficacy in some pain types, and the development of significant side effects. However, few alternate efficacious pain management treatments exist, leaving opioids as often the only choice. In addition, high prescription rates in the 2000s have driven an opioid boom in the United States, with a concomitant increase in opioid addiction and abuse, as well as overdose and death (Warner, Trinidad, Bastian, Minino, & Hedegaard, 2016). New treatments are desperately needed to manage pain without the side effects of opioids, as well as treat opioid addiction and overdose. Hsp90 represents an attractive target to investigate in the opioid system, due to the centrality of this protein in signaling regulation as described above.

A few studies suggested that Hsp90 interacts with and regulates opioid signaling, antinociception, and side effects. The research group of Fujimura and colleagues showed that Hsp90 inhibition blocked morphine-induced cyclic AMP (cAMP) sensitization in neuroblastoma cells (Koshimizu et al., 2010). cAMP sensitization and overshoot is an in vitro consequence of chronic opioid treatment, and has been hypothesized to be the molecular basis of opioid dependence and withdrawal (Ananthan et al., 2012; Stevenson et al., 2015). These results thus suggested that Hsp90 inhibition could block opioid dependence and withdrawal. This hypothesis was supported by a further study from Lakshmi Devi's group, who found through proteomic analysis that Hsp90 was increased in the brain synaptic fraction after chronic morphine treatment. They further found that acute treatment with an Hsp90 inhibitor reduced morphine withdrawal behavior, supporting this hypothesis, and suggesting that increased Hsp90 in the brain mediates at least one opioid side effect (Abul-Husn et al., 2011). In line with these findings, several groups found that chronic morphine treatment increased the levels of Hsp90 and Hsp70 mRNA in the brain, although they did not determine the function of Hsp90 in these studies (Ammon-Treiber et al., 2004; Salas et al., 2007). One of the studies cited above in the pain section also described how Hsp90 mediates in part the role of morphine in producing hyperalgesia in neuropathic pain, suggesting another role for Hsp90 in regulating opioid side effects (Grace et al., 2018). Lastly, some of the pain studies above found that Hsp90 inhibition potentiated the efficacy of morphine in these pain models, although these effects are likely through decreasing inflammation and not due to interaction with the opioid system (Hutchinson et al., 2009; Lewis et al., 2010).

A few other studies also described a role for Hsp90 in regulating opioid signaling outside of the pain modulatory systems. Wong and colleagues found that morphine promoted Hsp90 cleavage, which further promoted chemotaxis in microglia; this effect was reversed by the opioid receptor antagonist naloxone (Tsai, Cheng, & Wong, 2015). Another group found that Hsp90 was responsible for the beneficial preconditioning effect caused by opioid treatment in models of myocardial infarction (Small, Lu, Hsu, Gross, & Gross, 2015). These studies show that Hsp90 could have a broad role in regulating opioid physiology outside of the pain system.

These combined studies are suggestive, but are limited, and do not investigate the role of the opioid receptor system in regulating pain. We thus set out to determine the role of Hsp90 in regulating opioid signaling and acute antinociception. We began our studies by investigating Hsp90 in the brain. We found that injection of Hsp90 inhibitor into the brain strongly blocked morphine antinocieption in models of post-surgical paw incision pain and HIV peripheral neuropathy; we observed a 131-fold shift in potency with Hsp90 inhibition in the paw incision model, demonstrating the impact of this regulation (Lei et al., 2017). We found widespread changes in kinase signaling in the brain, including the loss of ERK MAPK activation in response to opioid drugs. We further found that this loss of ERK MAPK signaling was responsible for the observed loss of morphine efficacy in these pain models, suggesting a molecular mechanism by which Hsp90 promotes opioid-induced ERK signaling in the brain to promote anti-nociception (Lei et al., 2017).

In further as-yet unpublished studies, we extended this investigation to the spinal cord, and found that Hsp90 inhibition *enhanced* opioid anti-nociception (*data not shown*). As in the brain, the role of Hsp90 was to regulate opioid-induced ERK MAPK signaling; however, in the spinal cord, we found that Hsp90 inhibition *enhanced* ERK MAPK signaling which was responsible for the enhanced anti-nociception. These studies establish an impactful and central role for Hsp90 in regulating opioid-induced ERK signaling, which then impacts opioid-mediated anti-nociception, which is opposite in the brain vs. the spinal cord (Fig. 1). In further studies we will explore the differential molecular mechanisms in brain vs. spinal cord that leads to these effects; using selective inhibitors and CRISPR/Cas9 editing in adult brain, we have already determined that the Hsp90α isoform and the co-chaperones p23 and Cdc37 are responsible for the brain effects observed (Fig. 1; *data not shown*).

Our studies have begun to illuminate the molecular signaling mechanisms by which Hsp90 regulates opioid signaling and anti-nociception in the CNS. Combined with the above studies, this work suggests that Hsp90 inhibition could be used to reduce opioid side effects, or enhance opioid efficacy when given in the spinal cord. We have already begun these studies. Further understanding the molecular cascades by which Hsp90 regulates ERK MAPK and opioid anti-nociception could also suggest additional targets that could be successfully manipulated to improve therapeutic outcomes; analogous approaches are being pursued for Hsp70 and Hsp90 in other disease contexts (Assimon, Gillies, Rauch, & Gestwicki, 2013; Liu et al., 2015; Mishra et al., 2017).

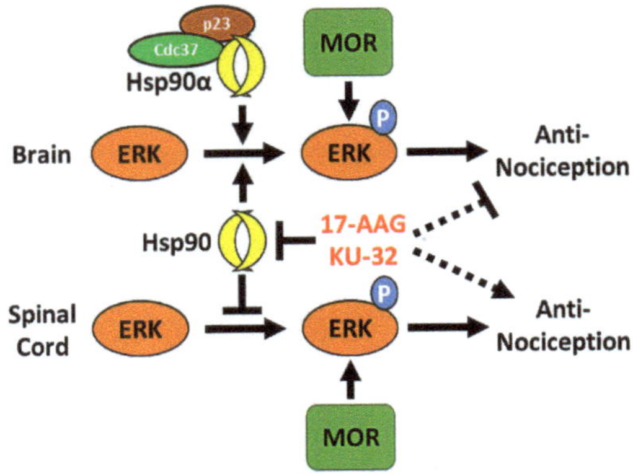

Fig. 1 Summary of the Role of Hsp90 in Regulating ERK MAPK in the Brain vs. Spinal Cord. The mu opioid receptor (MOR) promotes ERK MAPK phosphorylation and activation in response to opioid drugs in both the brain and spinal cord. Activated ERK MAPK then goes on to promote antinociception in both tissues. Hsp90 differentially regulates this signaling activation in the brain vs. spinal cord, promoting in the brain, and repressing in the spinal cord. Hsp90 inhibitors (17-AAG, KU-32) or co-chaperone inhibitors (gedunin, celastrol) reverse this relationship, leading to reduced ERK MAPK stimulation in the brain and enhanced in the spinal cord, thus blocking antinociception in the brain and enhancing it in the spinal cord (dashed lines). Using isoform and co-chaperone selective inhibitors and CRISPR constructs, we further showed that the Hsp90α isoform and the co-chaperones p23 and Cdc37 mediate this regulation in the brain. *This figure summarizes data from Lei, W., Mullen, N., McCarthy, S., Brann, C., Richard, P., Cormier, J., et al. (2017). Heat-shock protein 90 (Hsp90) promotes opioid-induced anti-nociception by an ERK mitogen-activated protein kinase (MAPK) mechanism in mouse brain.* The Journal of Biological Chemistry, 292*(25), 10414–10428. https://doi.org/10.1074/jbc.M116.769489 (web archive link) and unpublished data from the Streicher lab.*

4. Conclusion

The field of Hsp90 regulation of pain and opioid response is still small and limited. However, the studies that have been performed suggest that Hsp90 powerfully regulates pain and opioid response, via several different molecular mechanisms, including the regulation of inflammation, neuronal survival, and ERK MAPK signaling. These studies further suggest that Hsp90 inhibitors could block the development or extent of the pain state, or could enhance opioid anti-nociception and therapy if given in the spinal cord. Further studies are needed to extend these findings in multiple clinically relevant pain models, including an evaluation of the limits of therapy, including side effects. Further understanding of the molecular mechanism could permit more selective and efficacious targeting of the molecular machinery to improve outcomes; novel approaches to manipulate signaling such as biased agonism represent a potential way forward once this molecular signaling is understood (Schmid et al., 2017). Numerous Hsp90-selective tools, including isoform and co-chaperone selective inhibitors, are available, which will enhance these efforts (Anyika et al., 2016). The work has just begun for this exciting, impactful, and novel field of Hsp90 signaling regulation.

Acknowledgments

This work was supported by an Arizona Biomedical Research Commission New Investigator Award, as well as institutional funds from the University of Arizona. The author has no relevant conflicts of interest to disclose.

References

Abul-Husn, N. S., Annangudi, S. P., Ma'ayan, A., Ramos-Ortolaza, D. L., Stockton, S. D., Jr., Gomes, I., et al. (2011). Chronic morphine alters the presynaptic protein profile: Identification of novel molecular targets using proteomics and network analysis. *PLoS One*, *6*(10), e25535. https://doi.org/10.1371/journal.pone.0025535.

Ammon-Treiber, S., Grecksch, G., Stumm, R., Riechert, U., Tischmeyer, H., Reichenauer, A., et al. (2004). Rapid, transient, and dose-dependent expression of hsp70 messenger RNA in the rat brain after morphine treatment. *Cell Stress & Chaperones*, *9*(2), 182–197.

Ananthan, S., Saini, S. K., Dersch, C. M., Xu, H., McGlinchey, N., Giuvelis, D., et al. (2012). 14-Alkoxy- and 14-acyloxypyridomorphinans: Mu agonist/delta antagonist opioid analgesics with diminished tolerance and dependence side effects. *Journal of Medicinal Chemistry*, *55*(19), 8350–8363. https://doi.org/10.1021/jm300686p.

Ansar, S., Burlison, J. A., Hadden, M. K., Yu, X. M., Desino, K. E., Bean, J., et al. (2007). A non-toxic Hsp90 inhibitor protects neurons from Abeta-induced toxicity. *Bioorganic & Medicinal Chemistry Letters*, *17*(7), 1984–1990. https://doi.org/10.1016/j.bmcl.2007.01.017.

Anyika, M., McMullen, M., Forsberg, L. K., Dobrowsky, R. T., & Blagg, B. S. (2016). Development of Noviomimetics as C-terminal Hsp90 inhibitors. *ACS Medicinal Chemistry Letters*, 7(1), 67–71. https://doi.org/10.1021/acsmedchemlett.5b00331.

Assimon, V. A., Gillies, A. T., Rauch, J. N., & Gestwicki, J. E. (2013). Hsp70 protein complexes as drug targets. *Current Pharmaceutical Design*, *19*(3), 404–417.

Berthelot, J. M., Darrieutort-Lafitte, C., Le Goff, B., & Maugars, Y. (2015). Strong opioids for noncancer pain due to musculoskeletal diseases: Not more effective than acetaminophen or NSAIDs. *Joint, Bone, Spine*, *82*(6), 397–401. https://doi.org/10.1016/j.jbspin.2015.08.003.

Boles, R. G., Hornung, H. A., Moody, A. E., Ortiz, T. B., Wong, S. A., Eggington, J. M., et al. (2015). Hurt, tired and queasy: Specific variants in the ATPase domain of the TRAP1 mitochondrial chaperone are associated with common, chronic "functional" symptomatology including pain, fatigue and gastrointestinal dysmotility. *Mitochondrion*, *23*, 64–70. https://doi.org/10.1016/j.mito.2015.05.002.

Breivik, H., Cherny, N., Collett, B., de Conno, F., Filbet, M., Foubert, A. J., et al. (2009). Cancer-related pain: A pan-European survey of prevalence, treatment, and patient attitudes. *Annals of Oncology*, *20*(8), 1420–1433. https://doi.org/10.1093/annonc/mdp001.

Burlison, J. A., Neckers, L., Smith, A. B., Maxwell, A., & Blagg, B. S. (2006). Novobiocin: Redesigning a DNA gyrase inhibitor for selective inhibition of hsp90. *Journal of the American Chemical Society*, *128*(48), 15529–15536. https://doi.org/10.1021/ja065793p.

Busconi, L., Guan, J., & Denker, B. M. (2000). Degradation of heterotrimeric Galpha(o) subunits via the proteosome pathway is induced by the hsp90-specific compound geldanamycin. *The Journal of Biological Chemistry*, *275*(3), 1565–1569.

Charles, A., & Pradhan, A. A. (2016). Delta-opioid receptors as targets for migraine therapy. *Current Opinion in Neurology*, *29*(3), 314–319. https://doi.org/10.1097/WCO.0000000000000311.

Chen, M., Sato, P. Y., Chuprun, J. K., Peroutka, R. J., Otis, N. J., Ibetti, J., et al. (2013). Prodeath signaling of G protein-coupled receptor kinase 2 in cardiac myocytes after ischemic stress occurs via extracellular signal-regulated kinase-dependent heat shock protein 90-mediated mitochondrial targeting. *Circulation Research*, *112*(8), 1121–1134. https://doi.org/10.1161/circresaha.112.300754.

Gaskin, D. J., & Richard, P. (2012). The economic costs of pain in the United States. *The Journal of Pain*, *13*(8), 715–724. S1526-5900(12)00559-7 [pii]. https://doi.org/10.1016/j.jpain.2012.03.009.

Grace, P. M., Strand, K. A., Galer, E. L., Rice, K. C., Maier, S. F., & Watkins, L. R. (2018). Protraction of neuropathic pain by morphine is mediated by spinal damage associated molecular patterns (DAMPs) in male rats. *Brain, Behavior, and Immunity*, *72*, 45–50. https://doi.org/10.1016/j.bbi.2017.08.018.

Hutchinson, M. R., Ramos, K. M., Loram, L. C., Wieseler, J., Sholar, P. W., Kearney, J. J., et al. (2009). Evidence for a role of heat shock protein-90 in toll like receptor 4 mediated pain enhancement in rats. *Neuroscience*, *164*(4), 1821–1832. https://doi.org/10.1016/j.neuroscience.2009.09.046.

Jhaveri, K., Ochiana, S. O., Dunphy, M. P., Gerecitano, J. F., Corben, A. D., Peter, R. I., et al. (2014). Heat shock protein 90 inhibitors in the treatment of cancer: Current status and future directions. *Expert Opinion on Investigational Drugs*, *23*(5), 611–628. https://doi.org/10.1517/13543784.2014.902442.

Kang, B. H., Plescia, J., Song, H. Y., Meli, M., Colombo, G., Beebe, K., et al. (2009). Combinatorial drug design targeting multiple cancer signaling networks controlled by mitochondrial Hsp90. *The Journal of Clinical Investigation*, *119*(3), 454–464. https://doi.org/10.1172/JCI37613.

Kim, N., Kim, J. Y., & Yenari, M. A. (2015). Pharmacological induction of the 70-kDa heat shock protein protects against brain injury. *Neuroscience*, *284*, 912–919. https://doi.org/10.1016/j.neuroscience.2014.11.010.

Kim, H., Yang, J., Kim, M. J., Choi, S., Chung, J. R., Kim, J. M., et al. (2016). Tumor necrosis factor receptor-associated protein 1 (TRAP1) mutation and TRAP1 inhibitor Gamitrinib-triphenylphosphonium (G-TPP) induce a Forkhead box O (FOXO)-dependent cell protective signal from mitochondria. *The Journal of Biological Chemistry*, *291*(4), 1841–1853. https://doi.org/10.1074/jbc.M115.656934.

Koshimizu, T. A., Tsuchiya, H., Tsuda, H., Fujiwara, Y., Shibata, K., Hirasawa, A., et al. (2010). Inhibition of heat shock protein 90 attenuates adenylate cyclase sensitization after chronic morphine treatment. *Biochemical and Biophysical Research Communications*, *392*(4), 603–607. https://doi.org/10.1016/j.bbrc.2010.01.089.

Kroenke, K., Krebs, E. E., & Bair, M. J. (2009). Pharmacotherapy of chronic pain: A synthesis of recommendations from systematic reviews. *General Hospital Psychiatry*, *31*(3), 206–219. S0163-8343(09)00003-6 [pii]. https://doi.org/10.1016/j.genhosppsych.2008.12.006.

Lei, W., Mullen, N., McCarthy, S., Brann, C., Richard, P., Cormier, J., et al. (2017). Heat-shock protein 90 (Hsp90) promotes opioid-induced anti-nociception by an ERK mitogen-activated protein kinase (MAPK) mechanism in mouse brain. *The Journal of Biological Chemistry*, *292*(25), 10414–10428. https://doi.org/10.1074/jbc.M116.769489.

Lewis, S. S., Hutchinson, M. R., Rezvani, N., Loram, L. C., Zhang, Y., Maier, S. F., et al. (2010). Evidence that intrathecal morphine-3-glucuronide may cause pain enhancement via toll-like receptor 4/MD-2 and interleukin-1beta. *Neuroscience*, *165*(2), 569–583. https://doi.org/10.1016/j.neuroscience.2009.10.011.

Li, J., & Buchner, J. (2013). Structure, function and regulation of the hsp90 machinery. *Biomedical Journal*, *36*(3), 106–117. https://doi.org/10.4103/2319-4170.113230.

Liu, W., Vielhauer, G. A., Holzbeierlein, J. M., Zhao, H., Ghosh, S., Brown, D., et al. (2015). KU675, a concomitant heat-shock protein inhibitor of Hsp90 and Hsc70 that manifests isoform selectivity for Hsp90alpha in prostate Cancer cells. *Molecular Pharmacology*, *88*(1), 121–130. https://doi.org/10.1124/mol.114.097303.

Lu, Y., Ansar, S., Michaelis, M. L., & Blagg, B. S. (2009). Neuroprotective activity and evaluation of Hsp90 inhibitors in an immortalized neuronal cell line. *Bioorganic & Medicinal Chemistry*, *17*(4), 1709–1715. https://doi.org/10.1016/j.bmc.2008.12.047.

Ma, J., Pan, P., Anyika, M., Blagg, B. S., & Dobrowsky, R. T. (2015). Modulating molecular chaperones improves mitochondrial bioenergetics and decreases the inflammatory transcriptome in diabetic sensory neurons. *ACS Chemical Neuroscience*, *6*(9), 1637–1648. https://doi.org/10.1021/acschemneuro.5b00165.

Mishra, S. J., Ghosh, S., Stothert, A. R., Dickey, C. A., & Blagg, B. S. (2017). Transformation of the non-selective Aminocyclohexanol-based Hsp90 inhibitor into a Grp94-Seletive scaffold. *ACS Chemical Biology*, *12*(1), 244–253. https://doi.org/10.1021/acschembio.6b00747.

Mitra, S., Ghosh, B., Gayen, N., Roy, J., & Mandal, A. K. (2016). Bipartite role of heat shock protein 90 (Hsp90) keeps CRAF kinase poised for activation. *The Journal of Biological Chemistry*, *291*(47), 24579–24593. https://doi.org/10.1074/jbc.M116.746420.

Montgomery, E. R., Temple, B. R., Peters, K. A., Tolbert, C. E., Booker, B. K., Martin, J. W., et al. (2014). Galpha12 structural determinants of Hsp90 interaction are necessary for serum response element-mediated transcriptional activation. *Molecular Pharmacology*, *85*(4), 586–597. https://doi.org/10.1124/mol.113.088443.

Nascimento, D. S. M., Potes, C. S., Soares, M. L., Ferreira, A. C., Malcangio, M., Castro-Lopes, J. M., et al. (2017). Drug-induced HSP90 inhibition alleviates pain in Monoarthritic rats and alters the expression of new putative pain players at the DRG. *Molecular Neurobiology*. https://doi.org/10.1007/s12035-017-0628-x.

National Collaborating Centre for Cancer.. (2012). *Opioids in palliative care: Safe and effective prescribing of strong opioids for pain in palliative care of adults*. National Institute for Health and Care Excellence, Cardiff (UK).

Raehal, K. M., & Bohn, L. M. (2011). The role of beta-arrestin2 in the severity of antinociceptive tolerance and physical dependence induced by different opioid pain therapeutics. *Neuropharmacology*, *60*(1), 58–65. doi:S0028-3908(10)00202-9 [pii]. https://doi.org/10.1016/j.neuropharm.2010.08.003.

Raehal, K. M., Walker, J. K., & Bohn, L. M. (2005). Morphine side effects in beta-arrestin 2 knockout mice. *The Journal of Pharmacology and Experimental Therapeutics*, *314*(3), 1195–1201. doi:jpet.105.087254 [pii]10.1124/jpet.105.087254.

Rathore, A. P., Haystead, T., Das, P. K., Merits, A., Ng, M. L., & Vasudevan, S. G. (2014). Chikungunya virus nsP3 & nsP4 interacts with HSP-90 to promote virus replication: HSP-90 inhibitors reduce CHIKV infection and inflammation in vivo. *Antiviral Research*, *103*, 7–16. https://doi.org/10.1016/j.antiviral.2013.12.010.

Sabath, E., Negoro, H., Beaudry, S., Paniagua, M., Angelow, S., Shah, J., et al. (2008). Galpha12 regulates protein interactions within the MDCK cell tight junction and inhibits tight-junction assembly. *Journal of Cell Science*, *121*, 814–824. https://doi.org/10.1242/jcs.014878. Pt 6.

Salas, E., Alonso, E., Sevillano, J., Herradon, G., Bocos, C., Morales, L., et al. (2007). Morphine differentially regulates hsp90beta expression in the nucleus accumbens of Lewis and Fischer 344 rats. *Brain Research Bulletin*, *73*(4–6), 325–329. https://doi.org/10.1016/j.brainresbull.2007.04.007.

Salim, S., & Eikenburg, D. C. (2007). Role of 90-kDa heat shock protein (Hsp 90) and protein degradation in regulating neuronal levels of G protein-coupled receptor kinase 3. *The Journal of Pharmacology and Experimental Therapeutics*, *320*(3), 1106–1112. https://doi.org/10.1124/jpet.106.114835.

Samadi, A. K., Zhang, X., Mukerji, R., Donnelly, A. C., Blagg, B. S., & Cohen, M. S. (2011). A novel C-terminal HSP90 inhibitor KU135 induces apoptosis and cell cycle arrest in melanoma cells. *Cancer Letters*, *312*(2), 158–167. https://doi.org/10.1016/j.canlet.2011.07.031.

Schmid, C. L., Kennedy, N. M., Ross, N. C., Lovell, K. M., Yue, Z., Morgenweck, J., et al. (2017). Bias factor and therapeutic window correlate to predict safer opioid analgesics. *Cell*, *171*(5), 1165–1175. e1113. https://doi.org/10.1016/j.cell.2017.10.035.

Shervington, A., Cruickshanks, N., Lea, R., Roberts, G., Dawson, T., & Shervington, L. (2008). Can the lack of HSP90alpha protein in brain normal tissue and cell lines, rationalise it as a possible therapeutic target for gliomas? *Cancer Investigation*, *26*(9), 900–904. https://doi.org/10.1080/07357900802087259.

Sidera, K., & Patsavoudi, E. (2014). HSP90 inhibitors: Current development and potential in cancer therapy. *Recent Patents on Anti-Cancer Drug Discovery*, *9*(1), 1–20.

Small, B. A., Lu, Y., Hsu, A. K., Gross, G. J., & Gross, E. R. (2015). Morphine reduces myocardial infarct size via heat shock protein 90 in rodents. *BioMed Research International*, *2015*, 129612. https://doi.org/10.1155/2015/129612.

Stevenson, G. W., Luginbuhl, A., Dunbar, C., LaVigne, J., Dutra, J., Atherton, P., et al. (2015). The mixed-action delta/mu opioid agonist MMP-2200 does not produce conditioned place preference but does maintain drug self-administration in rats, and induces in vitro markers of tolerance and dependence. *Pharmacology, Biochemistry, and Behavior*, *132*, 49–55. https://doi.org/10.1016/j.pbb.2015.02.022.

Tsai, R. Y., Cheng, Y. C., & Wong, C. S. (2015). (+)-naloxone inhibits morphine-induced chemotaxis via prevention of heat shock protein 90 cleavage in microglia. *Journal of the Formosan Medical Association*, *114*(5), 446–455. https://doi.org/10.1016/j.jfma.2014.12.004.

Urban, M. J., Li, C., Yu, C., Lu, Y., Krise, J. M., McIntosh, M. P., et al. (2010). Inhibiting heat-shock protein 90 reverses sensory hypoalgesia in diabetic mice. *ASN Neuro*, *2*(4), e00040. https://doi.org/10.1042/AN20100015.

Urban, M. J., Pan, P., Farmer, K. L., Zhao, H., Blagg, B. S., & Dobrowsky, R. T. (2012). Modulating molecular chaperones improves sensory fiber recovery and mitochondrial function in diabetic peripheral neuropathy. *Experimental Neurology*, *235*(1), 388–396. https://doi.org/10.1016/j.expneurol.2012.03.005.

Vaiskunaite, R., Kozasa, T., & Voyno-Yasenetskaya, T. A. (2001). Interaction between the G alpha subunit of heterotrimeric G(12) protein and Hsp90 is required for G alpha(12) signaling. *The Journal of Biological Chemistry*, *276*(49), 46088–46093. https://doi.org/10.1074/jbc.M108711200.

Waheed, A. A., & Jones, T. L. (2002). Hsp90 interactions and acylation target the G protein Galpha 12 but not Galpha 13 to lipid rafts. *The Journal of Biological Chemistry*, *277*(36), 32409–32412. https://doi.org/10.1074/jbc.C200383200.

Walsby, E. J., Lazenby, M., Pepper, C. J., Knapper, S., & Burnett, A. K. (2013). The HSP90 inhibitor NVP-AUY922-AG inhibits the PI3K and IKK signalling pathways and synergizes with cytarabine in acute myeloid leukaemia cells. *British Journal of Haematology*, *161*(1), 57–67. https://doi.org/10.1111/bjh.12215.

Warner, M., Trinidad, J. P., Bastian, B. A., Minino, A. M., & Hedegaard, H. (2016). Drugs most frequently involved in drug overdose deaths: United States, 2010–2014. *National Vital Statistics Reports*, *65*(10), 1–15.

Wu, Z., Chen, Y., Yang, T., Gao, Q., Yuan, M., & Ma, L. (2012). Targeted ubiquitination and degradation of G-protein-coupled receptor kinase 5 by the DDB1-CUL4 ubiquitin ligase complex. *PLoS One*, 7(8), e43997. https://doi.org/10.1371/journal.pone.0043997.

Zhao, H., Anyika, M., Girgis, A., & Blagg, B. S. (2014). Novologues containing a benzamide side chain manifest anti-proliferative activity against two breast cancer cell lines. *Bioorganic & Medicinal Chemistry Letters*, *24*(15), 3633–3637. https://doi.org/10.1016/j.bmcl.2014.05.020.

CHAPTER FIVE

Enkephalinase regulation

Manuel Ramírez-Sánchez[a,*], Isabel Prieto[a], Ana-Belén Segarra[a], Magdalena Martínez-Cañamero[a], Inmaculada Banegas[a], Marc de Gasparo[b]

[a]Department of Health Sciences, University of Jaén, Jaén, Spain
[b]Cardiovascular & Metabolic Syndrome Adviser, Rossemaison, Switzerland
*Corresponding author: e-mail address: msanchez@ujaen.es

Contents

Abstract

After millennia of knowledge of opium, it was only recently that endogenous substances called opioids with similar properties to opium and derivatives were discovered. The first to be discovered were enkephalins. In addition to the regulation of their synthesis and expression of receptors, an important mechanism for the regulation of their functions carried out by multiple proteolytic enzymes acting at all levels of their structure is described. The action of such enzymes, known as enkephalinases, is also regulated by endogenous and exogenous factors which ultimately affect the control of the enkephalins's action. For therapeutic purposes, it is not only necessary to develop specific inhibitors but also to acquire a deep knowledge of the influence that such factors exert on their activities. This knowledge could help us to establish adapted therapeutic strategies in the treatment of pain or other processes in which enkephalinases are involved. In this chapter, some of these regulatory factors are discussed, such as regional and subcellular distribution, developmental changes, diurnal variations, hormonal influences, stress, dietary factors or interactions with other neurotransmitters.

Vitamins and Hormones, Volume 111
ISSN 0083-6729
https://doi.org/10.1016/bs.vh.2019.05.007

1. Enkephalins and enkephalinases

As the active form of morphine is the levo-isomer while the dextro-isomer has no biological activity, it was speculated on the existence of specific receptors for one form that did not recognize the other. It was Candace Pert and Snyder (1973) who first identified such receptors. Their discovery, along with the subsequent search for endogenous substances that bind such receptors, constituted a true revolt in the field of neurobiology. Finally, thanks to the work of the team lead by Hughes and Kosterlitz in Aberdeen (Hughes et al., 1975), these substances were discovered. These authors described two small penta-peptides in the brain with a high affinity for the receptors described by Snyder and called them "enkephalins" (inside the encephalon). These peptides, depending on the nature of the carboxy-terminal amino acid, were designated as Met-enkephalin and Leu-enkephalin (Fig. 1).

Almost simultaneously, several enzymatic activities capable of hydrolyzing both enkephalins were identified. Thus, a *soluble* aminopeptidase was described in brain of rat (Schnebli, Phillipps, & Barclay, 1979), monkey (Hayashi, 1978), bovine (Hersh, Smith, & McKelvy, 1980) and human (Traficante, Rotrosen, Siekierski, Tracer, & Gershon, 1980). Furthermore, at least two *membrane-bound* aminopeptidases from rat brain (Hersh, 1981)

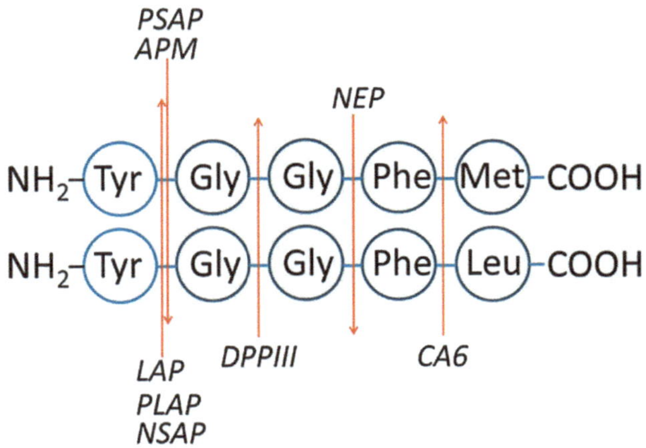

Fig. 1 Enkephalinases. Proteolytic enzymes acting on Leu- and Met-enkephalin. Arrows indicate the peptide bond on which they act. LAP, leucyl aminopeptidase; APM, alanyl aminopeptidase; PLAP, placental leucyl aminopeptidase; PSAP, Puromycin sensitive aminopeptidase; NSAP, neuron specific aminopeptidase; DPPIII, Dipeptidyl peptidase III, Carboxypeptidase A6; NEP, neutral endopeptidase (Table 1).

and a *membrane* dipeptidyl aminopeptidase that released the dipeptide Tyr-Gly (Gorenstein & Snyder, 1980; Hersh, 1986) were also identified. Currently, several proteolytic enzymes (enkephalinases) capable of acting on all the peptide bonds of both penta-peptides are identified (Table 1, Fig. 1). They gives an idea of the complex regulatory mechanisms that can be exerted on these peptides. In addition, the enzymes themselves are subject to the influence of multiple endogenous and exogenous factors, which will ultimately affect the control of the function of their endogenous substrates (Ramírez et al., 2008).

Enkephalins and enkephalinases, due to their heterogeneous substrate specificity, have been involved in multiple processes such as analgesia, cognitive processes, cardiovascular functions or feeding behavior (Bodnar, 2018; Mendez, Ostlund, Maidment, & Murphy, 2015; Henry, Gendron, Tremblay, & Drolet, 2017). Therefore, considering their important

Table 1 Enkephalinases.

Enzyme	Abbreviation	EC number	References
Leucyl aminopeptidase	LAP	EC 3.4.11.1	Gibson, Biggins, Lauffart, Mantle, and McDermott (1991)
Alanyl aminopeptidase	APM	EC 3.4.11.2	Hersh (1985)
Placental leucyl aminopeptidase	PLAP	EC 3.4.11.3	Gibson et al. (1991) and Matsumoto et al. (2001)
Puromycin sensitive aminopeptidase	PSAP	EC 3.4.11.14	Johnson and Hersh (1990) and Thompson and Hersh (2004)
Neuron specific aminopeptidase	NSAP	EC 3.4.11.-	Hui, Saito, and Hui, (1998)
Dipeptidyl peptidase III	DPPIII	EC 3.4.14.4	Lee and Snyder (1982) and Inaoka and Tamaoki (1987)
Carboxypeptidase A6	CA6	EC 3.4.17.1	Lyons, Callaway, and Fricker (2008)
Neutral endopeptidase, Dipeptidyl carboxypeptidase, Neprilysin	NEP	EC 3.4.24.11	Li, Booze, and Hersh (1996)

Proteolytic enzymes with enkephalinase activity.

pathophysiologic involvements and in order to prolong their beneficial effects, a large number of studies were quickly directed toward the identification and development of inhibitors of their activities (Fournie-Zaluski, 1988; Llorens et al., 1980) for the treatment, among others especially, of pain (Thanawala, Kadam, & Ghosh, 2008), anxiety (Zozulya et al., 2001) depressive disorders (de Felipe, Jiménez, Castro, & Fuentes, 1989; Tejedor-Real, Micó, Maldonado, Roques, & Gibert-Rahola, 1993) or hypertension and heart failure (Andersen, Simonsen, Wehland, Pietsch, & Grimm, 2016; Maki et al., 2001).

The currently available data on the function of enkephalinases is still partial, due to their broad substrate specificity but also to the influence that other multiple endogenous or exogenous factors exert on them. Therefore, in order to know more appropriately the function of these enzymes, it is necessary to know their regional or subcellular distribution as well as the influence of various endogenous factors such as gender, age, hormonal levels as well as exogenous factors such as environmental light–dark conditions, dietary habits or drug treatments as well as their response under pathological conditions such as stress or hypertension. This chapter focuses on the analysis of the regulation of enkephalinases in the nervous system.

2. Brain distribution and development (regional and subcellular)

Since the synthesis of neuropeptides is presumably carried out in the cellular soma and as it was demonstrated that some of them such as the enkephalins accumulate in the synaptosomes (Demmer & Brand, 1983), it is unlikely that the speed of synthesis serves as an essential control mechanism of their activity. It is more probable that the rate of release and the efficiency of their metabolism are the main mechanisms that control the biological activity of these neuropeptides. The existence of regional differences of a given substance in the central nervous system has been classically used as an argument for attributing a specific role for the function of the cells that contain it and separating it from other substances that fulfill a common function in all cells (Graham, Shank, Werman, & Aprison, 1967). This reasoning is also applicable to enkephalinase activity.

Regarding the regional and subcellular distribution of enkephalinases, the results are inclusive as they are heterogeneous, probably due to the different techniques used, experimental conditions, age of the animals and so on. As an example, Alba, Ramirez, Cantalejo, and Iribar (1988) and Alba, Arenas, Iribar, and Ramirez (1993) analyzed bilaterally membrane-bound and soluble enkephalinase activities in 11 brain regions, including cortical and subcortical ones. While the soluble activity was lower in the cortical regions than in the subcortical ones, the membrane-bound activity predominated in the cortical brain areas. In addition, the soluble activity was more homogeneous between brain regions than the membrane-bound one that showed a much more heterogeneous distribution (Fig. 2).

Interestingly, this distribution pattern of Sol vs MB activities is just the opposite that the one exhibited by enkephalin-convertase (carboxypeptidase E, EC 3.4.17.10) which demonstrated a higher heterogeneous distribution of soluble activity than the membrane one between brain regions (Supattapone, Fricker, & Snyder, 1984). Also, in contrast to Alba et al. (1993) and Hui, Wang, Wong, Tsai, and Lajtha (1981) described little difference in the degradation of Met- and Leu-enkephalin between several brain regions. However, the most interesting aspect of Alba's study (1993) was the significant difference between the left and right sides of the prefrontal cortex (Alba et al., 1988). This observation was the first neurochemical asymmetry of an enzymatic activity described in the literature. Subsequent studies, not only with enkephalinase activity but also with other proteolytic activities, showed that the brain asymmetry is not a static but a dynamic element subject to modulation of endogenous and exogenous factors (Ramírez, Prieto, Vives, de Gasparo, & Alba, 2004). In the case of enkephalinase activity, the existence of interhemispheric differences in the frontal cortex are obvious and change between normotensive and hypertensive animals, which are modified under hypotensive and hypertensive treatments, are observed (Fig. 3) (Prieto et al., 2019). The authors also reported a tendency to change the side of predominance from left to right and to increase the right predominance in hypertensive rats in comparison with normotensive ones. Hypertension has been related to mood disorders such as depression and this has been linked to changes in the basal brain asymmetry. Considering the influence of drug treatments on the bilateral pattern of neuropeptidase activities, including enkephalinase activity, Prieto et al. (2019) have suggested a possible beneficial influence of captopril over the treatment with propranolol.

To correctly understand the mechanisms that regulate the functional role of enkephalins, it was also necessary to analyze the subcellular distribution of enkephalinase activities as well as the possible modification during development and aging. For this purpose, Arechaga et al., (1996) determined the

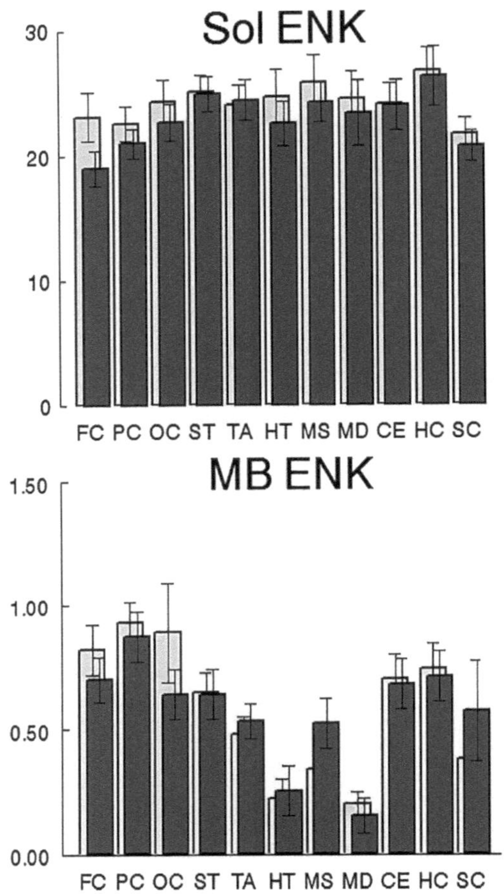

Fig. 2 Regional distribution of enkephalinase activity. Bilateral distribution of soluble (Sol) and membrane-bound (MB) enkephalinase (ENK) activity (leucyl aminopeptidase, LAP) in the left (light gray) and right (dark gray) selected zones of adult rat brain. FC, frontal cortex; PC, parietal cortex; OC, occipital cortex; ST, striatum; TA, thalamus; HY, hypothalamus; ME, mesencephalon; MD, medulla; CE, cerebellum; HC, hippocampus; SC, cervical spinal cord. Values represent mean ± SEM levels of nmol of leu-2-naphthylamide hydrolyzed per min per mg of proteins. *Modified from Alba, F., Ramirez, M., Cantalejo, E. S., & Iribar, C. (1988). Aminopeptidase activity is asymmetrically distributed in selected zones of rat brain.* Life Sciences, 43*(11), 935–939; Alba, F., Arenas, J. C., Iribar, C., & Ramirez, M. (1993). Regional distribution of soluble and membrane-bound aminopeptidase activities in rat brain.* Brain Research Bulletin, 31*(3–4), 393–396.*

subcellular distribution of enkephalinase activity at several stages of postnatal development of male rats: one-week-old rats, young adults (1-month-old), adults (5-month olds) and aged rats (2-years-old) (Fig. 4). They found a heterogeneous distribution of soluble and membrane-bound enkephalinase

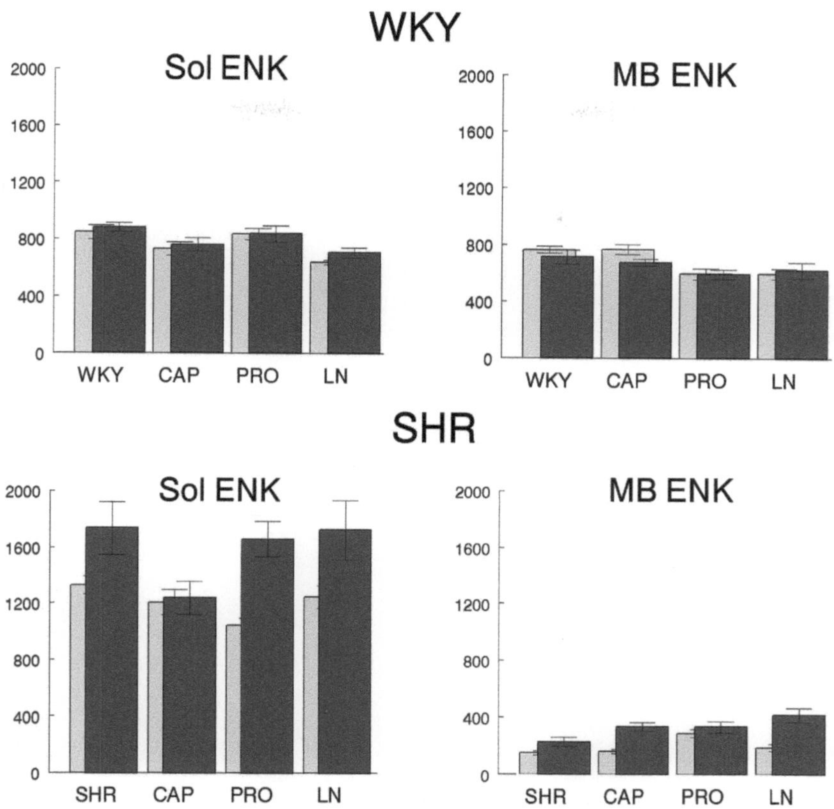

Fig. 3 Brain enkephalinase activity and blood pressure. Soluble (Sol) and membrane-bound enkephalinase (ENK) activity (Alanyl aminopeptidase, APM) in the in the left (light gray) and right (dark gray) frontal cortex of non-treated normotensive Wistar Kyoto (WKY) and spontaneously hypertensive rats (SHR) treated with captopril (CAP), propranolol (PRO) and L-NAME (LN). Values represent mean $\pm$ SEM levels of pmol of ala-2-naphthylamide hydrolyzed per min per mg of proteins. *Modified from Prieto, I., Segarra, A.B., Villarejo, A.B., de Gasparo, M., Martínez-Cañamero, M.M., & Ramírez-Sánchez, M. (2019). Neuropeptidase activity in the frontal cortex of Wistar-Kyoto and spontaneously hypertensive rats treated with vasoactive drugs: A bilateral study. Journal of Hypertension, 37(3), 612–628, https://doi.org/10.1097/HJH.0000000000001884.*

activities between fractions at all the stages analyzed. It was particularly interesting to notice the high levels of enkephalinase activity in the synaptosomal fraction of 5-months old adult in comparison with 1-week-old and 2-years-old aged rats: this suggested an active role in synaptic events that decreased in the advanced age of these animals. The high membrane-bound activity in the microsomal fraction that was evident at all ages is also remarkable especially its increase from 1-week to 1-month, which may be linked to the synthesis, catabolism or axoplasmic transport.

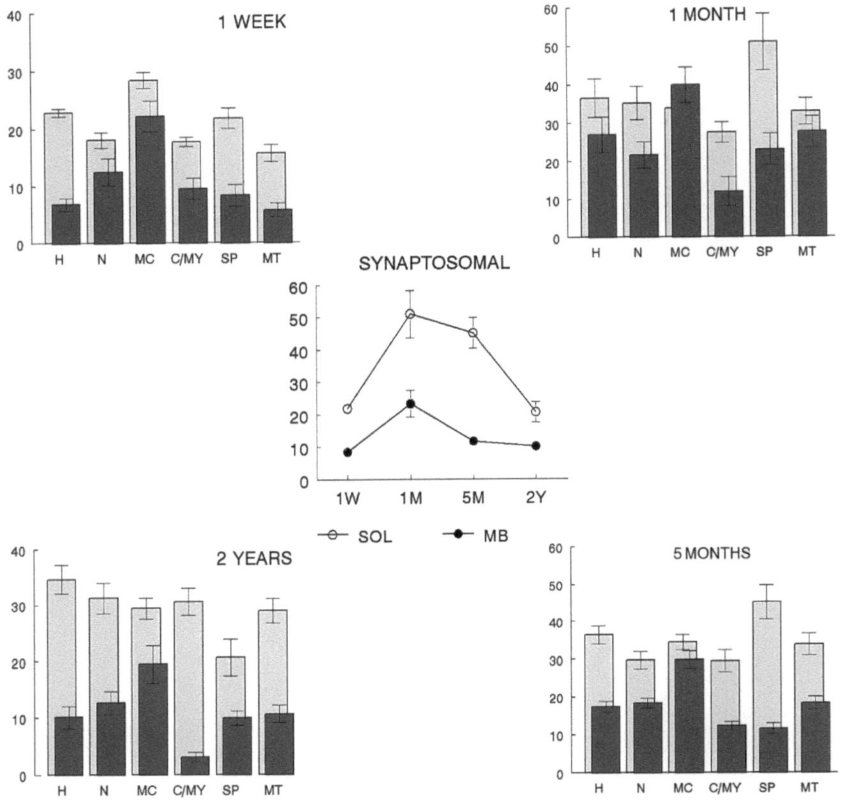

Fig. 4 Brain subcellular distribution of enkephalinase activity during development and aging. The *peripheral* figures represent the subcellular distribution of soluble (light gray) and membrane-bound (dark gray) enkephalinase activity (leucyl aminopeptidase, LAP) in 1-week-old (1 WEEK), 1-month-old (1 MONTH), 5-month-old (5 MONTHS) and 2-years-old (2 YEARS) male rats in the homogenate (H), nuclear (N), microsomal (MC), cytosol (C), myelin (MY), synaptosomal (SP) and mitochondrial (MT) fractions. The *central* figure represents the development of soluble (SOL) and membrane-bound (MB) fractions in the synaptosomal fraction of 1-week-old (1 W), 1-month-old (1 M), 5-month-old (5 M) and 2-years-old (2Y) male rats. Values represent mean $\pm$ SEM levels of nmol of ala-2-naphthylamide hydrolyzed per min per mg of proteins. *Modified from Arechaga, G., Sánchez, B., Prieto, I., Martínez, J.M., Alba, F., & Ramírez, M. (1996). Subcellular distribution of leucine aminopeptidase during the development and aging of rat brains.* Revista de Neurología, *24(136), 1503–1506.*

3. Endocrine regulation

Owing to the existence of gender differences in enkephalin-mediated analgesia (Kest, Sarton, & Dahan, 2000) and significant changes in the brain content of enkephalins during the estrous cycle of rats (Roman, Ploj, Gustafsson, Meyerson, & Nylander, 2006), there may also be gender differences and changes in the enkephalinases during the estrous cycle. In fact, changes of enkephalinase activity during the estrous cycle of the rat have been already reported in cortex, hypothalamus and pituitary of female rats, showing the highest levels in the proestrous phase, just when hormone release peaks (de Gandarias, Casis, Irazusta, Echevarria, & Ramirez, 1988) (Fig. 5).

In addition, gender differences in enkephalinase activity in hypothalamus were described, demonstrating higher levels in female than in male rats. However, gonadectomy inverted the difference to a predominance of gonadectomyzed males over gonadectomyzed females (de Gandarias, Ramirez, Zulaica, & Casis, 1989) (Fig. 5). These results suggested the influence of steroid hormones on enkephalinase activities, possibly linked to a change of the biochemical environment that regulates, at least in part, the activity of these enzymes. To analyze this hypothesis, the in vitro response of serum enkephalinase activity was evaluated in the

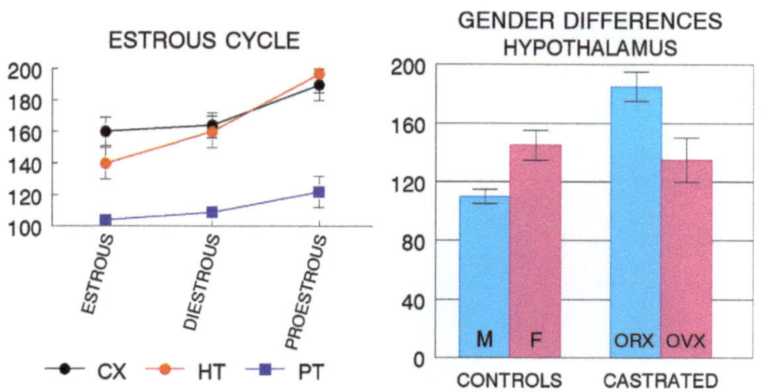

Fig. 5 Gender differences and change during the estrous cycle of brain enkephalinase activity. Left figure indicates the levels of enkephalinase activity (leucyl aminopeptidase, LAP) during the estrous cycle in the cortex (CX), hypothalamus (HT) and pituitary (PT) of adult female rats. Right figure indicates the levels of enkephalinase activity (leucyl aminopeptidase, LAP) in adult male (blue bars) (M), female (rose bars) (F), orchidectomyzed male (ORX) and ovariectomyzed (OVX) female rats. Values represent mean ± SEM levels of nmol of leu-2-naphthylamide hydrolyzed per min per mg of proteins. *Modified from de Gandarias, J.M., Casis, L., Irazusta, J., Echevarria, E., & Ramirez, M. (1988). Changes of aminopeptidase activity levels in serum and brain during the estrous cycle of the rat.* Hormone & Metabolic Research, 20*(12), 776; de Gandarias, J.M., Ramirez, M., Zulaica, J., & Casis, L. (1989). Aminopeptidase (arylamidase) activity in discrete areas of the rat brain: sex differences.* Hormone & Metabolic Research, 21*(5), 285–286.*

presence of cholesterol, estradiol, testosterone, progesterone, and hydrocortisone in control male and female and in gonadectomized male and female balb-C-mice (Martinez, Prieto, Ramirez, Alba, & Ramirez, 1997; Martinez, Ramirez, Prieto, Alba, & Ramirez, 1998) (Fig. 6).

This study clearly showed highly significant gender differences in enkephalinase activity. In addition, an influence of cholesterol and steroid hormones on enzyme activity was also demonstrated. Enkephalinase activity responded differently to the presence of these substances and also in a different way following gonadectomy: indeed, gonadectomy increased enkephalinase activity in male and female mice. On the other hand, whereas cholesterol did not modify enkephalinase activity, steroid hormones markedly stimulated

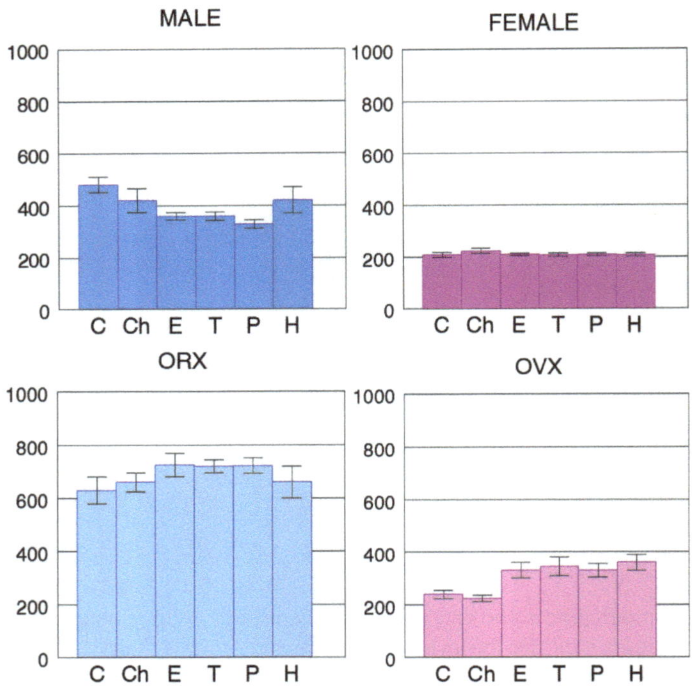

Fig. 6 Cholesterol and steroids influence on enkephalinase activity. Mean ± SEM levels of enkephalinase activity (Alanyl aminopeptidase, APM) in sera obtained from non-gonadectomized males, non-gonadectomized females, orchiectomized (ORX), and ovariectomized (OVX) rats, the four groups considered as controls (C), and in the same sera after incubation with cholesterol (Ch), 17-b-estradiol (E), testosterone (T), progesterone (P), and hydrocortisone (H). Specific activity was expressed as pmol of alanyl-2-naphthylamide hydrolyzed per min per mg of protein. *Modified from Martinez, J. M., Ramirez, M. J., Prieto, I., Alba, F., & Ramirez, M. (1998). Sex differences and in vitro effects of steroids on serum aminopeptidase activities.* Peptides, *19(9), 1637–1640.*

or inhibited it. Indeed, there was a clear difference between males and females: whereas in control females the activity did not change after incubation with steroid hormones, it increased significantly in ovariectomited rats. In contrast, whereas in control males enkephalinase decreased after incubation with steroid hormones, it did not change in orchiectomized males. These results not only demonstrated a direct or an indirect influence of steroid hormones on enkephalinase activity (stimulating in females and inhibiting in males), but also suggested that the enzyme may be different in males and in females or that factors other than sex steroids influence the different response between sexes. The increased activity after gonadectomy in male and female mice was not reversed by steroid treatment. In contrast, incubation with estradiol or testosterone increased the activity after gonadectomy in females. This implies that, apart from sex steroids, other factors must be involved (Martinez et al., 1997; Martinez et al., 1998) (Fig. 6). These results strongly suggest that steroid hormones, direct or indirectly, modulate enkephalinase activities and, as a consequence, may also regulate the functions in which their endogenous substrates, such as enkephalins, are involved in. In addition to steroid hormones, an influence of thyroid hormones on enkephalinase activity was also suggested. Indeed a significant increase in this enzyme activity was demonstrated in the posterior pituitary of hyperthyroid rats compared to euthyroid controls (Prieto et al., 2003).

4. Diet

The type of diet may be an important external factor influencing enkephalinase activity. The orexigenic character of enkephalins (Poon, Alam, Karatayev, Barson, & Leibowitz, 2015) could indeed be modulated by the type of fat predominant in the diet. In fact, studies in animals have shown that high-fat diets stimulate the level of enkephalins which, increase food intake (Huang, Han, South, & Storlien, 2003; Jo, Su, Gutierrez-Juarez, & Chua, 2009). Poon et al. (2015) have shown that the saturated fatty acid palmitic and the monounsaturated oleic acid had a differential effect on the brain transcription of enkephalins: while the palmitic acid stimulated the levels of mRNA and enkephalins in hypothalamic and forebrain neurons, the oleic acid had no effects on enkephalin mRNA but also increased enkephalin levels. The authors concluded that a saturated fatty acid has a stronger stimulatory effect on brain enkephalins compared to a monounsaturated acid.

In this context, Chang, Karatayev, Davydova, and Leibowitz (2004) studied the possible effect of an increased circulating triglycerides on the hypothalamic levels of orexigenic peptides, including enkephalins. Their results showed that the increase in triglycerides produced a marked elevation in the expression of enkephalins, specifically in the arcuate nucleus. In apparent contrast with these results, Kaneko et al. (2014) have shown that central activation of the delta-opioid receptor stimulated the intake of a standard diet but suppressed the intake of a high-fat diet. In any case, there are numerous reports that relate the type of fat in the diet with the variation of brain enkephalin levels. There was indeed a significant modifications of frontal cortex enkephalinase activity depending on the type of fat (saturated or unsaturated) used in the diet (Segarra et al., 2019). These results demonstrated that the profile of fatty acids in frontal cortex changed depending on the type of fat used in the diet and also indicated that the higher the degree of saturation, the lower the level of soluble enkephalinase activity: polyunsaturated (fish oil) > monounsaturated (olive oil) > saturated (coconut oil). In contrast, the membrane-bound enkephalinase activity showed a reverse behavior: polyunsaturated > monounsaturated = saturated (Fig. 7).

In addition, these results demonstrated the existence of significant correlations between the type of fatty acids and enkephalinase activities depending on the diet. While no correlations were observed in animals fed the diet enriched with fish oil, significant negative correlations did appear in the groups fed olive or coconut oil. For example, in the group of rats fed olive oil, the higher the level of docoxohesanoic acid (p22: 6w3), the lower the level of soluble enkephalinase activity or in the coconut oil group, the higher the level of mead acid (p20:3w9) the lower the level of membrane-bound enkephalinase activity (Fig. 8).

The authors emphasize the importance of such results since changes in the level of activity of enkephalinases of the frontal cortex, which depend on the type of fat used in the diet, could be indicative of changes in the levels of their endogenous substrates (enkephalins) and as a consequence, of changes in the functions that they exert at that location as analgesic or cognitive functions including anxiety or depression. Such influence of the type of fat in the diet on enkephalinase activity was also demonstrated in other tissues than brain, such as in kidney (Villarejo, Ramírez-Sánchez, Segarra, Martínez-Cañamero, & Prieto, 2015). There was an increased activity under a diet enriched with olive oil in comparison with a standard diet.

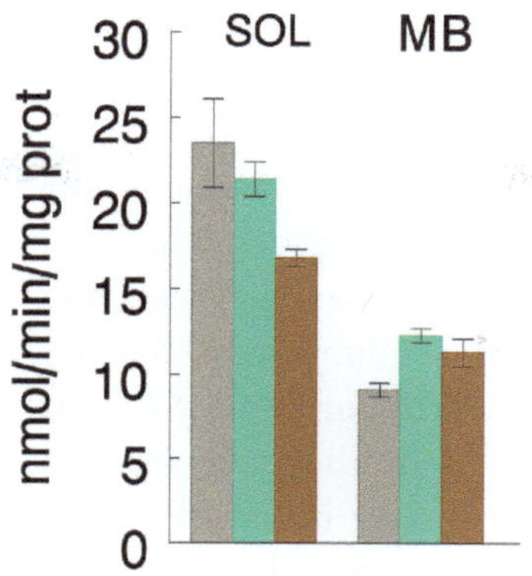

Fig. 7 Dietary fat influence on brain enkephalinase activity. Soluble (SOL) and membrane-bound (MB) enkephalinase activities (Alanyl aminopeptidase, APM), expressed as nmol of Ala-β-naphthylamide hydrolyzed per min per mg of proteins in frontal cortex of rats fed diets enriched with fish oil (gray bars), olive oil (cyan bars) or coconut oil (brown bars). *Modified from Segarra, A.B., Prieto, I., Martínez-Cañamero, M., Ruíz-Sanz, J.I., Ruíz-Larrea, M.B., de Gasparo, M., et al. (2019) Enkephalinase activity is modified and correlates with fatty acids in frontal cortex depending on the type of fat used in the diet.* Endocrine Regulations 53(2), 59–64, *https://doi.org/10.2478/enr-2019-0007.*

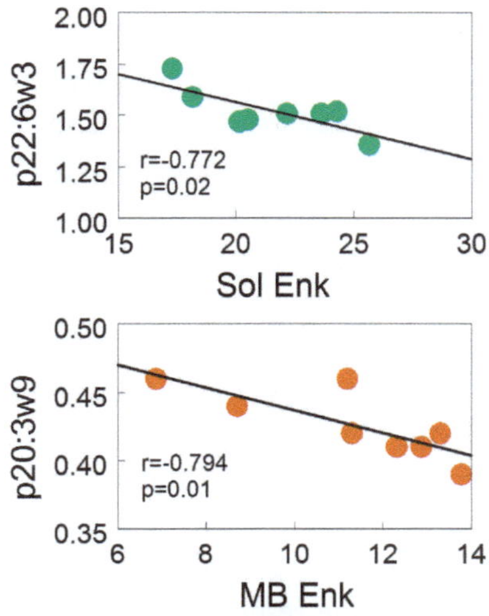

Fig. 8 Brain fatty acids and enkephalinase activity. Example of significant correlations (r/p values) observed between individual fatty acids such as p22:6w3 and p20:3w9 versus, respectively, soluble (Sol Enk) or membrane-bound (MB Enk) enkephalinase activities in frontal cortex of adult male rats fed diets enriched with olive (cyan) or coconut (brown) oils. *Modified from Segarra, A.B., Prieto, I., Martínez-Cañamero, M., Ruíz-Sanz, J.I., Ruíz-Larrea, M.B., de Gasparo, M., et al. (2019) Enkephalinase activity is modified and correlates with fatty acids in frontal cortex depending on the type of fat used in the diet. Endocrine Regulations 53(2), 59–64, https://doi.org/10.2478/enr-2019-0007.*

5. Daily rhythm

An important mechanism in the neuroendocrine regulation involves the circadian behavior of neuropeptides, which will affect the circadian variation of their functions and will be important for a possible therapeutic strategies that imply activators or inhibitors of these neuropeptides (Smolensky, Hermida, Reinberg, Sackett-Lundeen, & Portaluppi, 2016; Urbanski, 2011). Circadian variation of the enkephalins levels as well as of the enkephalinases activities should be considered. Indeed, enkephalins are under the influence of environmental light and dark conditions and exhibit a circadian rhythm (Kumar et al., 1982) which can therefore influence the factors and functions that control these neuropeptides. Modifications in this rhythm could lead to important pathophysiological consequences (Valdés-Tovar, Escobar, Solís-Chagoyán, Asai, & Benítez-King, 2015).

Previous reports described the circadian variation of angiotensinases involved in blood pressure control (Domínguez-Vías et al., 2017). Based also on the effect of chronotherapy of conventional hypertension medications (Hermida et al., 2013), it was speculated that more knowledge on angiotensinase and enkephalinase activities might have implications for the treatment of hypertension (Domínguez-Vías et al., 2018). Indeed, the results of this study demonstrated a specific rhythm of enkephalinase activity in hypothalamus and pituitary gland under standard light and dark conditions. While the zenith is observed in conditions of darkness and the nadir under light conditions in the hypothalamus, this rhythm is reversed in the pituitary gland. This study also showed a direct influence of light and dark conditions which alter the standard rhythm observed under constant light conditions (Fig. 9).

These results also call attention about some role of enkephalins on analgesia or on the immune response. Although the chronobiological aspects of pain are not clear (Junker & Wirz, 2010), alterations in the standard light and dark cycle, such as constant light conditions, cause an increase in pain response and also an increase in the analgesic effect of morphine (Oliverio, Castellano, & Puglisi-Allegra, 1982). In addition, alterations in the daily rhythm of the immune response can be reflected in the behavior of the enkephalins (Finn, Agren, Bjellerup, Vedin, & Lundeberg, 2004). Therefore, the use of inhibitors of enkephalinase activity could help in various types of therapies in which it would be beneficial to prolong the effect of the enkephalins (Bonnard, Poras, Fournié-Zaluski, & Roques, 2016).

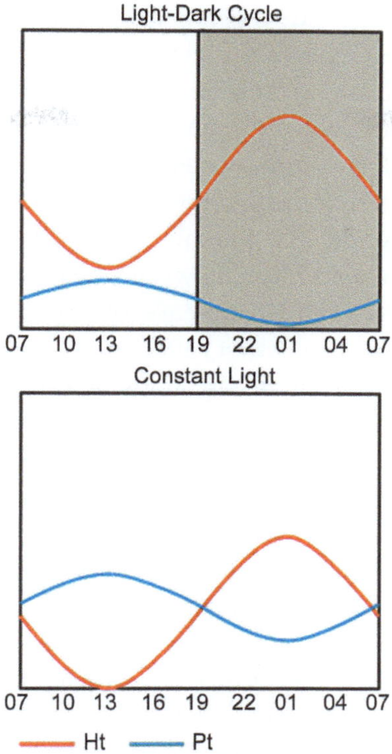

Fig. 9 Circadian rhythm in enkephalinase activity. Schematic representation of enkephalinase activity (Alanyl aminopeptidase, APM) under 12:12h light: dark conditions (upper figure) and under constant light conditions in hypothalamus (HT, red lines) and pituitary (Pt, blue lines). *Modified from Domínguez-Vías, G., Aretxaga-Maza, G., Prieto, I., Segarra, A. B., Luna, J. D., de Gasparo, M. et al. (2018). Light-dark influence on enkephalinase activity in hypothalamus and pituitary.* Neuroendocrinology Letters, *39, 277–280.*

6. Stress

In order to increase the chances of survival when faced to adverse conditions, the organism responds in a coordinated manner trying to maintain homeostasis (Goldstein & Kopin, 2007). For this response, the enkephalins play in an essential role. The high content of these peptides in the limbic system is suggestive of their important function in emotional processes and memory (Drolet et al., 2001), importance that may be extended to the involvement of enkephalinases in this process (reviewed in Prieto, Segarra, de Gasparo, & Ramírez-Sánchez, 2016). In this sense, a change in the level of the enzyme activity was described and a possible interaction between enkephalinase and oxytocinase activity in cortico-limbic regions under acute stress conditions was proposed (Hernández et al., 2009; Hernández et al., 2015). The study showed an increase of enkephalinase activity in hippocampus and a decrease in amygdala after acute stress due to immobilization. In addition, in control animals, a high level of correlation between amygdala and prefrontal cortex was observed, together with the absence of correlation of these regions with the hippocampus. In contrast, after stress, amygdala and prefrontal cortex reduced their correlation, but both established a high correlation with the hippocampus (Fig. 10). Changes in the level of connectivity between corticolimbic regions under stress conditions may be important for the relationship between emotions and memory (Maroun & Richter-Levin, 2003).

As a stress situation implies an integral response in which the establishment of a bidirectional connection between the brain and (virtually) the whole organism through reciprocal mechanisms of regulation, (neurovisceral integration model) (Prieto et al., 2017), the existence of a possible interaction between cerebral and plasma enkephalinases was suggested. The results confirmed such interaction in control animals which changed significantly after stress conditions. This supported the aforementioned neurovisceral integrative response between brain and the periphery which probably involve the autonomic nervous system (Segarra, Hernández, Prieto, de Gasparo, & Ramírez-Sánchez, 2016).

In addition, Prieto et al. (1998) reported a significant positive correlation between neurohypophysis and adrenal gland for enkephalinase activity under a hyperosmotic stimulus induced by a reduced renal mass in a model of hypertension (Fig. 11). The posterior lobe of the pituitary contains important amounts of enkephalins (Zamir, 1985) and the enkephalins coexist with adrenaline in the adrenal medulla (Livett, Day, Elde, & Howe, 1982). Further, an hypertonic saline stress induces proenkephalin gene expression in the paraventricular nucleus (Borsook, Konradi, Falkowski, Comb, & Hyman, 1994). The results of Prieto et al. (1998) support therefore a coordinated role for enkephalins between the neurohypophysis and adrenals under this specific type of stress. It would be interesting to verify whether these results indicate a role for such unspecific coordinated response of the body under general stress conditions.

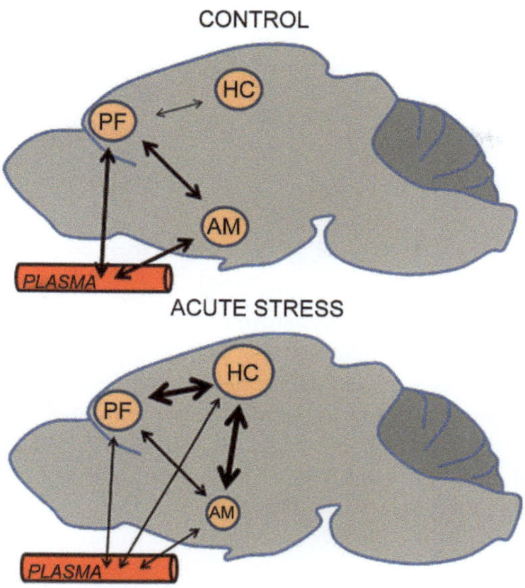

Fig. 10 Stress and enkephalinase activity. Schematic representation of the influence of acute stress on the levels enkephalinase activity (Placental leucyl aminopeptidase, PLAP) in the amygdala (AM), prefrontal cortex (PF), hippocampus (HC) and plasma of adult male rats as well as the significant correlations observed between these locations. The size of the circles is proportional to the level of enzyme activity. Line thickness is proportional to the number of correlations. *Modified from Hernández, J., Segarra, A. B., Ramírez, M., Banegas, I., de Gasparo, M., Alba, F., et al. (2009). Stress influences brain enkephalinase, oxytocinase and angiotensinase activities: a new hypothesis.* Neuropsychobiology. *59(3), 184–189; Hernández, J., Prieto, I., Segarra, A. B., de Gasparo, M., Wangensteen, R., Villarejo, A. B., et al. (2015). Interaction of neuropeptidase activities in cortico-limbic regions after acute restraint stress.* Behavioural Brain Research, 287, *42–48. doi: 10.1016/j.bbr.2015.03.036;; Segarra, A. B., Hernández, J., Prieto, I., de Gasparo, M., & Ramírez-Sánchez M. (2016). Neuropeptidase activities in plasma after acute restraint stress. Interaction with cortico-limbic areas.* Acta Neuropsychiatrica. 28*(4), 239–243. doi: 10.1017/neu.2016.2; Prieto, I., Segarra, A. B., de Gasparo, M., & Ramírez-Sánchez, M. (2016). Neuropeptidases, Stress, and Memory—A Promising Perspective.* AIMS Neuroscience 3*(4), 487–501.*

7. Neurotransmitters interactions

Finally, it is necessary to highlight the importance of the interaction between classical neurotransmitters and neuropeptides in the regulation of enkephalin function and the role of enkephalinases in this mechanism. In this sense, the discovery that classical neurotransmitters and neuropeptides could coexist in the same neurons is of special significance (reviewed in Banegas, Prieto, Segarra, de Gasparo, & Ramírez-Sánchez, 2017;

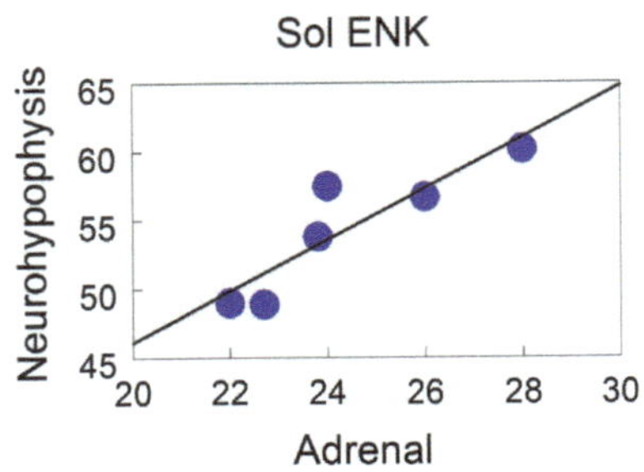

Fig. 11 Enkephalinase activity in a hyperosmotic model of hypertension. Significant positive correlation of soluble enkephalinase activity between neurohypophysis and adrenal gland in the reduced renal mass model of hypertension. Values are expressed as nanomoles of Ala-β-naphthylamide hydrolyzed per minute per milligram of protein. *Modified from* Prieto, I., Martinez, A., Martinez, J.M., Ramírez, M.J., Vargas, F., Alba, F., et al., (1998). Activities of aminopeptidases in a rat saline model of volume hypertension. Hormone & Metabolic Research, 30(5), 246–248.

Banegas, Prieto, Segarra, Vives, et al., 2017). If several neurotransmitters coexist in the same neuron, the alteration of one of them can affect the others and, as a consequence, the functions they exert. Considering enkephalins, these peptides have been described as coexisting, among others, with glutamate (Van Bockstaele, Saunders, Commons, Liu, & Peoples, 2000), gamma-aminobutyric acid (Van Bockstaele & Chan, 1997), serotonin (Tanaka, Okamura, Yanaihara, Tanaka, & Ibata, 1993), cholecystokinin (Gall, Lauterborn, Burks, & Seroogy, 1987) or catecholamines (Charnay et al., 1982). In relation to enkephalinases, Cicin-Sain, Simaga, Froebe, and Abramić (2008) studied enkephalinase activity in two sublines of rats, one with upregulated and the other with downregulated serotonin transporter activity. They observed that the animals with high transporter activity exhibited lower enkephalinase activity in brain.

Taking into account that modifications in the basal bilateral brain distribution profile of peptides may be, at least in part, responsible for some brain alterations that involve disorders in the mood, and as there is alteration of dopamine in Parkinson's disease, Banegas, Prieto, Segarra, de Gasparo, and Ramírez-Sánchez (2017) and Banegas, Prieto, Segarra, Vives, et al. (2017) determined the enkephalinase activity in the left and right prefrontal cortex after selective lesions of the left or right nigrostriatal system, in two strains that clearly differ in their anxiety-related behaviors: spontaneously hypertensive rats (SHR) and Wistar Kyoto (WKY). The authors compared these two lesioned strains with their respective controls of simulated (sham) left or right lesions (Fig. 12).

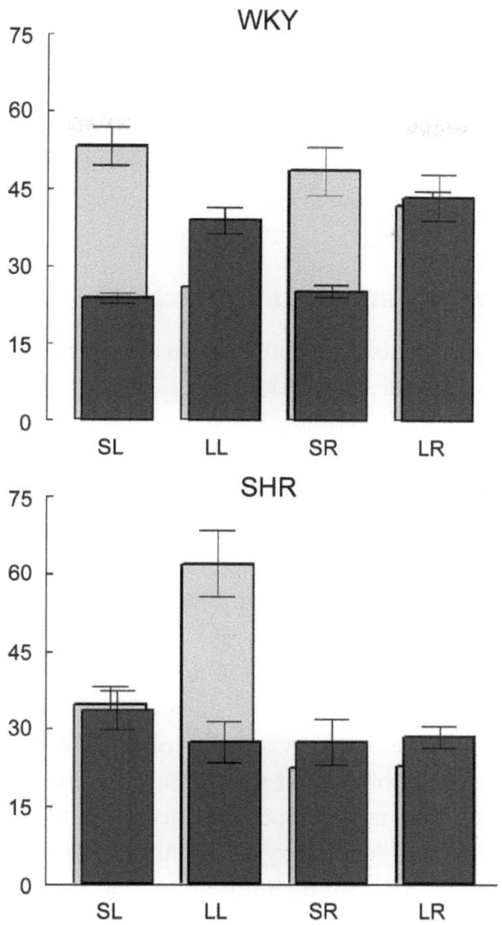

Fig. 12 Enkephalinase activity and dopamine interaction. Enkephalinase activity (Alanyl aminopeptidase, APM) in the left (light gray bars) and right (dark gray bars) prefrontal cortex of left (SL) and right (SR) sham operated and left (LL) or right (LR) 6-hydroxydopamine-lesioned Wistar-Kyoto (WKY) and spontaneously hypertensive rats (SHR). Values represent the mean ± SEM of specific enkephalinase activity expressed as nanomoles of Ala-β-naphthylamide hydrolyzed per minute per milligram of protein. *Modified from Banegas, I., Prieto, I., Segarra, A. B., Vives, F., de Gasparo, M., Duran, R., de Dios Luna, J., & Ramírez-Sánchez M. (2017) Bilateral distribution of enkephalinase activity in the medial prefrontal cortex differs between WKY and SHR rats unilaterally lesioned with 6-hydroxydopamine.* Progress in Neuropsychopharmacology & Biological Psychiatry, *75, 213–218. doi: 10.1016/j.pnpbp.2017.02.015.*

The results showed that the left and right WKY controls had a clear left predominance but, while the left lesion reversed the predominance, the right one led to a symmetric distribution. In contrast, the left or right control animals of the SHR strain showed no asymmetry in the enkephalinase activity.

However, the left injury induced a high left predominance and the right one a slight but significant right predominance. These results indicated that unilateral lesions of the nigrostriatal system altered the bilateral distribution of enkephalinase activity depending on both the side of the lesion and the strain used. This may indicate that dopamine interacts asymmetrically with enkephalins in the prefrontal cortex and that enkephalinase activity modulates this interaction.

8. Conclusions and future directions

Opium, a drug capable of producing analgesia and euphoria, has been known since immemorial time. Although Sumerian writings mention it 4000 years before Christ (Krikorian, 1975), their properties were most probably known much earlier. Currently, opium derivatives are one of the main tools for the treatment of pain and are an essential part of the protocols designed for such treatment (Beal & Wallace, 2016). The discovery of the existence of endogenous opiates opened immense possibilities to improve the strategies in the fight against pain and other pathologies in which these peptides are involved. Today there is a vast literature concerning the enkephalins and enkephalinases and there are excellent reviews that systematically address the new contributions (Bodnar, 2018). However, in relation to enkephalinases, there is a wide variability of reported data due in part to broad substrate specificity of these enzymes, various experimental conditions, differences in the species analyzed and influence of multiple endogenous and exogenous factors that modulate enzymatic activity and consequently endogenous substrates. The challenge for future directions would be to achieve a deep knowledge of the role of such factors in the regulation of these enzymes, which would optimize our therapeutic strategies in the treatment of pain and the pathologies in which enkephalins are involved.

Acknowledgments

The results represented in Figs. 2–12 are part of projects performed at the University of Granada, University of the Basque Country and University of Jaén.

References

Alba, F., Arenas, J. C., Iribar, C., & Ramirez, M. (1993). Regional distribution of soluble and membrane-bound aminopeptidase activities in rat brain. *Brain Research Bulletin*, *31*(3–4), 393–396.

Alba, F., Ramirez, M., Cantalejo, E. S., & Iribar, C. (1988). Aminopeptidase activity is asymmetrically distributed in selected zones of rat brain. *Life Sciences*, *43*(11), 935–939.

Andersen, M. B., Simonsen, U., Wehland, M., Pietsch, J., & Grimm, D. (2016). LCZ696 (valsartan/Sacubitril)—A possible new treatment for hypertension and heart failure. *Basic & Clinical Pharmacology & Toxicology*, *118*(1), 14–22. https://doi.org/10.1111/bcpt.12453.

Arechaga, G., Sánchez, B., Prieto, I., Martínez, J. M., Alba, F., & Ramírez, M. (1996). Subcellular distribution of leucine aminopeptidase during the development and aging of rat brains. *Revista de Neurología*, *24*(136), 1503–1506.

Banegas, I., Prieto, I., Segarra, A. B., de Gasparo, M., & Ramírez-Sánchez, M. (2017). Study of the neuropeptide function in Parkinson's disease using the 6-hydroxydopamine model of experimental hemiparkinsonism. *AIMS Neuroscience*, *4*(4), 223–237. https://doi.org/10.3934/Neuroscience.2017.4.223.

Banegas, I., Prieto, I., Segarra, A. B., Vives, F., de Gasparo, M., Duran, R., et al. (2017). Bilateral distribution of enkephalinase activity in the medial prefrontal cortex differs between WKY and SHR rats unilaterally lesioned with 6-hydroxydopamine. *Progress in Neuropsychopharmacology & Biological Psychiatry*, *75*, 213–218. https://doi.org/10.1016/j.pnpbp.2017.02.015.

Beal, B. R., & Wallace, M. S. (2016). An overview of pharmacologic management of chronic pain. *The Medical Clinics of North America*, *100*(1), 65–79.

Bodnar, R. J. (2018). Endogenous opiates and behavior: 2016. *Peptides*, *101*, 167–212. https://doi.org/10.1016/j.peptides.2018.01.011.

Bonnard, E., Poras, H., Fournié-Zaluski, M. C., & Roques, B. P. (2016). Preventive and alleviative effects of the dual enkephalinase inhibitor (Denki) PL265 in a murine model of neuropathic pain. *European Journal of Pharmacology*, *788*, 176–182. https://doi.org/10.1016/j.ejphar.2016.05.041.

Borsook, D., Konradi, C., Falkowski, O., Comb, M., & Hyman, S. E. (1994). Molecular mechanisms of stress-induced proenkephalin gene regulation: CREB interacts with the proenkephalin gene in the mouse hypothalamus and is phosphorylated in response to hyperosmolar stress. *Molecular Endocrinology*, *8*(2), 240–248.

Chang, G. Q., Karatayev, O., Davydova, Z., & Leibowitz, S. F. (2004). Circulating triglycerides impact on orexigenic peptides and neuronal activity in hypothalamus. *Endocrinology*, *145*(8), 3904–3912.

Charnay, Y., Léger, L., Dray, F., Bérod, A., Jouvet, M., Pujol, J. F., et al. (1982). Evidence for the presence of enkephalin in catecholaminergic neurones of cat locus coeruleus. *Neuroscience Letters*, *30*(2), 147–151.

Cicin-Sain, L., Simaga, S., Froebe, A., & Abramić, M. (2008). Central aminopeptidase and serotonin system activities: Possible relationship. *Neuropeptides*, *42*(4), 435–440. https://doi.org/10.1016/j.npep.2008.04.007.

de Felipe, M. C., Jiménez, I., Castro, A., & Fuentes, J. A. (1989). Antidepressant action of imipramine and iprindole in mice is enhanced by inhibitors of enkephalin-degrading peptidases. *European Journal of Pharmacology*, *159*(2), 175–180.

de Gandarias, J. M., Casis, L., Irazusta, J., Echevarria, E., & Ramirez, M. (1988). Changes of aminopeptidase activity levels in serum and brain during the estrous cycle of the rat. *Hormone & Metabolic Research*, *20*(12), 776.

de Gandarias, J. M., Ramirez, M., Zulaica, J., & Casis, L. (1989). Aminopeptidase (arylamidase) activity in discrete areas of the rat brain: Sex differences. *Hormone & Metabolic Research*, *21*(5), 285–286.

Demmer, W., & Brand, K. (1983). Processing and degradation of met-enkephalin by peptidases associated with rat brain cortical synaptosomes. *Archives of Biochemistry & Biophysics*, *227*(1), 310–320.

Domínguez-Vías, G., Aretxaga-Maza, G., Prieto, I., Luna, J. D., de Gasparo, M., & Ramírez-Sánchez, M. (2017). Diurnal opposite variation between angiotensinase activities in photo-neuro-endocrine tissues of rats. *Chronobiology International*, *34*(9), 1180–1186.

Domínguez-Vías, G., Aretxaga-Maza, G., Prieto, I., Segarra, A. B., Luna, J. D., de Gasparo, M., et al. (2018). Light-dark influence on enkephalinase activity in hypothalamus and pituitary. *Neuroendocrinology Letters*, *39*, 277–280.

Drolet, G., Dumont, E. C., Gosselin, I., Kinkead, R., Laforest, S., & Trottier, J. F. (2001). Role of endogenous opioid system in the regulation of the stress response. *Progress in Neuropsychopharmacology & Biological Psychiatry*, *25*(4), 729–741.

Finn, A., Agren, G., Bjellerup, P., Vedin, I., & Lundeberg, T. (2004). Production and characterization of antibodies for the specific determination of the opioid peptide Met5-Enkephalin-Arg6-Phe7. *Scandinavian Journal of Clinical & Laboratory Investigation*, *64*(1), 49–56.

Fournie-Zaluski, M. C. (1988). Design and evaluation of inhibitors of encephalin degrading enzymes. *Neurochemistry International*, *12*(3), 375–382.

Gall, C., Lauterborn, J., Burks, D., & Seroogy, K. (1987). Co-localization of enkephalin and cholecystokinin in discrete areas of rat brain. *Brain Research*, *403*(2), 403–408.

Gibson, A. M., Biggins, J. A., Lauffart, B., Mantle, D., & McDermott, J. R. (1991). Human brain leucyl aminopeptidase: Isolation, characterization and specificity against some neuropeptides. *Neuropeptides*, *19*, 163–168.

Goldstein, D. S., & Kopin, I. J. (2007). Evolution of concepts of stress. *Stress*, *10*(2), 109–120.

Gorenstein, C., & Snyder, S. H. (1980). Characterization of Enkephalinases. In E. L. Way (Ed.), *Endogenous and exogenous opiate agonist and antagonists* (pp. 345–348). Pergamon Press.

Graham, L. T., Jr., Shank, R. P., Werman, R., & Aprison, M. H. (1967). Distribution of some synaptic transmitter suspects in cat spinal cord: Glutamic acid, aspartic acid, gamma-aminobutyric acid, glycine and glutamine. *Journal of Neurochemistry*, *14*(4), 465–472.

Hayashi, M. (1978). Monkey brain arylamidase. II. Further characterization and studies on mode of hydrolysis of physiologically active peptides. *Journal of Biochemistry*, *84*(6), 1363–1372.

Henry, M. S., Gendron, L., Tremblay, M. E., & Drolet, G. (2017). Enkephalins: Endogenous analgesics with an emerging role in stress resilience. *Neural Plasticity2017*. 1546125. https://doi.org/10.1155/2017/1546125.

Hermida, R. C., Ayala, D. E., Fernández, J. R., Mojón, A., Smolensky, M. H., Fabbian, F., et al. (2013). Administration-time differences in effects of hypertension medications on ambulatory blood pressure regulation. *Chronobiology International*, *30*, 280–314.

Hernández, J., Prieto, I., Segarra, A. B., de Gasparo, M., Wangensteen, R., Villarejo, A. B., et al. (2015). Interaction of neuropeptidase activities in cortico-limbic regions after acute restraint stress. *Behavioural Brain Research*, *287*, 42–48. https://doi.org/10.1016/j.bbr.2015.03.036.

Hernández, J., Segarra, A. B., Ramírez, M., Banegas, I., de Gasparo, M., Alba, F., et al. (2009). Stress influences brain enkephalinase, oxytocinase and angiotensinase activities: A new hypothesis. *Neuropsychobiology*, *59*(3), 184–189.

Hersh, L. B. (1981). Solubilization and characterization of two rat brain membrane-bound aminopeptidases active on Met-enkephalin. *Biochemistry*, *20*(8), 2345–2350.

Hersh, L. B. (1985). Characterization of membrane-bound aminopeptidases from rat brain: Identification of the enkephalin-degrading aminopeptidase. *Journal of Neurochemistry*, *44*, 1427–1435.

Hersh, L. B. (1986). Nomenclature for enkephalin degrading peptidases. *Life Sciences*, *38*(13), 1151–1153.

Hersh, L. B., Smith, T. E., & McKelvy, J. F. (1980). Cleavage of endorphins to des-Tyr endorphins by homogeneous bovine brain aminopeptidase. *Nature*, *286*(5769), 160–162.

Huang, X. F., Han, M., South, T., & Storlien, L. (2003). Altered levels of POMC, AgRP and MC4-R mRNA expression in the hypothalamus and other parts of the limbic system of mice prone or resistant to chronic high-energy diet-induced obesity. *Brain Research*, *992*, 9–19.

Hughes, J., Smith, T. W., Kosterlitz, H. W., Fothergill, L. A., Morgan, B. A., & Morris, H. R. (1975). Identification of two related pentapeptides from the brain with potent opiate agonist activity. *Nature*, *258*(5536), 577–580.

Hui, K. S., Saito, M., & Hui, M. (1998). A novel neuron-specific aminopeptidase in rat brain synaptosomes. Its identification, purification, and characterization. *Journal of Biological Chemistry*, *273*(47), 31053–31060.

Hui, K. S., Wang, Y. J., Wong, K. H., Tsai, H., & Lajtha, A. (1981). Regional and subcellular distribution of Met-and Leu-enkephalin-degrading activity in rat brain. *Neuropeptides*, *1*, 167–173.

Inaoka, Y., & Tamaoki, H. (1987). Purification and characterization of enkephalinase B from rat brain membrane. *Biochimica et Biophysica Acta*, *925*(1), 27–35.

Jo, Y. H., Su, Y., Gutierrez-Juarez, R., & Chua, S., Jr. (2009). Oleic acid directly regulates POMC neuron excitability in the hypothalamus. *Journal of Neurophysiology*, *101*, 2305–2316.

Johnson, G. D., & Hersh, L. B. (1990). Studies on the subsite specificity of the rat brain puromycin-sensitive aminopeptidase. *Archives of Biochemistry and Biophysics*, *276*(2), 305–309.

Junker, U., & Wirz, S. (2010). Review article: Chronobiology: Influence of circadian rhythms on the therapy of severe pain. *Journal of Oncology Pharmacy Practice*, *16*(2), 81–87. https://doi.org/10.1177/1078155209337665.

Kaneko, K., Mizushige, T., Miyazaki, Y., Lazarus, M., Urade, Y., Yoshikawa, M., et al. (2014). δ-Opioid receptor activation stimulates normal diet intake but conversely suppresses high-fat diet intake in mice. *American Journal of Physiology. Regulatory, Integrative and Comparative Physiology*, *306*(4), R265–R272. https://doi.org/10.1152/ajpregu.00405.2013.

Kest, B., Sarton, E., & Dahan, A. (2000). Gender differences in opioid-mediated analgesia: Animal and human studies. *Anesthesiology*, *93*(2), 539–547.

Krikorian, A. D. (1975). Were the opium poppy and opium known in the ancient near East? *Journal of the History of Biology*, *8*(1), 96–114.

Kumar, M. S., Chen, C. L., Sharp, D. C., Liu, J. M., Kalra, P. S., & Kalra, S. P. (1982). Diurnal fluctuations in methionine-enkephalin levels in the hypothalamus and preoptic area of the male rat: Effects of pinealectomy. *Neuroendocrinology*, *35*, 28–31.

Lee, C. M., & Snyder, S. H. (1982). Dipeptidyl-aminopeptidase III of rat brain. Selective affinity for enkephalin and angiotensin. *Journal of Biological Chemistry*, *257*(20), 12043–12050.

Li, C., Booze, R. M., & Hersh, L. B. (1996). Tissue-specific expression of rat neutral endopeptidase mRNAs. *Annals of the New York Academy of Sciences*, *780*, 145–155.

Livett, B. G., Day, R., Elde, R. P., & Howe, P. R. (1982). Co-storage of enkephalins and adrenaline in the bovine adrenal medulla. *Neuroscience*, 7(5), 1323–1332.

Llorens, C., Gacel, G., Swerts, J. P., Perdrisot, R., Fournie-Zaluski, M. C., Schwartz, J. C., et al. (1980). Rational design of enkephalinase inhibitors: Substrate specificity of enkephalinase studied from inhibitory potency of various dipeptides. *Biochemical and Biophysical Research Communications*, *96*(4), 1710–1716.

Lyons, P. J., Callaway, M. B., & Fricker, L. D. (2008). Characterization of carboxypeptidase A6, an extracellular matrix peptidase. *The Journal of Biological Chemistry*, *283*(11), 7054–7063. https://doi.org/10.1074/jbc.M707680200.

Maki, T., Nasa, Y., Yamaguchi, F., Yoshida, H., Mori, M., Takada, T., et al. (2001). Long-term treatment with neutral endopeptidase inhibitor improves cardiac function and reduces natriuretic peptides in rats with chronic heart failure. *Cardiovascular Research, 51*(3), 608–617.

Maroun, M., & Richter-Levin, G. (2003). Exposure to acute stress blocks the induction of long-term potentiation of the amygdala-prefrontal cortex pathway in vivo. *Journal of Neuroscience, 23*(11), 4406–4409.

Martinez, J. M., Prieto, I., Ramirez, M. J., Alba, F., & Ramirez, M. (1997). Cholesterol and steroids action on aminopeptidases. *Biochemical Society Transactions, 25*(1), 113S.

Martinez, J. M., Ramirez, M. J., Prieto, I., Alba, F., & Ramirez, M. (1998). Sex differences and in vitro effects of steroids on serum aminopeptidase activities. *Peptides, 19*(9), 1637–1640.

Matsumoto, H., Nagasaka, T., Hattori, A., Rogi, T., Tsuruoka, N., Mizutani, S., et al. (2001). Expression of placental leucine aminopeptidase/oxytocinase in neuronal cells and its action on neuronal peptides. *European Journal of Biochemistry, 268*, 3259–3266.

Mendez, I. A., Ostlund, S. B., Maidment, N. T., & Murphy, N. P. (2015). Involvement of endogenous Enkephalins and β-endorphin in feeding and diet-induced obesity. *Neuropsychopharmacology, 40*(9), 2103–2112. https://doi.org/10.1038/npp.2015.67.

Oliverio, A., Castellano, C., & Puglisi-Allegra, S. (1982). Opiate analgesia: Evidence for circadian rhythms in mice. *Brain Research, 249*(2), 265–270.

Pert, C. B., & Snyder, S. H. (1973). Opiate receptor: Demonstration in nervous tissue. *Science, 179*(4077), 1011–1014.

Poon, K., Alam, M., Karatayev, O., Barson, J. R., & Leibowitz, S. F. (2015). Regulation of the orexigenic neuropeptide, enkephalin, by PPARδ and fatty acids in neurons of the hypothalamus and forebrain. *Journal of Neurochemistry, 135*(5), 918–931.

Prieto, I., Martinez, A., Martinez, J. M., Ramírez, M. J., Vargas, F., Alba, F., et al. (1998). Activities of aminopeptidases in a rat saline model of volume hypertension. *Hormone & Metabolic Research, 30*(5), 246–248.

Prieto, I., Segarra, A. B., de Gasparo, M., & Ramírez-Sánchez, M. (2016). Neuropeptidases, stress, and memory—A promising perspective. *AIMS Neuroscience, 3*(4), 487–501.

Prieto, I., Segarra, A. B., Martinez-Canamero, M., de Gasparo, M., Zorad, S., & Ramirez-Sanchez, M. (2017). Bidirectional asymmetry in the neurovisceral communication for the cardiovascular control: New insights. *Endocrine Regulations, 51*(3), 157–167. https://doi.org/10.1515/enr-2017-0017.

Prieto, I., Segarra, A. B., Vargas, F., Alba, F., de Gasparo, M., & Ramírez, M. (2003). Angiotensinase activity in hypothalamus and pituitary of hypothyroid, euthyroid and hyperthyroid adult male rats. *Hormone & Metabolic Research, 35*(5), 279–281.

Prieto, I., Segarra, A. B., Villarejo, A. B., de Gasparo, M., Martínez-Cañamero, M. M., & Ramírez-Sánchez, M. (2019). Neuropeptidase activity in the frontal cortex of Wistar-Kyoto and spontaneously hypertensive rats treated with vasoactive drugs: A bilateral study. *Journal of Hypertension, 37*(3), 612–628. https://doi.org/10.1097/HJH.0000000000001884.

Ramírez, M., Prieto, I., Alba, F., Vives, F., Banegas, I., & de Gasparo, M. (2008). Role of central and peripheral aminopeptidase activities in the control of blood pressure: A working hypothesis. *Heart Failure Reviews, 13*(3), 339–353. https://doi.org/10.1007/s10741-007-9066-6.

Ramírez, M., Prieto, I., Vives, F., de Gasparo, M., & Alba, F. (2004). Neuropeptides, neuropeptidases and brain asymmetry. *Current Protein & Peptide Science, 5*(6), 497–506.

Roman, E., Ploj, K., Gustafsson, L., Meyerson, B. J., & Nylander, I. (2006). Variations in opioid peptide levels during the estrous cycle in Sprague-Dawley rats. *Neuropeptides, 40*(3), 195–206.

Schnebli, H. P., Phillipps, M. A., & Barclay, R. K. (1979). Isolation and characterization of an enkephalin-degrading aminopeptidase from rat brain. *Biochimica et Biophysica Acta, 569*(1), 89–98.

Segarra, A. B., Hernández, J., Prieto, I., de Gasparo, M., & Ramírez-Sánchez, M. (2016). Neuropeptidase activities in plasma after acute restraint stress. Interaction with corticolimbic areas. *Acta Neuropsychiatrica, 28*(4), 239–243. https://doi.org/10.1017/neu.2016.2.

Segarra, A. B., Prieto, I., Martínez-Cañamero, M., Ruíz-Sanz, J. I., Ruíz-Larrea, M. B., de Gasparo, M., et al. (2019). Enkephalinase activity is modified and correlates with fatty acids in frontal cortex depending on the type of fat used in the diet. *Endocrine Regulations, 53*(2), 59–64. https://doi.org/10.2478/enr-2019-0007.

Smolensky, M. H., Hermida, R. C., Reinberg, A., Sackett-Lundeen, L., & Portaluppi, F. (2016). Circadian disruption: New clinical perspective of disease pathology and basis for chronotherapeutic intervention. *Chronobiology International, 33*, 1101–1119.

Supattapone, S., Fricker, L. D., & Snyder, S. H. (1984). Purification and characterization of a membrane-bound enkephalin-forming carboxypeptidase, "enkephalin convertase" *Journal of Neurochemistry, 42*(4), 1017–1023.

Tanaka, M., Okamura, H., Yanaihara, N., Tanaka, Y., & Ibata, Y. (1993). Differential expression of serotonin and [Met]enkephalin-Arg6-Gly7-Leu8 in neurons of the rat brain stem. *Brain Research Bulletin, 30*(5–6), 561–570.

Tejedor-Real, P., Micó, J. A., Maldonado, R., Roques, B. P., & Gibert-Rahola, J. (1993). Effect of mixed (RB 38A) and selective (RB 38B) inhibitors of enkephalin degrading enzymes on a model of depression in the rat. *Biological Psychiatry, 34*(1–2), 100–107.

Thanawala, V., Kadam, V. J., & Ghosh, R. (2008). Enkephalinase inhibitors: Potential agents for the management of pain. *Current Drug Targets, 9*(10), 887–894. https://doi.org/10.2174/138945008785909356.

Thompson, M. W., & Hersh, L. B. (2004). The puromycin sensitive aminopeptidase. In N. M. Hooper & U. Lendeckel (Eds.), *Aminopeptidases in biology and disease* (pp. 1–15). New York: Kluwer Academic/Plenum.

Traficante, L. J., Rotrosen, J., Siekierski, J., Tracer, H., & Gershon, S. (1980). Enkephalin inactivation by N-terminal tyrosine cleavage: Purification and partial characterization of a highly specific enzyme from human brain. *Life Sciences, 26*(20), 1697–1706.

Urbanski, H. F. (2011). Role of circadian neuroendocrine rhythms in the control of behavior and physiology. *Neuroendocrinology, 93*, 211–222.

Valdés-Tovar, M., Escobar, C., Solís-Chagoyán, H., Asai, M., & Benítez-King, G. (2015). Constant light suppresses production of met-enkephalin-containing peptides in cultured splenic macrophages and impairs primary immune response in rats. *Chronobiology International, 32*, 164–177.

Van Bockstaele, E. J., & Chan, J. (1997). Electron microscopic evidence for coexistence of leucine5-enkephalin and catecholamines in a subpopulation of axon terminals in the rat locus coeruleus region. *Brain Research, 746*(1–2), 171–182.

Van Bockstaele, E. J., Saunders, A., Commons, K. G., Liu, X. B., & Peoples, J. (2000). Evidence for coexistence of enkephalin and glutamate in axon terminals and cellular sites for functional interactions of their receptors in the rat locus coeruleus. *Journal of Comparative Neurology, 417*(1), 103–114.

Villarejo, A. B., Ramírez-Sánchez, M., Segarra, A. B., Martínez-Cañamero, M., & Prieto, I. (2015). Influence of extra virgin olive oil on blood pressure and kidney angiotensinase activities in spontaneously hypertensive rats. *Planta Medica, 81*(8), 664–669.

Zamir, N. (1985). On the origin of Leu-enkephalin and met-enkephalin in the rat neurohypophysis. *Endocrinology, 117*(4), 1687–1692.

Zozulya, A. A., Kost, N. V., Yu Sokolov, O., Gabaeva, M. V., Grivennikov, I. A., Andreeva, L. N., et al. (2001). The inhibitory effect of Selank on enkephalin-degrading enzymes as a possible mechanism of its anxiolytic activity. *Bulletin of Experimental Biology and Medicine, 131*(4), 315–317.

CHAPTER SIX

Limbic circuit connectivity and the stress response: New insights into the mammalian nociceptin peptide system

Allison Jane Fulford[a,*], Sarah Keskes[a,b]
[a]Centre for Applied Anatomy, University of Bristol, Bristol, United Kingdom
[b]Keele University Medical School, Keele, United Kingdom
*Corresponding author: e-mail address: a.j.fulford@bristol.ac.uk

Contents

Abstract

Considerable progress has been made in understanding the neurobehavioral effects of nociceptin peptide (N/OFQ) and its NOP receptor. Recent interest has focused on its role in stress and cognition, with consideration of therapeutic potential in regard to anxiety and mood disorders. Research has interrogated the mechanisms of action of N/OFQ peptide in the context of stress-related behavior. We are interested in the endogenous role of N/OFQ and NOP receptor in terms of adaptation to chronic stress. Our research has highlighted the importance of associated limbic regions including the bed nucleus, extended amygdala, in addition to thalamic reticular nucleus as important sites for long-term adaptations in endogenous N/OFQ function in chronic stress. Such research raises interest in appreciation of extended limbic connections and novel pathways which allow us to reevaluate current understanding of stress neurocircuitry. Examination of endogenous N/OFQ-NOP receptor modulation of monoaminergic and amino acid transmitter systems in this extended limbic architecture will facilitate deeper understanding of the tonic control of behavior. Application of *in vivo* experimental approaches to models of abnormal neurodevelopment and heightened stress vulnerability in adulthood will enable the role of N/OFQ in complex neuropsychiatric disorders including schizophrenia and post-traumatic stress disorder to be defined.

Vitamins and Hormones, Volume 111
ISSN 0083-6729
https://doi.org/10.1016/bs.vh.2019.05.002

Abbreviations

BNST	bed nucleus of the stria terminalis
CRF	corticotropin releasing factor
CNS	central nervous system
ERK	extracellular signal regulated kinase
GABA	gamma-aminobutyric acid
GPCR	G-proten-coupled receptor
HPA	hypothalamo-pituitary-adrenal
MAP	mitogen-activated protein kinase
N/OFQ	nociceptin/orphanin FQ
NOP	nociceptin/orphanin FQ peptide receptor
PFC	prefrontal cortex
pOFC	posterior orbitofrontal cortex
PPI	prepulse inhibition
ppN/OFQ	prepronociceptin
PTSD	post-traumatic stress disorder
RT-PCR	reverse transcriptase polymerase chain reaction
TRN	thalamic reticular nucleus

1. Mechanisms underlying the stress modulatory effects of the neuropeptide nociceptin

Nociceptin (N/OFQ) is a multifunctional peptide within the brain, particularly important in the neuromodulation of stress and anxiety (see Fulford, 2015; Toll, Bruchas, Calo', Cox, & Zaveri, 2016). N/OFQ affects anxiety, depressive behavior, learning and memory performance in rodents in a range of tasks, although the mechanism of action of N/OFQ remains incompletely understood (Gavioli & Calo', 2006; Higgins et al., 2002; Holanda et al., 2016; Rekik et al., 2017). Highly specific pharmacological tools have been used to investigate the integrative role of N/OFQ and NOP receptors in major limbic sites including the basolateral amygdala and bed nucleus of the stria terminalis (BNST), two major limbic sites that control emotional stress responses and associated behavior. N/OFQ effects are dependent on interactions with limbic neurotransmitter and neuropeptide mediators (Werthwein, Bauer, Nakazi, Kathmann, & Schlicker, 1999), although the precise basis for this functional interaction requires further evaluation. The objectives remain to advance understanding of the role of N/OFQ in the neuromodulation of stress responses, anxiety and depressive behavior (Asth et al., 2016; Post et al., 2016) in order to interrogate the possibility that this peptide may have therapeutic potential.

Despite considerable evidence showing that N/OFQ reduces anxiety-related behaviors the underlying mechanisms remain poorly understood. Rodent models of stress and conditioned fear have been employed to investigate the relevance of hippocampal and amygdala NOP receptors to MAP kinase activity, neuroplasticity and expression of emotional behavior. Interactions with the corticotrophin-releasing factor (CRF) system in the limbic forebrain is involved in N/OFQ effects on behavior and stress responses (Ciccocioppo et al., 2003; Rodi et al., 2008), however, the intracellular mechanisms modulated by activation of N/OFQ (NOP) receptors are little understood. Activation of the ERK signaling pathway has been implicated in the N/OFQ-induced nociceptive effect in nucleus accumbens (Chen, Huang, & Yu, 2008). Studies show that N/OFQ differentially activates isoforms of MAP kinases *in vitro* (Wang et al., 2010) raising the question of the role of N/OFQ and NOP receptors in the modulation of ERK-MAP kinase signaling *in vivo*. N/OFQ agonists have been shown to impair recognition memory in the mouse hippocampus *in vivo via* inhibition of ERK activation (Goeldner et al., 2008). However, ERK signaling pathways and ERK phosphorylation are required for contextual fear learning in mice and retention of trace fear memory in rats (Villarreal & Barea-Rodriguez, 2006). These findings raise the possibility that N/OFQ impairment in contextual fear conditioning involves ERK signaling pathways. The wider relevance of hippocampal and amygdala NOP receptors to MAP kinase activity, neuroplasticity and expression of emotional behavior requires scrutiny.

Varied stimuli alter the activity of G protein-coupled kinase signaling pathways including cytokines, inflammatory mediators and environmental stressors. Strong evidence suggests that central N/OFQ peptide and its NOP receptor regulate the physical, immunological and psychological stress response (Leggett, Dawe, Jessop, & Fulford, 2009; Leggett, Harbuz, Jessop, & Fulford, 2006; Leggett, Jessop, & Fulford, 2007). The opioid-like peptide N/OFQ differentially activates isoforms of MAP kinases *in vitro,* implicating MAP kinase cascade activation in diverse cellular and behavioral actions of N/OFQ peptide (Buzas, 2002; Collins et al., 2016). One question worthy of investigation is whether activation of ERK pathways underlies N/OFQ-induced anxiolysis in the "extended amygdala," where NOP interactions with the CRF system underlie robust anxiolytic effects. Site-specific injections of selective kinase inhibitors may be employed to dissect neuronal pathways involved in NOP regulation of behavior. In addition, the behavioral consequences of stress-induced regulation of ERK 1 and 2 activation and dependence on NOP receptors is of interest. A further objective must be to identify limbic neurotransmitter substrates for N/OFQ and NOP regulation, involving N/OFQ-monoamine interactions coupled with functional *in vivo* approaches such as behavioral microdialysis. An advance in understanding neuropeptide specific G protein-coupled kinases in behavioral stress responses *in vivo* will progress our understanding of the mechanisms *via* which N/OFQ induces its cellular effects.

2. Neuropharmacological basis of nociceptin effects in stress and anxiety

The neuropharmacology of the N/OFQ system in stress and anxiety regulation has been a significant topic of study of recent years (see Fulford, 2015; Toll et al., 2016). We have undertaken detailed studies to investigate the endogenous N/OFQ system *in vivo* in an attempt to discriminate the functional roles of this peptide system with regard to stress, anxiety and inflammation (Leggett et al., 2006, 2007, 2009). The NOP receptor has a hypothalamic and limbic distribution similar to CRF suggesting that its peptide ligand, N/OFQ, and the NOP receptor system are involved in the regulation of hypothalamo-pituitary-adrenal (HPA) axis activity and stress responses. We have investigated the involvement of N/OFQ peptide precursor (ppNOC) or NOP in the regulation of hypothalamic function during stress. ppNOC and NOP mRNA expression levels were measured by *in situ* hybridization histochemistry in limbic regions and hypothalamic nuclei in rats subjected to models of acute and repeated psychological (restraint) stress. We have previously shown that the acute stressor, restraint (60 min), regulates expression of NOP receptor using RT-PCR (Delaney et al., 2012). Our studies have enabled site-specific changes in limbic-hypothalamic gene expression to be determined, and to address the involvement of ppN/OFQ and NOP mRNA in chronic stress adaptation. Our in-house assays for ppN/OFQ and NOP mRNA expression using *in situ* assay were validated previously. The protocols for highly sensitive radioactive *in situ* hybridization using full-length hydrolysed riboprobes to detect subregional changes in ppN/OFQ and NOP mRNA expression in brain in response to aversive challenge have been previously published (Delaney et al., 2012).

The degree to which the N/OFQ system expression in the HPA axis is influenced by circadian variation in circulating glucocorticoids is a valid question. In addition, it is important to determine effects of immunological stress on N/OFQ-NOP system expression throughout the limbic system.

We provided the first integrated analysis of relationships between N/OFQ, NOP receptor and HPA axis activity under basal and stress-stimulated conditions (Leggett et al., 2006, 2007, 2009). Employing the selective peptidic NOP antagonist, UFP-101, we identified that blockade of NOP receptors differentially modulates the corticosterone response to restraint in rats dependent upon the phase of the circadian rhythm (Leggett et al., 2007).

An additional aim was to investigate glucocorticoid-involvement in the regulation of N/OFQ functions in the BNST. We characterized the role of steroid hormones in stress-induced changes in BNST activity by analyzing the acute and repeated restraint stress effects on hypothalamic and bed nucleus NOP mRNA levels. N/OFQ and NOP transcripts are densely expressed throughout the amygdala, although the functional response of the subregional endogenous amygdala N/OFQ system is incompletely understood (Neal et al., 1999). As the amygdala is a key site of plasticity underlying fear conditioning, it is of interest to consider limbic N/OFQ system expression in association with unconditioned and conditioned fear. The effect of site-specific injections of N/OFQ ligands into the amygdala on expression of unconditioned and conditioned fear would be helpful in this regard. N/OFQ-NOP receptor function in animal models that have previously shown to elicit aberrant conditioned behavior (e.g., social isolation models in rats, see Lapiz et al., 2003) are warranted.

Our consideration is that integration & interaction between physiological & psychological systems relies upon a systems-based and coordinated multi-level analysis. This necessitates the examination of neural circuits, endocrine & behavioral analysis to fully understand system complexity. A coordinated analysis of effects of central site-specific drug infusion, sampling from indwelling *iv* catheters (hormone analysis) in conjunction with behavioral recording provides a robust and integrative experimental approach. *Ex-vivo* analysis through gene expression analysis, receptor pharmacology and electrophysiology studies allow a comprehensive study of N/OFQ-NOP function.

In our research, specific limbic sites have emerged as major sites of interest in stress neurobiology. The BNST is closely anatomically and functionally connected to the amygdala, and apparently serves to integrate stress and fear inputs (Choi et al., 2007; Dong, Petrovich, & Swanson, 2001; Walker, Miles, & Davis, 2009). N/OFQ, an opioid-like peptide implicated in the regulation of the HPA axis and anxiety, is thought to exert its actions, in part, *via* interaction with CRF neurones in the bed nucleus (Ciccocioppo et al., 2003; Rodi et al., 2008). We characterized the organization and synchronization of neuronal firing responses in defined subregions of the BNST and examined the impact of corticosterone on BNST neurones (Dawe, Wakerley, & Fulford, 2010). We identified sub-regional variations in firing patterns of different cell types in the BNST, dose-related inhibitory effects of N/OFQ and sensitivity to NOP receptor antagonists and corticosterone (Dawe et al., 2010). The neuronal inhibitory response to N/OFQ application was similar across dorsal and ventral BNST subregions, irrespective of firing pattern. Our demonstration that pre-treatment with corticosterone attenuates the response of neurones in the bed nucleus to N/OFQ (under conditions of NMDA stimulation), but not baseline firing *per se*, indicates that glucocorticoids interact with N/OFQ to modulate BNST neuronal activity. This is of significant interest given the presence of multiple glucocorticoid response elements in the human ppN/OFQ gene promoter (Xie et al., 1999) emphasizing the wider implications for glucocorticoid-mediated regulation of endogenous N/OFQ in the limbic system. These collective findings reinforce the importance of the BNST as a locus for the action of N/OFQ in chronic adaptation to stress, however, in order for the extent of the N/OFQ system involvement in stress adaptation to be determined, a critical analysis of the functional limbic network is required.

3. Stress and the thalamic reticular nucleus

Analysis of integration of stress neurocircuits focused on classic components of the limbic system including amygdala, hippocampus and prefrontal cortex. The BNST emerged as an integrative hub for gateway control of limbic-hypothalamic inputs to the paraventricular nucleus. However, our recent studies have identified additional structures which may influence limbic responses providing a greater understanding of limbic stress processing. Thalamic structures, in particular the thalamic reticular nucleus (TRN), have been a focus of interest given that they have been implicated in response to stress and limbic functions *via* its connections with thalamo-hippocampal circuitry (Cavdar et al., 2008).

The TRN is uniquely targeted by organized projections from widespread areas of neocortex including cingulate and infralimbic cortices, retrosplenial and frontal cortices, in addition to medial thalamic subregions including paraventricular, anteromedial, centromedial and mediodorsal thalamic nuclei (Carman, Cowan, & Powell, 1964; Cavdar et al., 2008; Guillery, 2003). These extensive projections advance our current understanding of the limbic circuitry and extend the role of the TRN to emotional responses. The role of discrete subregions of thalamus in limbic regulation have been studied previously. The neuronal populations of the thalamic nucleus reuniens, for instance, are shown to modulate hippocampal projections from entorhinal cortex (Dolleman-Van der Weel & Witter, 1996; Dolleman-Van der Weel, Lopes da Silva, & Witter, 1997). In addition, posterior orbitofrontal cortex (pOFC) is widely connected to sensory cortex and has strong connections to mediodorsal thalamus and amygdala (Barbas, Zikopoulos, & Timbie, 2011). The rostral TRN, connecting the hippocampal formation with the thalamus, highlights a novel and additional locus (and associated pathway) that reinforces the established functional limbic loop. Characterization of the TRN in functional stress responses and anxiety will advance our understanding of the precise organization of cortical- thalamic projections that integrate limbic responses.

The TRN has been suggested to be involved in prioritizing gating for selective attention (Crick, 1984; Guillery, Feig, & Lozsadi, 1998). The relevance of TRN thalamo-hippocampal connections are debated, however. These provide a theoretical route for integration of sensory stimuli with emotional salience. The TRN itself is implicated in dysfunction in neuropsychiatric and pathological states that characteristically involve susceptibility to heightened emotionality. Imbalance in the TRN functional system has been implicated in anxiety and behavioral disorders (Barbas et al., 2011). Interest has also focused on TRN function in neurodevelopmental conditions, e.g., schizophrenia (Krol, Wimmer, Halassa, & Feng, 2018; Manoach & Stickgold, 2019). TRN regulation is complex given that the TRN, as the "pacemaker" of sleep spindle oscillations, is crucial to gating and attention (Crabtree, 2018). Positive symptoms of schizophrenia including hallucinations, primarily arise from imbalance between reciprocal corticothalamic-thalamocortical pathways which encompass the major pathways of the TRN (Barch & Ceaser, 2012).

Although numerous functional studies have investigated the roles of the TRN, there has been little direct effort paid to investigating its roles in stress and fear. As the TRN functions as a "modulator" of projections from all cortical regions and associated thalamic nuclei (Zikopoulos & Barbas, 2012), it may be a critical integrator of attentional stimuli in emotional states. The TRN receives numerous inputs from regions with recognized roles in emotional regulation, such as the hippocampus and amygdala. TRN neurones display distinctive firing characteristics and coordinate sleep spindle activity. These neuronal networks may be dysregulated by stress and in stress disorders, a number of which are associated with altered sleep patterns (Selvi et al., 2018; Vargas et al., 2018).

With regard to preclinical research, rats exposed to an adverse early environment (maternal separation) display anxiety behavior in the open field test compared to controls (Ali et al., 2013). In addition, early maternal separation is associated with a reduction in tonic TRN neuronal activity but enhanced burst firing *in vivo* at 7 weeks of age (Ali et al., 2013). Furthermore rats exposed to early maternal separation demonstrate greater susceptibility to amygdala kindling of TRN neurones than controls, providing evidence that early environmental stress impacts on development of firing characteristics of TRN neuronal populations (Ali et al., 2013). Further investigation is required to characterize the role of the TRN in regard to the effects of environmental stressors. This may aid understanding of disorders such as schizophrenia or childhood PTSD and associated impacts on thalamocortical networks.

The TRN itself has complex neuropharmacological regulation, combined with its rich and varied network of projections. It follows that the TRN may be a critical regulatory region influencing activity in the limbic system. The relevance of the TRN to attentional regulation depends on TRN subregions and their specific site projections. In primate brain, the amygdala projects to the posterior and central subregions of TRN, directing attention to salient external stimuli (Zikopoulos & Barbas, 2012). Disrupted attention toward exteroceptive stimuli in PTSD may precipitate the stereotypical state of high anxiety. PTSD, associated with reinforced amygdala-prefrontal cortical (PFC) coupling, could therefore be dependent upon aberrant TRN integration. Clarification of the role of TRN in PTSD may therefore enlighten our understanding of the activation in amygdala-PFC networks.

With regard to the regulation of the TRN it is interesting to reflect upon the role of the opioid-like peptide, N/OFQ in this limbic network. Our evidence indicated that ppN/OFQ mRNA is greater in the dorsal reticular thalamic nucleus of chronic restraint rats, with no change seen in acute stress only rats (Delaney et al., 2012). Our *in situ* hybridization studies identified that ppN/OFQ mRNA expression is particularly concentrated in the TRN structure, indicating that N/OFQ is potentially able to modulate inputs arising from widespread regions encroaching on limbic circuits, including ventral striatum and brainstem. Neurones of the reticular thalamic nucleus are GABAergic, indicating that N/OFQ regulation of intrinsic GABAergic control participates in integration of dorsal thalamic functions. Inhibitory neurones of the TRN govern behavioral state and attentional response, therefore studies identifying the dorsal TRN region in anxiety and fear-related behavior (Picazo et al., 2009) may suggest that N/OFQ in TRN is functionally relevant. 6-hydroxydopamine lesions of the dorsal reticular thalamus also reduce anxiety in behavioral tests in rats (e.g., elevated plus maze). This is not related to an impairment in motor function (Picazo et al., 2009). Our findings revealed a role for N/OFQ neurones of the dorsal reticular thalamus in adaptation to chronic psychological stress. Other opioid peptide systems have also been implicated in functional responses in the TRN, for example delta opioid receptor expression is reduced in TRN in monoarthritis (Neto, Carvalhosa, Ferreira-Gomes, Reguenga, & Castro-Lopes, 2008). Regarding other peptide modulators, somatostatin mRNA in mouse TRN is upregulated in the brain dopamine-deficient weaver mouse mutant model (Zavitsanou et al., 2002). As repeated restraint can heighten anxiety behavior, it is conceivable that chronic restraint induced increase in ppN/OFQ mRNA in the TRN may be related to the heightened anxiety level of chronically stressed animals. It is possible that chronic stress increases N/OFQ transcription in the TRN to compensate for the behavioral activation associated with persistent stress. In theory, N/OFQ inhibition of GABA release in the dorsal thalamic nucleus, could increase anxiety in rats. Clearly further research will need to elucidate the role of N/OFQ in chronic stress and behavior, expanding upon its limbic functions to extend understanding of its role to thalamic reticular nucleus.

4. Nociceptin and neurodevelopmental disorders

Neurodevelopmental conditions like schizophrenia are associated with enhanced presynaptic nigrostriatal dopaminergic function which is associated with behavioral hyperactivity (Grace, 2016; Robbins, 2005). More recently, N/OFQ has been implicated in the pathophysiology of schizophrenia (Khan, Boileau, Kolla, & Mizrahi, 2018), with preclinical research indicating that there are multiple examples of functional interactions between N/OFQ and dopamine systems in the midbrain and forebrain regions. Dopaminergic neurones of both the midbrain tegmentum and striatum express NOP receptor mRNA (Norton, Neal, Jumar, Akil, & Watson, 2002). Furthermore, N/OFQ suppresses mesolimbic dopamine efflux (Murphy & Maidment, 1999), presynaptic dopamine release in striatum (Flau, Redmer, Liedtke, Kathmann, & Schlicker, 2002) and decreases morphine-stimulated dopamine efflux in nucleus accumbens *in vivo* (Di Giannuario, Pieretti, Catalani, & Loizzo, 1999). N/OFQ also inhibits dopamine neurons at the level of D_1 receptor signaling and inhibition of tyrosine hydroxylase, an indicator of reduced dopamine synthesis in striatal neurons (Olianas, Dedoni, Boi, & Onali, 2008). N/OFQ has well-documented effects in terms of inducing a reduction in locomotor behavior at higher peptide doses (Devine et al., 1996; Rizzi et al., 2001). N/OFQ-induced suppression of dopaminergic transmission is consistent with N/OFQ-mediated locomotor inhibition.

Post-weaning social isolation rearing in rats is a model with behavioral and neurochemical sequelae bearing similarities to human schizophrenia, a neurodevelopmental disorder (Robbins, 2016). Both sensorimotor gating and selective attention are significantly disrupted in schizophrenia. Isolation-reared in rats and mice also display persistent deficits in prepulse inhibition of the startle response (PPI) indicative of sensorimotor gating impairments (Geyer, Wilkinson, Humby, & Robbins, 1993; Varty, Paulus, Braff, & Geyer, 2000; Varty, Powell, Lehmann-Masten, Buell, & Geyer, 2006) and in isolation-reared rats this has been associated with elevated anxiety and enduring changes in enhanced presynaptic monoaminergic function in the dorsal and ventral striatum (Fulford & Marsden, 1998; Howes, Dalley, Morrison, Robbins, & Everitt, 2000). From the perspective of functional interactions between neuromodulatory peptides and monoamine neurotransmitters, it is interesting to consider the role of N/OFQ in the integration of monoamine neurotransmitter systems. N/OFQ agonists suppress PPI in tests of sensorimotor gating in mice (Ces et al., 2012). This NOP-dependent effect is suppressed by the dopamine receptor antagonist, haloperidol, indicating a direct interaction between dopamine and N/OFQ mechanisms in functional attention. These findings raise the question of the role of N/OFQ in neurodevelopmental conditions associated with attentional deficits. Preliminary evidence supports an effect of isolation-housing in rats on central N/OFQ levels *ex-vivo* (Granholm, Roman, & Nylander, 2015) warranting an investigation of the long-term impact of early environmental manipulation on this peptide system. Furthermore, the integration of attentional responses by N/OFQ in contexts such as abnormal neurodevelopment, provides an opportunity to address the role of N/OFQ in wider limbic regions, such as TRN, given its implicit role in attention and stress adaptation.

5. Conclusions

In summary, there has been good progress made in delineation of the effect of N/OFQ ligands in behavioral regulation, in particular with regard to anxiety, mood and cognitive states. These studies have raised questions regarding the translational basis of N/OFQ function to human conditions. Furthermore, there are well-characterized examples of human conditions associated with abnormal neurodevelopment (e.g., schizophrenia; childhood PTSD) which are also associated with heightened stress vulnerability. Established rodent models of adverse social environments (e.g., postweaning isolation-rearing in rats) result in permanent behavioral and neurochemical alterations in the brain, underpinning disturbed stress responsiveness in limbic forebrain. There are clear opportunities for extending understanding of the N/OFQ-NOP receptor function to the context of aberrant neurodevelopment. In addition, recent characterization of brain regions with novel roles in emotionality (e.g., thalamic reticular nucleus) present opportunities to address the extent of the endogenous N/OFQ system more widely in chronic stress adaptation in adulthood.

References

Ali, I., O'Brien, P., Kumar, G., Zheng, T., Jones, N. C., Pinault, D., et al. (2013). Enduring effects of early life stress on firing patterns of hippocampal and thalamocortical neurons in rats: Implications for limbic epilepsy. *PLoS One*, *8*(6), 1–12.

Asth, L., Ruzza, C., Malfacin, D., Medeiros, I., Guerrini, R., Zaveri, N. T., et al. (2016). Beta-arrestin rather than G protein efficacy determines the anxiolytic-versus antidepressant-like effects of nociception/orphanin FQ receptor ligands. *Neuropharmacology*, *105*, 434–442.

Barbas, H., Zikopoulos, B., & Timbie, C. (2011). Sensory pathways and emotional context for action in primate prefrontal cortex. *Biological Psychiatry*, *69*, 1133–1139.

Barch, D. M., & Ceaser, A. (2012). Cognition in schizophrenia: Core psychological and neural mechanisms. *Trends in Cognitive Sciences*, *16*, 27–34.

Buzas, B. (2002). Regulation of nociceptin/orphanin FQ gene expression in astrocytes by ceramide. *Neuroreport*, *13*, 1707–1710.

Carman, J. B., Cowan, W. M., & Powell, T. P. S. (1964). Cortical connexions of the thalamic reticular nucleus. *Journal of Anatomy*, *98*, 587–598.

Cavdar, S., Onat, F. Y., Cakmak, Y. O., Yananli, H. R., Gulcebi, et al. (2008). The pathways connecting the hippocampal formation, the thalamic reuniens nucleus and the thalamic reticular nucleus in the rat. *Journal of Anatomy*, *212*, 249–256.

Ces, A., et al. (2012). Activation of nociceptin/orphaninFQ peptide receptors disrupts visual but not auditory sensorimotor gating in BALB/cByJ mice: Comparison to dopamine receptor agonists. *Neuropsychopharmacology*, *37*, 378–389.

Chen, L. Y., Huang, J. X., & Yu, L. C. (2008). Involvement of ORL1 receptor and ERK kinase in the orphanin FQ-induced nociception in the nucleus accumbens of rats. *Regulatory Peptides*, *151*, 43–47.

Choi, D. C., Furay, A. R., Evanson, N. K., Ostrander, M. M., Ulrich-Lai, Y. M., & Herman, J. P. (2007). Bed nucleus of the stria terminalis subregions differentially regulate hypothalamo-pituitary-adrenal axis activity: Implications for the integration of limbic inputs. *The Journal of Neuroscience*, *27*, 2025–2034.

Ciccocioppo, R., Fedeli, A., Economidou, D., Policani, F., Weiss, F., & Massi, M. (2003). The bed nucleus is a neuroanatomical substrate for the anorectic effect of corticotropin-releasing factor and for its reversal by nociceptin/orphanin FQ. *The Journal of Neuroscience*, *23*, 9445–9451.

Collins, L. M., Dal Bo, G., Calcagno, M., Monzón-Sandoval, J., Sullivan, A. M., Gutierrez, H., et al. (2016). Nociceptin/orphanin FQ inhibits the survival and axon growth of midbrain dopaminergic neurons through a p38-MAPK dependent mechanism. *Molecular Neurobiology*, *53*, 7284–7297.

Crabtree, J. W. (2018). Functional diversity of thalamic reticular subnetworks. *Frontiers in Systems Neuroscience*, *12*, 41.

Crick, F. (1984). Function of the thalamic reticular complex: The searchlight hypothesis. *Proceedings of the National Academy of Sciences of the United States of America*, *81*, 4586–4590.

Dawe, K. L., Wakerley, J. B., & Fulford, A. J. (2010). Nociceptin/orphanin FQ and the regulation of neuronal excitability in the rat bed nucleus of the stria terminalis: Interaction with glucocorticoids. *Stress*, *13*, 516–527.

Delaney, G., Dawe, K. L., Hogan, R., Hunjan, T., Roper, J., Hazell, G., et al. (2012). Role of nociceptin/orphanin FQ and NOP receptors in the response to acute and repeated restraint stress in rats. *Journal of Neuroendocrinology*, *24*, 1527–1541.

Devine, D. P., et al. (1996). Rats rapidly develop tolerance to the locomotor-inhibiting effects of the novel neuropeptide orphanin FQ. *Neurochemical Research*, *21*, 1387–1396.

Di Giannuario, A., Pieretti, S., Catalani, A., & Loizzo, A. (1999). Orphanin FQ reduces morphine-induced dopamine release in the nucleus accumbens: A microdialysis study in rats. *Neuroscience Letters*, *272*, 183–186.

Dolleman-Van der Weel, M. J., Lopes da Silva, F. H., & Witter, M. P. (1997). Nucleus reuniens thalami modulates activity in hippocampal field CA1 through excitatory and inhibitory mechanisms. *The Journal of Neuroscience*, *17*, 5640–5650.

Dolleman-Van der Weel, M. J., & Witter, M. P. (1996). Projections from the nucleus reuniens thalami to the entorhinal cortex, hippocampal field CA1, and the subiculum in the rat arise from different populations of neurons. *The Journal of Comparative Neurology*, *364*, 4.

Dong, H. W., Petrovich, G. D., & Swanson, L. W. (2001). Topography of projections from amygdala to bed nuclei of the stria terminalis. *Brain Research Reviews*, *38*, 192–246.

Flau, K., Redmer, A., Liedtke, S., Kathmann, M., & Schlicker, E. (2002). Inhibition of striatal and retinal dopamine release via nociceptin/orphaninFQ receptors. *British Journal of Pharmacology*, *137*, 1355–1361.

Fulford, A. J. (2015). Endogenous nociceptin system involvement in stress responses and anxiety behavior. *Vitamins and Hormones*, *97*, 267–293.

Fulford, A. J., & Marsden, C. A. (1998). Effect of isolation-rearing on conditioned dopamine release in vivo in the nucleus accumbens of the rat. *Journal of Neurochemistry*, *70*, 384–390.

Gavioli, E. C., & Calo', G. (2006). Antidepressant- and anxiolytic-like effects of nociceptin/orphanin FQ receptor ligands. *Naunyn-Schmiedeberg's Archives of Pharmacology*, *372*, 319–330.

Geyer, M. A., Wilkinson, L. S., Humby, T., & Robbins, T. W. (1993). Isolation rearing of rats produces a deficit in prepulse inhibition of acoustic startle similar to that in schizophrenia. *Biological Psychiatry*, *34*, 361–372.

Goeldner, C., Reiss, D., Wichmann, J., Meziane, H., Kieffer, B. L., & Ouagazzal, A. M. (2008). Nociceptin receptor impairs recognition memory via interaction with NMDA receptor-dependent mitogen-activated protein kinase/extracellular signal-regulated kinase signalling in the hippocampus. *The Journal of Neuroscience*, *28*, 2190–2198.

Grace, A. A. (2016). Dysregulation of the dopamine system in the pathophysiology of schizophrenia and depression. *Nature Reviews Neuroscience*, *17*, 524–532.

Granholm, L., Roman, E., & Nylander, I. (2015). Single housing during early adolescence causes time-, area- and peptide-specific alterations in endogenous opioids of rat brain. *British Journal of Pharmacology*, *172*, 606–614.

Guillery, R. W. (2003). Structure and connections of the thalamic reticular nucleus: Advancing views over half a century. *The Journal of Comparative Neurology*, *463*, 360–371.
Guillery, R. W., Feig, S. L., & Lozsadi, D. A. (1998). Paying attention to the thalamic reticular nucleus. *Trends in Neurosciences*, *21*, 28–32.
Higgins, G. A., Kew, J. N., Richards, J. G., Takeshima, H., Jenck, F., Adam, G., et al. (2002). A combined pharmacological and genetic approach to investigate the role of orphanin FQ in learning and memory. *The European Journal of Neuroscience*, *15*, 911–922.
Holanda, V. A., Medeiros, I. U., Asth, L., Guerrini, R., Calo', G., & Gavioli, E. C. (2016). Antidepressant activity of nociceptin/orphanin FQ receptor antagonists in the mouse learned helplessness. *Psychopharmacology*, *233*, 2525–2532.
Howes, S. R., Dalley, J. W., Morrison, C. H., Robbins, T. W., & Everitt, B. J. (2000). Leftward shift in the acquisition of cocaine self-administration in isolation-reared rats: Relationship to extracellular levels of dopamine, serotonin and glutamate in the nucleus accumbens and amygdala-striatal FOS expression. *Psychopharmacology*, *151*, 55–63.
Khan, M. S., Boileau, I., Kolla, N., & Mizrahi, R. (2018). A systematic review of the role of the nociceptin receptor system in stress, cognition, and reward: Relevance to schizophrenia. *Translational Psychiatry*, *8*, 38.
Krol, A., Wimmer, R. D., Halassa, M. M., & Feng, G. (2018). Thalamic reticular dysfunction as a circuit endophenotype in neurodevelopmental disorders. *Neuron*, *98*, 282–295.
Lapiz, M. D., Fulford, A., Muchimapura, S., Mason, R., Parker, T., & Marsden, C. A. (2003). Influence of postweaning social isolation in the rat on brain development, conditioned behavior, and neurotransmission. *Neuroscience and Behavioral Physiology*, *33*, 13–29.
Leggett, J. D., Dawe, K. L., Jessop, D. S., & Fulford, A. J. (2009). Endogenous nociceptin/orphanin FQ system involvement in hypothalamic–pituitary–adrenal axis responses: Relevance to models of inflammation. *Journal of Neuroendocrinology*, *21*, 888–897.
Leggett, J. D., Harbuz, M. S., Jessop, D. S., & Fulford, A. J. (2006). The nociceptin receptor antagonist [Nphe1,Arg14,Lys15]nociceptin/orphanin FQ-NH2 blocks the stimulatory effects of nociceptin/orphanin FQ on the HPA axis in rats. *Neuroscience*, *141*, 2051–2057.
Leggett, J. D., Jessop, D. S., & Fulford, A. J. (2007). The nociceptin/orphanin FQ antagonist UFP-101 differentially modulates the glucocorticoid response to restraint stress in rats during the peak and nadir phases of the hypothalamo-pituitary–adrenal axis circadian rhythm. *Neuroscience*, *147*, 757–764.
Manoach, D. S., & Stickgold, R. (2019). Abnormal sleep spindles, memory consolidation, and schizophrenia. *Annual Review of Clinical Psychology*, *15*, 451–479.
Murphy, N. P., & Maidment, N. T. (1999). Orphanin FQ/nociceptin modulation of mesolimbic dopamine transmission determined by microdialysis. *Journal of Neurochemistry*, *73*, 179–186.
Neal, C. R., Mansour, A., Reinscheid, R., Nothacker, H. P., Civelli, O., & Watson, S. J. (1999). Localization of orphanin FQ (nociceptin) peptide and messenger RNA in the central nervous system of the rat. *The Journal of Comparative Neurology*, *406*, 503–547.
Neto, F. L., Carvalhosa, A. R., Ferreira-Gomes, J., Reguenga, C., & Castro-Lopes, J. M. (2008). Delta opioid receptor mRNA expression is changed in the thalamus and brainstem of monoarthritic rats. *Journal of Chemical Neuroanatomy*, *36*, 122–127.
Norton, C. S., Neal, C. R., Jumar, S., Akil, H., & Watson, S. J. (2002). Nociceptin/orphaninFQ and opioid receptor-like receptor mRNA expression in dopamine systems. *The Journal of Comparative Neurology*, *444*, 358–368.
Olianas, M. C., Dedoni, S., Boi, M., & Onali, P. (2008). Activation of nociception/orphanin FQ-NOP recptor system inhibits tyrosine hydroxylase phosphorylation, dopamine synthesis, and dopamine D_1 receptor signaling in rat nucleus accumbens and dorsal striatum. *Journal of Neurochemistry*, *107*, 544–556.

Picazo, O., Chuc-Meza, E., Anaya-Martinez, V., Jimenez, I., Aceves, J., & Garcia-Ramirez, M. (2009). 6-Hydroxydopamine lesion in thalamic reticular nucleus reduces anxiety behaviour in the rat. *Behavioural Brain Research, 197*, 317–322.

Post, A., Smart, T. S., Krikke-Workel, J., Dawson, G. R., Harmer, C. J., Browning, M., et al. (2016). A selective nociceptin receptor antagonist to treat depression: Evidence from preclinical and clinical studies. *Neuropsychopharmacology, 41*, 1803–1812.

Rekik, K., Faria da Silva, R., Colom, N., Pacifico, S., Zaveri, N. T., Calo, G., et al. (2017). Activation of nociceptin/orphanin FQ receptors inhibits contextual fear memory consolidation. *Neuropharmacology, 125*, 39–49.

Rizzi, A., Bigoni, R., Marzola, G., Guerrini, R., Salvadori, S., Regoli, D., et al. (2001). Characterization of the locomotor activity-inhibiting effect of nociceptin/orphanin FQ in mice. *Naunyn-Schmiedeberg's Archives of Pharmacology, 363*, 161–165.

Robbins, T. W. (2005). Synthesizing schizophrenia: A bottom-up, symptomatic approach. *Schizophrenia Bulletin, 31*, 854–864.

Robbins, T. W. (2016). Neurobehavioral sequelae of social deprivation in rodents revisited: Modeling social adversity for developmental neuropsychiatric disorders. *Journal of Psychopharmacology, 30*, 1082–1089.

Rodi, D., Zucchini, S., Simonato, M., Cifani, C., Massi, M., & Polidori, C. (2008). Functional antagonism between nociceptin/orphanin FQ (N/OFQ) and corticotropin-releasing factor (CRF) in the rat brain: Evidence for involvement of the bed nucleus of the stria terminalis. *Psychopharmacology, 196*, 523–531.

Selvi, Y., Boysan, M., Kandeger, A., Uygur, O. F., Sayin, A. A., Akbaba, N., et al. (2018). Heterogeneity of sleep quality in relation to circadian preferences and depressive symptomatology among major depressive patients. *Journal of Affective Disorders, 235*, 242–249.

Toll, L., Bruchas, M. R., Calo', G., Cox, B. M., & Zaveri, N. T. (2016). Nociceptin/orphanin FQ receptor structure, signaling, ligands, functions, and interactions with opioid systems. *Pharmacological Reviews, 68*, 419–457.

Vargas, I., Vgontzas, A. N., Abelson, J. L., Faghih, R. T., Morales, K. H., & Perlis, M. L. (2018). Altered ultradian cortisol rhythmicity as a potential neurobiologic substrate for chronic insomnia. *Sleep Medicine Reviews, 41*, 234–243.

Varty, G. B., Paulus, M. P., Braff, D. L., & Geyer, M. A. (2000). Environmental enrichment and isolation rearing in the rat: Effects on locomotor behavior and startle response plasticity. *Biological Psychiatry, 47*, 864–873.

Varty, G. B., Powell, S. B., Lehmann-Masten, V., Buell, M. R., & Geyer, M. A. (2006). Isolation rearing of mice induces deficits in prepulse inhibition of the startle response. *Behavioural Brain Research, 169*, 162–167.

Villarreal, J. S., & Barea-Rodriguez, E. J. (2006). ERK phosphorylation is required for retention of trace fear memory. *Neurobiology of Learning and Memory, 85*, 44–57.

Walker, D. L., Miles, L. A., & Davis, M. (2009). Selective participation of the bed nucleus of the stria terminalis and CRF in sustained anxiety-like *versus* phasic fear-like responses. *Progress in Neuro-Psychopharmacology & Biological Psychiatry, 33*, 1291–1308.

Wang, W., Cui, Q., Li, Y., Yang, X., Cui, L., Jin, H., et al. (2010). The role of ERK-1/2 in the N/OFQ-induced inhibition of delayed rectifier potassium currents. *Biochemical and Biophysical Research Communications, 394*, 1058–1062.

Werthwein, S., Bauer, U., Nakazi, M., Kathmann, M., & Schlicker, E. (1999). Further characterization of the ORL1 receptor-mediated inhibition of noradrenaline release in the mouse brain in vitro. *British Journal of Pharmacology, 127*, 300–308.

Xie, G. X., Ito, E., Maruyama, K., Suzuki, Y., Sugano, S., Sharma, M., et al. (1999). The promoter region of human prepro-nociceptin gene and its regulation by cyclic AMP and steroid hormones. *Gene, 238*, 427–436.

Zavitsanou, K., Triarhou, L. C., Kouvelas, E. D., Mitsacos, A., Palacios, J. M., & Mengod, G. (2002). Somatostatin, cholecystokinin and neuropeptide Y mRNAs in normal and weaver mouse brain. *Journal of Neural Transmission, 109*, 1337–1351.
Zikopoulos, B., & Barbas, H. (2012). Pathways for emotions and attention converge on the thalamic reticular nucleus in primates. *The Journal of Neuroscience, 32*, 5338–5350.

Further reading

Fulford, A. J., & Marsden, C. A. (2007). An intact dopaminergic system is required for context-conditioned release of 5-HT in the nucleus accumbens of postweaning isolation-reared rats. *Neuroscience, 149*, 392–400.

CHAPTER SEVEN

Enkephalins and ACTH in the mammalian nervous system

Ewing Duque-Díaz[a,*], Olga Alvarez-Ojeda[b], Rafael Coveñas[c]
[a]Universidad de Santander UDES, Laboratory of Neurosciences, School of Medicine, Bucaramanga, Colombia
[b]Universidad Industrial de Santander, Department of Pathology, School of Medicine, Bucaramanga, Colombia
[c]Laboratory of Neuroanatomy of the Peptidergic Systems, Institute of Neurosciences of Castilla y León (INCYL), University of Salamanca, Salamanca, Spain
*Corresponding author: e-mail address: ew.duque@mail.udes.edu.co

Contents

Abstract

The pentapeptides methionine-enkephalin and leucine-enkephalin belong to the opioid family of peptides, and the non-opiate peptide adrenocorticotropin hormone (ACTH) to the melanocortin peptide family. Enkephalins/ACTH are derived from pro-enkephalin, pro-dynorphin or pro-opiomelanocortin precursors and, via opioid and

Vitamins and Hormones, Volume 111
ISSN 0083-6729
https://doi.org/10.1016/bs.vh.2019.05.001

melanocortin receptors, are responsible for many biological activities. Enkephalins exhibit the highest affinity for the δ receptor, followed by the μ and κ receptors, whereas ACTH binds to the five subtypes of melanocortin receptor, and is the only member of the melanocortin family of peptides that binds to the melanocortin-receptor 2 (ACTH receptor). Enkephalins/ACTH and their receptors exhibit a widespread anatomical distribution. Enkephalins are involved in analgesia, angiogenesis, blood pressure, embryonic development, emotional behavior, feeding, hypoxia, limbic system modulation, neuroprotection, peristalsis, and wound repair; as well as in hepatoprotective, motor, neuroendocrine and respiratory mechanisms. ACTH plays a role in acetylcholine release, aggressive behavior, blood pressure, bone maintenance, hyperalgesia, feeding, fever, grooming, learning, lipolysis, memory, nerve injury repair, neuroprotection, sexual behavior, sleep, social behavior, tissue growth and stimulates the synthesis and secretion of glucocorticoids. Enkephalins/ACTH are also involved in many pathologies. Enkephalins are implicated in alcoholism, cancer, colitis, depression, heart failure, Huntington's disease, influenza A virus infection, ischemia, multiple sclerosis, and stress. ACTH plays a role in Addison's disease, alcoholism, cancer, Cushing's disease, dermatitis, encephalitis, epilepsy, Graves' disease, Guillain-Barré syndrome, multiple sclerosis, podocytopathies, and stress. In this review, we provide an updated description of the enkephalinergic and ACTH systems.

1. Introduction

Enkephalins (methionine (met)-enkephalin (Tyr-Gly-Gly-Phe-Met) and leucine (leu)-enkephalin (Tyr-Gly-Gly-Phe-Leu)) belong to the opioid peptide family (Fig. 1). Opioid peptides contain in their N-terminals the Tyr-Gly-Gly-Phe motif (referred to as the opioid core sequence), and do not possess a C-terminal amide (Larhammar, Sundström, & Dores, 2013). The non-opiate peptide adrenocorticotropin hormone (ACTH) belongs to the melanocortin family of peptides, a group which also includes α-, β- and γ-melanocyte-stimulating hormones (Jégou, Cone, Eberle, & Vaudry, 2013). Melanocortin peptides share the core amino acid motif His-Phe-Arg-Trp (Larhammar et al., 2013) (Fig. 1). ACTH contains 39 amino acids (Fig. 1), although fragments of the hormone (e.g., $ACTH_{1-14}$, $ACTH_{1-17}$, $ACTH_{1-24}$, $ACTH_{1-34}$, $ACTH_{7-21}$, $ACTH_{18-39}$ or CLIP (corticotropin-like intermediate lobe peptide)) have been also reported.

2. Historical background

In 1973, opioid receptors were identified in the gastrointestinal tract and brain (Pert & Snyder, 1973; Simon, Hiller, & Edelman, 1973; Terenius, 1973).

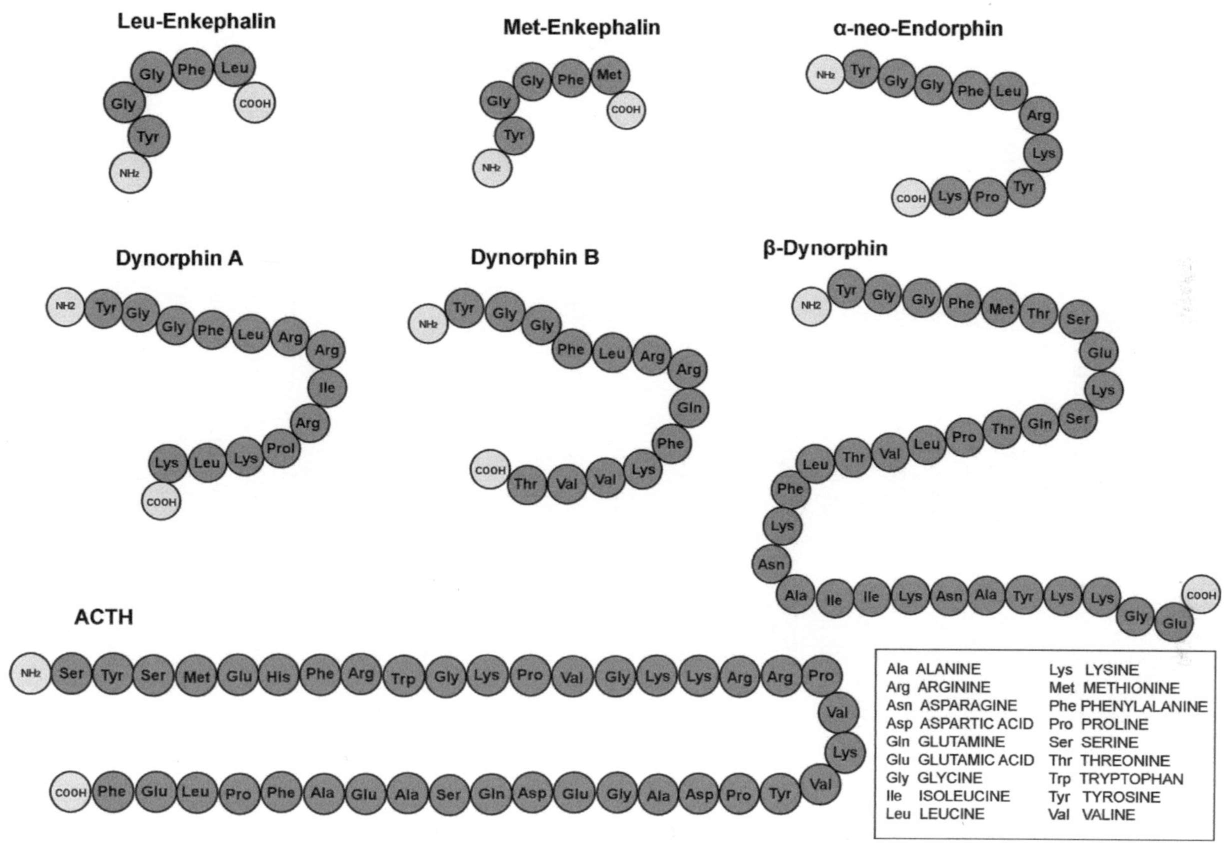

Fig. 1 Enkephalins, endorphins, dynorphins and ACTH chemical models and amino acid sequences.

In 1976, the opioid receptors μ (for morphine) and κ (for ketocyclazocine) were discovered and named based upon the drugs used in the pioneering study; followed by the δ receptor (found in the vas deferens) (Lord, Waterfield, Hughes, & Kosterlitz, 1977; Martin, Eades, Thompson, Gilbert, & Huppler, 1976). In 1992, the *Oprd1* (which encodes the δ receptor) was the first opioid receptor gene cloned, followed by the *Oprm1* and *Oprk1* genes that, respectively, encode the μ and κ receptors (Kieffer & Evans, 2009). In 1975, natural ligands (the pentapeptides met-enkephalin and leu-enkephalin that were isolated from the porcine brain) of the δ receptor were discovered; and shortly thereafter the natural ligands of the μ (β-endorphins) and κ (dynorphins) receptors (Akil et al., 1984; Hughes et al., 1975). Both pentapeptides mimic the action of morphine and this action is inhibited by naloxone, an opioid antagonist (Snyder, 2004). In the 1990s, Endomorphins 1 and 2 (natural ligands of the μ receptor) would be discovered in the 1990s (Zadina, Hackler, Ge, & Kastin, 1997).

ACTH was first reported in 1933 (Collip, Anderson, & Thomson, 1933). In 1955, the researches isolated the ACTH polypeptide from sheep pituitary glands, and demonstrated that the molecule stimulated adrenocortical activity (Li, Geschwind, Dixon, Levy, & Harris, 1955). ACTH's biological activity occurs by means of the melanocortin receptors, which were first cloned in 1992 (Mountjoy, Robbins, Mortrud, & Cone, 1992).

3. Structure of the enkephalin/ACTH genes

3.1 Enkephalins

The pro-enkephalin gene is comprised of three exons separated by two introns. The 5′-untranslated region is separated from the signal peptide by one of the introns, while the other intron separates the signal peptide from the sequence encoding the biologically active peptide. The pro-dynorphin gene contains four exons separated by three introns (Spampinato, Bedini, & Baiula, 2013). Exons 1 and 4 contain non-translated sequences, with translated regions found in exons 3 and 4. Pro-dynorphin mRNA contains, at the 3′-terminal end, a long (1550 nucleotides) non-translated region (Spampinato et al., 2013). In the promoter region of the pro-dynorphin gene, several transcription factor binding sites have been reported, which are involved in the enhancement of pro-dynorphin mRNA expression in response to protein kinase A or AMP (Douglass, McKinzie, & Pollock, 1994).

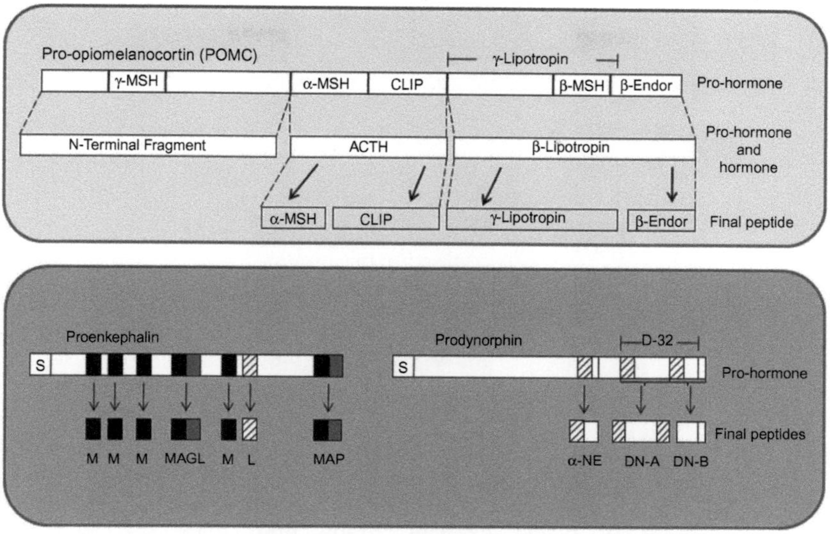

Fig. 2 Genomic structure of pro-opiomelanocortin (POMC). ACTH: adrenocorticotropin hormone; β-LPH: β-lipotropin.

3.2 ACTH

The pro-opiomelanocortin gene has 7665 base pairs, and is composed of three exons separated by two introns (Eberle, 1988). In mammal, the third exon encodes all biologically active pro-opiomelanocortin-derived peptides, while the second exon encodes the first 18 amino acid residues and the signal peptide (Jégou et al., 2013) (Fig. 2). In the species studied, the sequence of the pro-opiomelanocortin cDNA has been preserved in those regions coding for β-endorphin, β-melanocyte-stimulating hormone, ACTH/α-melanocyte-stimulating hormone and pro-γ-melanocyte-stimulating hormone/γ-melanocyte-stimulating hormone (Jégou et al., 2013). In general, dibasic motifs that are cleavage sites for pro-hormone convertases are also preserved, except for the Arg^{63} and Lys^{190} of human pro-opiomelanocortin that are not present in mouse and rat pro-opiomelanocortin sequences (Jégou et al., 2013). Consequently, in rodents, neither the γ3-melanocyte-stimulating hormone nor γ-lipotropin can be processed to generate γ- and β-melanocyte-stimulating hormones, respectively.

4. Processing of the precursors

4.1 Enkephalins

Three genes (pro-enkephalin, pro-opiomelanocortin and pro-dynorphin) generate the endogenous opioid peptides (Sternini, Duraffourd, & Anselmi,

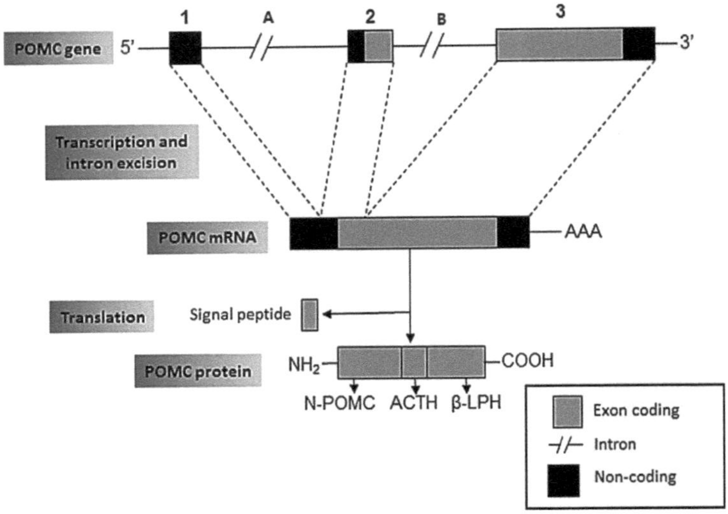

Fig. 3 Biosynthesis of opioid peptides arising from three precursors: pro-opiomelanocortin (POMC), pro-enkephalin and pro-dynorphin. The post-translational processing of POMC produces several peptides: α-, β- and γ- melanocyte stimulating hormone (MSH); β-endorphin, adrenocorticotropin hormone (ACTH) and corticotropin-like intermediate lobe peptide (CLIP). Enkephalins originate from pro-enkephalin (L: leu-enkephalin (Tyr-Gly-Gly-Phe-Leu); M: met-enkephalin (Tyr-Gly-Gly-Phe-Met); MAGL (M-Arg-Gly-Phe); MAP: (M-Arg-Phe)) and dynorphins from pro-dynorphin (dynorphin-A (DN-A); dynorphin-B (DN-B); dynorphin 32 (D-32); α-neo-endorphin (α-NE)).

2013) (Fig. 3). Through the cleavage of complex and large precursor proteins, met-enkephalin, leu-enkephalin (Fig. 1), methionine-enkephalin-Arg^6-Phe^7 and methionine-enkephalin-Arg^6-Phe^7-Leu^8 are produced from pro-enkephalin (Fig. 3); met-enkephalin, β-endorphin (Fig. 1), α-endorphin and γ-endorphin are produced from pro-opiomelanocortin (this precursor is not considered as a primary source for met-enkephalin) (McLaughlin & Zagon, 2013) (Fig. 3), and leu-enkephalin, dynorphin (dynorphin A_{1-32}), dynorphin A (dynorphin A_{1-17}), dynorphin B (dynorphin B_{1-13}), leumorphin (dynorphin B_{1-29}) and α-neo-endorphin (Fig. 1) are produced from pro-dynorphin (Fig. 3). The three precursors contain multiple repetitive units in the C-terminal region (indicating that the structural genes for the precursors evolved through a series of direct duplications of DNA segments), and a cysteine-containing N-terminal sequence (implying that the correct folding of this region, effected by disulfide bonds, is involved in the processing of precursors). The three precursors represent multi-peptide precursors with multiple biological activities. Biological activity can be regulated by both gene regulation and tissue-specific processing. In general, in the precursor, opioid sequences are flanked by paired basic amino acids (Larhammar et al., 2013). Each precursor contains a signal peptide to prompt its transfer into the endoplasmic reticulum during synthesis, followed by an N-terminal part containing cysteine residues (Larhammar et al., 2013). Post-translational events (e.g., glycosylation, palmitoylation, phosphorylation) regulate the final products, the pharmacological profiles, and the selectivity of the opioid peptides (Akil et al., 1984; Dietis, Rowbotham, & Lambert, 2011).

The pro-dynorphin precursor is highly basic; the processing sequences are cleaved on the carboxyl side of the basic amino acid (lysine-lysine, lysine-arginine and arginine-arginine), after which the C-terminal basic residue is removed (Spampinato et al., 2013). The first of these steps is carried out by the endopeptidase dynorphin-converting enzyme, and the second by carboxypeptidases. Thus, processing enzymes are crucial in the modulation of peptide gene expression. By these mechanisms, dynorphin A_{1-32} originated from the pro-dynorphin precursor is converted into dynorphin A_{1-17}, this into dynorphin A_{1-8}, this in leu-enkephalin-Arg and, finally, the latter is converted into leu-enkephalin (Spampinato et al., 2013).

After the processing of the pro-enkephalin precursor, six copies of met-enkephalin and one copy of leu-enkephalin are generated (McLaughlin & Zagon, 2013). As noted above, the cleavage occurs between pairs of basic amino acids. Enkephalins are biodegraded in a few minutes by hydrolysis and cleavage of the Tyr-Gly bond (McLaughlin & Zagon, 2013). Further degradation occurs when non-specific enkephalinases or amino peptidases (e.g., NAP and NAP-2 degrade the pentapeptides in tissue and CD10 and CD13 in plasma) carve up the pentapeptides into 2, 3 or 4 amino acids (Lukiw, 2006).

4.2 ACTH

ACTH is exclusively produced from pro-opiomelanocortin, and is thus a good marker for the pro-opiomelanocortin system (Duque-Díaz, Díaz-Cabiale, Narváez, & Coveñas, 2017; Manso et al., 2014) (Figs. 2, 3). Moreover, the melanocortins α-, β- and γ-melanocyte-stimulating hormones are also derived from pro-opiomelanocortin (Jégou et al., 2013) (Fig. 3). The sequence of the α-melanocyte-stimulating hormone is identical to the first third of ACTH (Jégou et al., 2013). Once the signal peptide has been removed, ACTH and pro-γ-melanocyte-stimulating hormone are glycosylated, and ACTH is phosphorylated (Eberle, 1988). Convertases 1 and 2 then catalyze the cleavage of pro-opiomelanocortin at dibasic or tetrabasic sites. The corticotroph cells located in the pars distalis of the pituitary gland only express the convertase 1 that generates the N-terminal fragment pro-γ-melanocyte-stimulating hormone, ACTH and the C-terminal peptide β-lipotropin (Jégou et al., 2013). However, in those cells (e.g., melanotroph cells of the pars intermedia of the pituitary gland and hypothalamus) containing convertases 1 and 2, the processing of the precursor generates $ACTH_{1-17}$, $ACTH_{18-39}$, β-endorphins, γ-lipotropin and (Lys-) γ-melanocyte-stimulating hormone (Jégou et al., 2013). From the cleaved peptides, basic residues placed at the C-terminal are trimmed by carboxypeptidase E. $ACTH_{1-14}$ and (Lys-) γ-melanocyte-stimulating hormone are C-terminally amidated by peptidylglycine-α-amidating-monooxygenase. Finally, Nα-acetylation of β-endorphin can occur, an important factor for the biological activity of the pro-opiomelanocortin-derived peptides. For example, the acetylation of the N-terminal tyrosine residue blocks the analgesic action of β-endorphin (Bertolini, Tacchi, & Vergoni, 2009).

5. Tissue distribution

5.1 Enkephalins

Opioid peptides have been discovered in endocrine cells, and in the enteric nervous system along the mammalian gastrointestinal tract (Kromer, 1988; Sternini, Patierno, Selmer, & Kirchgessner, 2004). Although met-enkephalin and leu-enkephalin show a different distribution, many clusters of neurons contain both pentapeptides (McLaughlin & Zagon, 2013). Fibers and perikarya containing enkephalins are widely distributed throughout the central and peripheral mammalian nervous systems. For example, met-enkephalin has been observed in the amygdala, hippocampus, globus pallidus, thalamus, hypothalamus, brainstem and spinal cord, whereas the expression of the pro-enkephalin gene has been reported in the posterior pituitary, sympathetic nervous system, pancreas, intestine, adrenal medulla, placenta and cardiovascular system (Duque, Mangas, Díaz-Cabiale, Narváez, & Coveñas, 2011; McLaughlin & Zagon, 2013). For example, the presence of both leu-enkephalin- and met-enkephalin-immunoreactive fibers has been noted in 82% of the brainstem nuclei/tracts of the squirrel monkey (*Saimiri sciureus*) (Duque-Díaz et al., 2017) (Fig. 4). Thus, a close neuroanatomical relationship exists between both enkephalins. Pro-enkephalin derived peptides have been observed in the myenteric

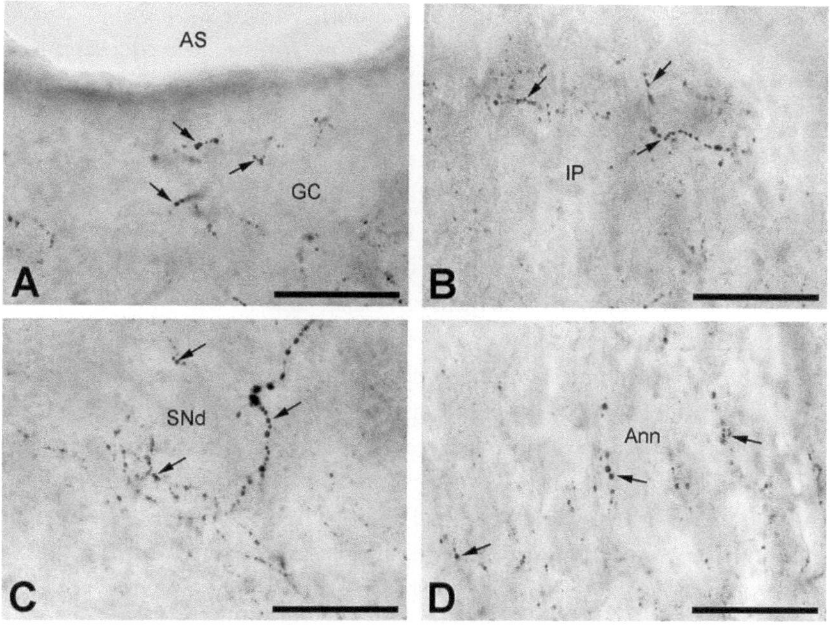

Fig. 4 Immunoreactive structures in the *Saimiri sciureus* brainstem containing enkephalins. A, B: Arrows indicate fibers containing leucine-enkephalin located in the substantia grisea centralis (GC) and interpeduncular nucleus (IP). AS: Aquaeductus Sylvii. C, D: Arrows indicate fibers containing methionine-enkephalin located in the substantia nigra, pars diffusa (SNd) and nucleus annularis (Ann). Scale bar: 50 μm.

neurons that send projections to the submucosal plexus and circular muscle, and in enterochromaffin and enteroendocrine cells located in the intestine and stomach (Furness, Costa, & Miller, 1983; Sternini et al., 2004). In descending neuronal pathways, enkephalins co-localize with other peptides (e.g., galanin, substance P, gastrin-releasing peptide, vasoactive intestinal polypeptide) or classical neurotransmitters (e.g., acetylcholine) (Furness et al., 1983; Steele & Costa, 1990).

Opioid receptors for enkephalins are widely distributed by the body; being found in the neocortex, limbic and olfactory systems, basal ganglia, hippocampus, amygdala, diencephalon, brainstem, dorsal horn of the spinal cord, and in the primary afferent terminals (McLaughlin & Zagon, 2013). μ receptors, located within the central and peripheral nervous systems, are involved in pain processes, while δ receptors have been detected in cancer cells and cardiovascular tissues (Snyder, 2004). Opioid receptors (mainly μ and δ) are widely distributed throughout the mammalian gastrointestinal tract, in particular the enteric nervous system (Bagnol, Mansour, Akil, &

Watson, 1997; Ho et al., 2003; Poonyachoti, Kulkarni-Narla, & Brown, 2002; Sternini et al., 2004). Thus, neurons in the submucosal plexus express more μ receptors than those located in the myenteric plexus, whereas the opposite occurs in the δ receptor system (Bagnol et al., 1997). In the rat myenteric plexus, but not in other species, μ or δ receptors have been located in the interstitial cells of Cajal (Bagnol et al., 1997; Gray, Coupar, & White, 2006; Ho et al., 2003). Moreover, μ receptor-immunoreactive fibers are abundant in the deep muscular plexus in close association with the interstitial cells of Cajal, and in the muscle layer. In the gastrointestinal tract, neurons expressing μ or δ receptors are more abundant in the proximal colon/stomach compared to other regions of the tract (Sternini et al., 2013), and fibers expressing μ or δ receptors have been observed in mucosal and muscular layers. Terminal fibers from myenteric neurons expressing the μ receptor are very abundant in the small intestine, followed by the stomach and proximal colon (Ho et al., 2003; Sternini et al., 2013). In the small intestine, δ receptors are mainly located in the submucous and myenteric (inhibitory and excitatory neurons) plexuses and are abundant in the inhibitory neurons located in the myenteric plexus of the colon (Poole et al., 2011). In the rat ileum, the coexistence of μ and κ receptors has been observed in the interstitial cells of the submucous plexus (Gray et al., 2006) and the colocalization of μ and δ receptors has been also reported (Poole et al., 2011). In humans, μ receptor immunoreactivity has been observed in the immune cells of the lamina propia, and in the submucosal and myenteric neurons, whereas δ receptor immunoreactivity has been reported in fibers located in the mucosa and muscle, and in both plexuses (Sternini et al., 2004).

5.2 ACTH

The expression of the pro-opiomelanocortin gene has been reported in the pars intermedia (melanotroph cells) and pars distalis (corticotroph cells) of the pituitary gland (Eberle, 1988). In rodents, ACTH-immunoreactive cell bodies have been observed in the hypothalamic arcuate nucleus, and in the nucleus of the solitary tract (Joseph, Plicher, & Bennett-Clarke, 1983). In dogs, cell bodies containing ACTH have been also found in the nucleus of the solitary tract and in the ventral mesencephalon (ventral tegmental area, substantia nigra compacta and interfascicular nucleus) (Pesini, Pego-Reigosa, Tramu, & Coveñas, 2004). In cats, dogs, monkeys and humans, ACTH-immunoreactive perikarya are located in the arcuate

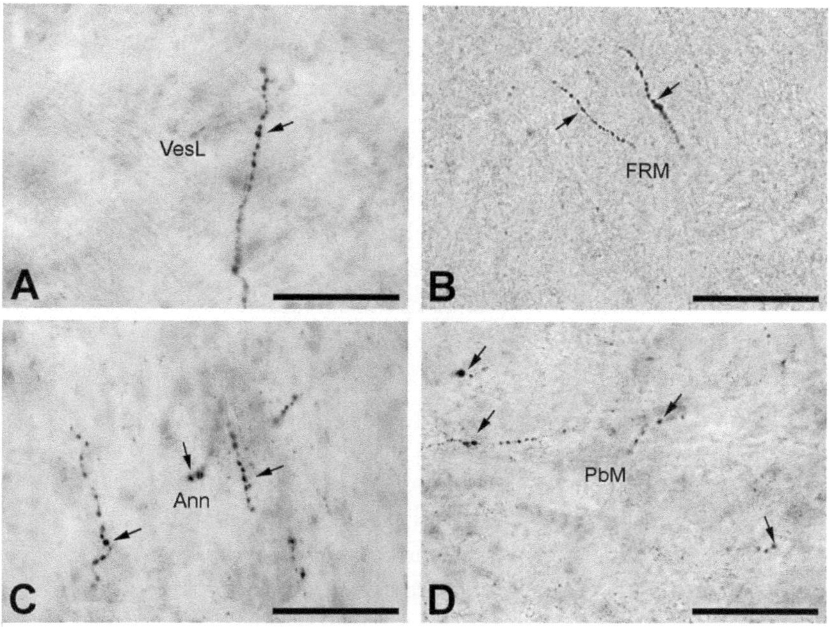

Fig. 5 Immunoreactive structures in the *Saimiri sciureus* brainstem containing ACTH. Arrows indicate fibers containing ACTH located in the nucleus vestibularis lateralis (VesL) (A), formatio reticularis myelencephalic (FRM) (B), nucleus annularis (Ann) (C) and nucleus parabrachialis medialis (PbM) (D). Scale bar: 50 μm.

nucleus and, for the first two species, in the ventromedial hypothalamic nucleus, as well (Abrams et al., 1980; Coveñas et al., 1996; Pesini et al., 2004). No immunoreactive cell bodies containing ACTH have been reported either in the alpaca diencephalon (Manso et al., 2014), or in the squirrel monkey brainstem (Duque et al., 2011; Duque-Díaz et al., 2017) (Fig. 5). It is also known that the ACTH-immunoreactive perikarya located in the nucleus of the solitary tract send projections to the spinal cord, nucleus reticularis gigantocellularis, and A_1 cell group (Jégou et al., 2013) and that those located in the arcuate nucleus send projections to the diencephalon (e.g., nucleus ventromedialis hypothalami, nucleus paraventricularis thalami, perifornical region) and locus coeruleus (Kitahama, Sakai, & Jouvet, 1984; Kitahama et al., 1986).

ACTH-immunoreactive fibers are widely distributed throughout the mammalian central nervous system (e.g., amygdala, thalamus, hypothalamus, brainstem, spinal cord) (Manso et al., 2014; Palkovits, 1988). For example, ACTH-immunoreactive fibers have been observed in regions of the central nervous system in which clusters of serotonergic (raphe nuclei), noradrenergic (Kölliker-Fuse/parabrachial nuclei, locus coeruleus) and dopaminergic (ventral tegmental area) neurons are located (Pesini et al., 2004). In 60.5% of the brainstem nuclei/tracts of the squirrel monkey (*Saimiri sciureus*), the presence of both ACTH- and met-enkephalin-immunoreactive fibers has been observed; this relationship being less marked for ACTH and leu-enkephalin (52.5%) (Duque-Díaz et al., 2017). In 42.5% of the brainstem nuclei/tracts of the same species in which ACTH- and met-enkephalin-immunoreactive fibers were observed, fibers containing leu-enkephalin were also found (Duque-Díaz et al., 2017). It is known that the release of ACTH is controlled by met-enkephalin (del Pozo, Martín-Pérez, Stadelmann, Girard, & Brownell, 1980).

ACTH binds to the five subtypes of the melanocortin receptor, and it is the only member of the melanocortin peptide family that binds to the melanocortin-receptor 2 (Jégou et al., 2013). The melanocortin-receptor 1 is expressed in melanocytes, keratinocytes, adipose tissue, leukocytes and macrophages; melanocortin-receptor 2 in adrenocortical cells and adipocytes; melanocortin-receptor 3 in the gut, kidney, testis, ovary, skeletal muscle, mammary gland, placenta and central nervous system (hypothalamus, thalamus, brainstem); melanocortin-receptor 4 in sympathetic ganglia and the central nervous system (forebrain, autonomic centers of the hindbrain, hypothalamus, brainstem, spinal cord); and melanocortin-receptor 5 in skeletal muscle, adipocytes, exocrine glands, liver, spleen, lung and central nervous system (cerebral cortex, striatum, olfactory bulb, medulla) (Jégou et al., 2013).

6. Receptors and signaling cascades

6.1 Enkephalins

Opioid receptors belong to the family of seven transmembrane G protein-coupled receptors (Dietis et al., 2011; Kieffer & Evans, 2009), and the sequences of the μ, κ and δ receptors have been found to exhibit approximately 60% similarity (Sternini et al., 2013). The transmembrane regions of these receptors show the highest similarity (73–76%), followed by the intracellular (63–66%) and extracellular (34–49%) regions. The extracellular domains are involved in the selectivity of the opioid receptors, and in their ability to bind antagonists and agonists (Sternini et al., 2013). Splicing of the μ, κ and δ receptors occurs, leading to multiple receptor subtypes μ_1, μ_2, μ_3, κ_1, κ_2, κ_3, δ_1 and δ_2 (Dietis et al., 2011). Genes encoding the opioid receptors have two introns in the corresponding position (Larhammar et al., 2013). Enkephalins exhibit the highest affinity for the δ receptor, followed for the μ receptor (high affinity), and κ receptor (low affinity) (Sternini et al., 2013). The binding of enkephalins to δ/μ receptors induced the hyperpolarization/inhibition of neurons, whereas agonists of the κ receptor lead to both activated and inhibitory signaling pathways (McLaughlin & Zagon, 2013).

All types of opioid receptors promote the dissociation of heterotrimeric G proteins (into separated βγ and α subunits) via the exchange of 5′-guanosine diphosphate by 5′-guanosine triphosphate. Main protein effectors of opioid action are adenylyl cyclase, phospholipase C, N and L type Ca^{++} channels, mitogen-activated protein kinases, and inward rectifying K^{+} channels (Rose, Shaw, Prendergast, & Little, 2010). Opioid receptors, via G proteins (Gi/Go), increase potassium currents, activate mitogen-activated protein kinase, block calcium channel activity, regulate inositol triphosphate turnover, and inhibit the formation of cyclic AMP and adenylyl cyclase (Kieffer & Evans, 2009; Simon, 1991). Opioid peptides, via the opioid receptors, initiates a signaling cascade that includes phosphorylation by G protein receptor kinases, promoting an interaction with β arrestins thatprevent further stimulation and downstream signaling pathway by uncoupling receptors from G proteins, thereby causing an acute desensitization (Bohm, Grady, & Bunnett, 1997). The complex (ligand and the phosphorylated opioid receptor) is then translocated from the cell surface into the cytoplasm via a clathrin-dependent mechanism, undergoes intracellular sorting into endosomes, dephosphorylation and is either recycled to the cell surface or degraded into lysosomes (Bohm et al., 1997). In enteric neurons, it has been reported that the endocytosis of μ receptors occurs in neuronal processes and perikarya and that this mechanism is inhibited by D-Phe-Cys-Tyr-D-Trp-Orn-Thr-Pen-Thr-NH_2 (a selective μ receptor antagonist) or by naloxone (an opioid receptor antagonist) (Minnis et al., 2003). Additionally, once internalized, μ receptors are recycled to the cell surface, and no synthesis of new μ receptors is required (Minnis et al., 2003). In enteric neurons, the internalization of μ receptors (e.g., by enkephalins) is concentration dependent, but morphine (an opiate drug) does not promote the internalization of these receptors (Patierno et al., 2011). However, in the enteric neurons of animals made morphine-tolerant through chronic administration of the drug, internalization of μ receptors occurred, a process which can be blocked by a dynamin inhibitor that prevents the formation of coated vesicles (Patierno et al., 2011).

The internalization of μ receptors has been described in stress conditions (e.g., abdominal surgery) (Patierno, Raybould, & Sternini, 2004). In enteric neurons, δ receptors are internalized in neuronal processes and perikarya and this is blocked when a δ receptor antagonist is used. δ receptors are degraded into lysosomes; thus, they do not return to the cell surface, and the synthesis of new δ receptors is required (Poole et al., 2011).

6.2 ACTH

In mammals, five subtypes of the melanocortin receptor (which belongs to the rhodopsin family of G protein-coupled receptors) have been cloned (Tatro, 2000). When activated by ligands, they promote the production of intracellular cyclic adenosine 3′, 5′-monophosphate (cAMP). ACTH and α-, β-, and γ-MSH are agonists for the melanocortin receptors, whereas agouti and agouti-related proteins (expressed in the adrenal gland and hypothalamus) act as antagonists (Yang, 2011). As indicated above, ACTH binds to the five subtypes of the melanocortin receptor, and this molecule is the only member of melanocortin peptide family that binds to the melanocortin-receptor 2. For this reason, this receptor is also called the ACTH receptor (Jégou et al., 2013). The melanocortin-receptor 2 is a hetero-hexameric structure former by four molecules of melanocortin-receptor 2 accessory protein (see below) and two molecules of the ACTH receptor (Fridmanis, Roga, & Klovins, 2017). The third transmembrane domain of the ACTH receptor is crucial for the ligand potency and selectivity (Yang et al., 2015). It has been also demonstrated that the replacement of short segments within the transmembrane domains of the melanocortin-receptor 2 disrupts the retention signal (Fridmanis, Petrovska, Pjanova, Schiöth, & Klovins, 2014). Additionally, it is known that residues 17–24 are crucial for the recognition of ACTH by its receptor (Ghaddhab, Vuissoz, & Deladoëy, 2017).

For the binding of ACTH, the interaction of the melanocortin-receptor 2 with its accessory protein is crucial for the trafficking of the receptor to the cell membrane of adrenal cells; thus, this protein is required for the formation of a functionally active melanocortin-receptor 2 (Babischkin, Aberdeen, Pepe, & Albrecht, 2016; Fridmanis et al., 2017). It has been reported that, in monkeys, the increasing levels of estradiol associated with advancing pregnancy block the interaction of the melanocortin-receptor 2 with its accessory protein, and hence the movement of the receptor to the cell membrane was diminished (Babischkin et al., 2016).

Melanocortin-receptor 2 accessory protein is expressed in the adipose tissue and adrenal gland and it is also known that Gαs is a novel interacting partner of this accessory protein. Melanocortin-receptor 2 accessory protein is crucial for the ACTH-induced activation, via the melanocortin-receptor 2, of the cAMP/protein kinase A pathway, leading to glucocorticoid synthesis and secretion (Zhang, Saarinen, Campbell, de Filippis, & Liu, 2018). Agonists can induce different conformation changes of melanocortin receptors, leading to the activation of various signaling pathways, even when the expression level of the receptor and the strength of the stimulus-response coupling are the same (Yang, 2011). The melanocortin-receptor 3 is also coupled to the inositol phospholipid/Ca^{++}-mediated signaling cascade and this receptor and the melanocortin-receptor 4 activate the mitogen-activated kinase pathway, whereas the latter inhibits the c-Jun N-terminal kinase (Chai, Li, Zhang, Wang, & Mulholland, 2009; Renquist, Lippert, Sebag, Ellacott, & Cone, 2011; Tao, 2010). It has been also reported that the melanocortin-receptor 4 activates pertussis toxin-sensitive Gi/o proteins and increases intracellular Ca^{++} via Gq/phospholipase C-dependent signaling pathways (Büch, Heling, Damm, Gudermann, & Breit, 2009).

ACTH receptor (mRNA and protein) levels were found to be higher in aldosterone-producing adenomas when compared with normal adrenal samples, but lower in cortisol-producing adenomas and non-functional adenomas (Hu, Liu, & Wang, 2017). The protein kinase C delta is known to restrain the growth of ACTH-secreting pituitary adenoma cells (Gentilin et al., 2016). It has been also reported that signals induced by ACTH included cAMP, TGF-β and ERK1/2 phosphorylation (Tourkova et al., 2017). ACTH promotes, via the activation of cAMP/protein kinase A, the synthesis of steroid hormones, eliciting a fast acute response and a long-term chronic response (Ruggiero & Lalli, 2016). In the acute response, cholesterol is mobilized from lipid stores to the inner membrane of the mitochondria; this mechanism being mediated by the steroidogenic acute regulatory protein. In the chronic response, an increase in the transcription of genes encoding steroidogenic enzymes occurs (Ruggiero & Lalli, 2016). ACTH, via the melanocortin-receptor 2, stimulates protein kinase A activation, adenylyl cyclase (promoting the synthesis of cAMP) and the phosphorylation of nuclear factors, which bind to promoters and elicit coactivator protein recruitment to direct steroidogenic gene transcription (Ruggiero & Lalli, 2016). Recently, it has been shown that ACTH binds to the orphan G protein-coupled receptor GPR139 (Nøhr et al., 2017).

Alterations in the ACTH receptor gene has been reported (Fridmanis et al., 2017) in terms of mutations that alter a single amino acid residue. These mutations are divided into those located within the central region of the transmembrane domains (affecting the overall conformation of the ACTH receptor) and those located in the extracellular (affecting ligand-receptor interaction) or intracellular (affecting G-protein interaction) parts of the ACTH receptor (Fridmanis et al., 2017). Glycosylation is often required for the transport of cell-surface proteins to the plasma membrane and, in the case of G protein-coupled receptors, glycosylation is also important for correct receptor folding and maturation (Fridmanis et al., 2017). In the ACTH receptor, two glycosylation sites have been found within the N-terminus (Fridmanis et al., 2017). Finally, it has been reported that the residues Thr^{131}, Thr^{143}, Thr^{147} and Ser^{280} are crucial for the ACTH receptor expression and functional regulation (Roy et al., 2011).

7. Biological actions

Due to the widespread anatomical distribution of met-enkephalin, leu-enkephalin and ACTH in the central and peripheral nervous systems, it seems that the three peptides are involved in many physiological actions, acting as neuromodulators and/or neurotransmitters. In this sense, for example, the presence of ACTH and/or enkephalins in the colliculus superior, colliculus inferior, nucleus of the solitary tract, substantia nigra and spinal trigeminal nucleus suggests that these neuroactive substances are involved in visual, auditory, cardiovascular, respiratory, gastric, motor and sensory mechanisms (Coveñas, Mangas, & Narváez, 2007; Pioro, Mai, & Claudio Cuello, 1990). Table 1 shows the biological actions mediated by enkephalins and ACTH.

7.1 Enkephalins

Opiate receptors are located in sensory pain terminal fibers and enkephalins block the release of neurotransmitters (e.g., dopamine, vasopressin, substance P), and also alter the Ca^{++} influx (Mudge et al., 1979). Attenuation of mechanical hypersensitivity and myogenic orofacial nociception by viral-mediated enkephalin overproduction has been reported (Kramer et al., 2015). The anatomical localization of enkephalins suggests that they are also involved in neuroendocrine, respiratory and motor activities; as well as peristalsis, emotional behavior and limbic system modulation (McLaughlin, 2013). Met-enkephalin controls microglia function and polarization (Xu et al., 2016), and has neuroprotective effects (Zhao et al., 2015).

Table 1 Biological actions of enkephalins/ACTH.

Enkephalins	References
Alter Ca^{++} influx	Mudge, Leeman, and Fischbach (1979)
Attenuation of mechanical hypersensitivity and myogenic orofacial nociception	Kramer, Umorin, and Bellinger (2015)
Block the release of neurotransmitters (dopamine, vasopressin, substance P)	Mudge et al. (1979)
Control microglia function and polarization	Xu et al. (2016)
Exert neuroprotective effects	Zhao et al. (2015)
Involved in secretion, electrolyte/fluid transport and motility in the digestive system	Kromer (1988) and Sternini et al. (2004)
Increase fat and carbohydrates intake	Mena and Sadeghian Baldo (2011), Taha (2010), Vucetic, Kimmel, and Reyes (2011) and Zhang and Kelley (2000)
Inhibit the activity of the interstitial cells of Cajal	Ho et al. (2003)
Inhibitory action on cancer cell proliferation	McLaughlin, 2013
Involved in angiogenesis, wound healing, embryonic development, emotional behavior and in normal homeostatic action	McLaughlin, 2013
Involved in energy balance, autonomic function and reward	Gosnell, Kotz, Billington, & Levine, 2013
Involved in the differentiation of mesenchymal stem cells into osteoblasts	Thakur, DeBoyace, & Margulies, 2016
ACTH	**References**
Alters the profile of blood neutrophil membrane receptors and increases the number of thymocytes and the expression of the delta-like ligand 4 factor	Machado et al. (2014) and Talaber, Tuckermann, and Okret (2015)
Controls learning, memory, attention and motivation	Bertolini et al. (2009) and Eberle (1988)
Controls the development and ovulatory process of ovarian follicles	Endo, Yamane, Rahayu, and Tanaka (2018)
Decreases social interaction	Vaudry and Eberle (1993)

Continued

Table 1 Biological actions of enkephalins/ACTH.—cont'd

ACTH	References
Exerts a hypoglycemic action	Endo et al. (2018)
Exerts anti-epileptic and anti-pyretic actions	Vaudry and Eberle (1993)
Favors calcium influx and increases the number of ACTH receptors	Penhoat, Jaillard, and Sáez (1989)
Increases aggressive behavior, reduces social interaction, delays copulatory behavior and reduces sexual performance	Argiolas, Melis, Murgia, and Schioth (2000), Jégou et al. (2013) and van der Ploeg et al. (2002)
Increases the activity of soluble protein tyrosine phosphatase, promotes the dephosphorylation of paxillin and activates the PTP115 protein (PTP-PEST)	Gorostizaga et al. (2016)
Increases adrenal blood flow	Shah, Kriska, Gauthier, Falck, and Campbell (2018)
Induces hyperalgesia and reverses the analgesic effect	Bertolini et al. (2009)
Involved in acetylcholine release (from presynaptic terminals at the neuromuscular junction) and tissue growth	Vaudry and Eberle (1993)
Involved in the physiological response to stress	Ghaddhab et al. (2017)
Mediates food intake behavior	Adan et al. (2006)
Promotes lipolysis and the secretion of gonadotropin	Aleknaviciute et al. (2016) and Zhang et al. (2018)
Promotes the protection of oligodendrocytes by astroglia	Lisak, Nedelkoska, and Benjamins (2016)
Promotes the stretching-yawning syndrome and excessive grooming	Argiolas et al. (2000) and Mul et al. (2011)
Protects neurons from inflammatory, excitotoxic and apoptotic insults	Lisak, Nedelkoska, Bealmear, and Benjamins (2015)
Stimulates brown adipose tissue and the expression of the thermogenin gene	van den Beukel et al. (2014)
Stimulates the synthesis and secretion of glucocorticoids and mineralocorticoids	Lindner et al. (2017)

Met-enkephalin (also called opioid growth factor) is known to inhibit the proliferation of normal and cancer cells. Met-enkephalin-Arg^6-Phe^7, met-enkephalin-Arg^6-Gly^7-Leu^8 and leu-enkephalin also exert an inhibitory action on cell proliferation, but to a lesser degree than met-enkephalin (McLaughlin, 2013). Met-enkephalin blocked DNA synthesis and the inhibition of the met-enkephalin/opioid growth factor receptor promoted accelerated cell proliferation (McLaughlin, 2013). The peptide uses the p16 and/or p21 cyclin-dependent inhibitory kinase pathways, and needs clathrin-mediated endocytosis for its inhibitory action (McLaughlin, 2013). In human tumor cells, mutations in the opioid growth factor receptor have been found to alter the function of the receptor (Kren, Zagon, & McLaughlin, 2015). Additionally, antagonism of the opioid growth factor receptor-signaling axis induced the differentiation of mesenchymal stem cells into osteoblasts (Thakur et al., 2016). The met-enkephalin/opioid growth factor receptor system has been involved in angiogenesis, wound healing, embryonic development, and in normal homeostatic action (McLaughlin, 2013).

Based on the presence of opioid receptors in enteric nerve cells (Ho et al., 2003), opioid peptides play an important role in secretion, electrolyte/fluid transport, and motility in the digestive system (Kromer, 1988). In fact, the activation of these receptors can inhibit the release of excitatory transmitters (e.g., acetylcholine), and thus the peristaltic contraction is inhibited, or this activation can block the release of inhibitory transmitters (e.g., nitric oxide), being the peristaltic distention inhibited (Sternini et al., 2004). It has been also reported that activation of the opioid receptors located in the myenteric neurons, inhibits longitudinal/circular muscle contraction. μ receptor-immunoreactive fibers have been observed surrounding the interstitial cells of Cajal that control intestinal motor activity (Ho et al., 2003). It has been suggested that the activation of opioid receptors could inhibit the activity of the interstitial cells of Cajal, because the release of transmitters from nerve cells innervating these cells is decreased.

The administration of D-Ala^2, NMe^4, $Glyol^5$-enkephalin (DAMGO, a selective μ receptor agonist) has been found to increase fat intake (Taha, 2010), but the administration of this μ receptor agonist into the ventromedial prefrontal cortex led to an observed increase in the intake of carbohydrates (Mena & Sadeghian Baldo, 2011). Animals lacking enkephalin/β-endorphin work less diligently for food than wild type animals (Hayward, Pintar, & Low, 2002), and a decrease in food-anticipatory activity is observed in animals lacking the μ receptor (Kas et al., 2004).

In humans, it has been reported that variability in the μ receptor gene is associated with variability in preference for fatty and sweet foods (Davis et al., 2011). These findings indicate that opioids mediate food reward and palatability. In the arcuate nucleus and ventromedial hypothalamus, an increase in the expression of the μ receptor is observed after food deprivation (Barnes, Primeaux, & Bray, 2008), and during a high-fat diet, the level of μ receptor mRNA decreases in the prefrontal cortex, nucleus accumbens and ventral tegmental area (Vucetic et al., 2011). Administration of DAMGO into the nucleus accumbens activates neurons (e.g., those located in the nucleus of the solitary tract, ventral tegmental area, lateral hypothalamus, substantia nigra), and increased the intake of a high-fat diet (Zhang & Kelley, 2000). According to the neuroanatomical distribution of these activated neurons, it seems that the nuclei in which they are located are involved in energy balance, autonomic function, and reward (Gosnell et al., 2013).

7.2 ACTH

ACTH is involved in the physiological response to stress (Ghaddhab et al., 2017). In the adrenal cortex, ACTH stimulates the synthesis and secretion of glucocorticoids and mineralocorticoids by stimulating the conversion of cholesterol to pregnenolone. The intravenous administration of ACTH increases the concentration of steroids (e.g., corticosterone, cortisol, 11-deoxycorticosterone, 17-OH-progesterone, 11-deoxycortisol) (Lindner et al., 2017). These effects are mediated by stimulation of the membrane-bound adenylate cyclase after the binding of ACTH to receptors located in the adrenal cortex. In adrenal cell membranes, ACTH increases the number of ACTH receptors and favors calcium influx (Penhoat et al., 1989). ACTH increases the adrenal blood flow, a process mediated by the epoxyeicosatrienoic acid (Shah et al., 2018). In the adrenal zona fasciculate of experimental animals, ACTH increases the activity of soluble protein tyrosine phosphatase (PTP) and promotes the dephosphorylation of paxillin (Gorostizaga et al., 2016). Moreover, the hormone activated the PTP115 protein (PTP-PEST), for which paxillin is a substrate (Gorostizaga et al., 2016). Thus, ACTH controls the activity of PTP-PEST, and also elicits a PTP-PEST-paxillin interaction (Gorostizaga et al., 2016).

In ACTH-treated animals, the prefrontal cortex mitochondrial function has been studied (Kim et al., 2016). Compared to controls, a decreased capacity for the production of adenosine triphosphate was reported. ACTH exerts a hypoglycemic action attributed to a direct insulin-releasing effect on pancreatic β cells; and the hormone also controls the development and ovulatory processes of ovarian follicles (Endo et al., 2018). Additionally, it is known that ACTH/ACTH fragments (e.g., $ACTH_{4-9}$, $ACTH_{4-10}$, $ACTH_{1-16}$, $ACTH_{1-24}$) are involved in epilepsy (exerting an anti-epileptic action), social behavior (decreasing social interaction), fever (inhibiting IL-1-induced fever and exerting an anti-pyretic effect), tissue growth (acting as growth factors), lipolysis (stimulation), and in acetylcholine release (from presynaptic terminals of the neuromuscular junction) (Vaudry & Eberle, 1993).

It has been published that a moderate exercise increases the secretion of ACTH-cortisol (Roelfsema et al., 2017) and that insufficient sleep is associated with a decrease of the overall ACTH response to evening corticotropin-releasing hormone stimulation (Guyon et al., 2017). In females, the ACTH/cortisol ratio increases in opioid-dependent patients (Reece & Hulse, 2016). This study also established that the ACTH/cortisol ratio is a biomarker for age. Additionally, in pre-menopausal women, ACTH elicits the secretion of gonadotropin (Aleknaviciute et al., 2016).

ACTH has been detected in B cells, indicating that the hormone could exert an autocrine regulatory action in the development of these cells (Pállinger, Kiss, & Csaba, 2016). ACTH increases the number of thymocytes and the expression of a factor (delta-like ligand 4) that supports T-cell development (Talaber et al., 2015). Additionally, ACTH alters the profile of blood neutrophil membrane receptors, and high levels of ACTH are known to mobilize granulocytes from the bone marrow into the blood (Machado et al., 2014).

ACTH has also been found to promote the protection of oligodendrocytes by means of astroglia (Lisak et al., 2016), and protects neurons from inflammatory, excitotoxic and apoptotic insults (Lisak et al., 2015). The administration of $ACTH_{1-24}$ into the hypothalamic periventricular regions (e.g., paraventricular nucleus, dorsomedial nucleus, ventromedial nucleus) has been found to promote the stretching-yawning syndrome and excessive grooming (Argiolas et al., 2000). Grooming is also induced when the peptide is administered into the substantia nigra or the periaqueductal gray, a mechanism apparently mediated by the melanocortin-receptor 4 (Argiolas et al., 2000; Mul et al., 2011). Local administration of $ACTH_{1-24}$ into the hypothalamic periventricular region promotes penile erection, this mechanism being mediated by melanocortin-receptors 3 and 4 (Argiolas et al., 2000). Additionally, the intracerebroventricular administration of $ACTH_{4-10}$ or $ACTH_{1-24}$ delays copulatory behavior, and the latter fragment also reduces sexual performance (Argiolas et al., 2000; van der Ploeg et al., 2002). The administration of $ACTH_{4-10}$ into the septum has been found to increase aggressive behavior and reduce social interaction (Jégou et al., 2013). ACTH facilitates acquisition, and delay extinction, of passive and active avoidance responses and of rewarded behavior (Bertolini et al., 2009; Eberle, 1988). This means that the peptide plays a role in learning, memory, attention and motivation. $ACTH_{4-7}$ is the smallest fragment exhibiting a full action on learning and memory, with $ACTH_{6-9}$ being the minimum sequence capable of stimulating melanocortin receptors (Jégou et al., 2013). The intracerebroventricular administration of $ACTH_{4-10}$ promotes a slight pressor effect that is not mediated by melanocortin receptors (Humphreys, Ni, & Pearce, 2011). Moreover, the prevalence of hypertension has been found to be lower in melanocortin-receptor 4-deficient patients than in control patients (Jégou et al., 2013).

Central administration of ACTH induces hyperalgesia and reverses the analgesic effect mediated by β-endorphin and morphine (Bertolini et al., 2009). It has been also reported that melanocortin receptor agonists antagonized the antinociceptive response of morphine, whereas melanocortin-receptor 4 antagonists potentiated it (Bertolini et al., 2009). Additionally, central administration of a melanocortin-receptor 4 antagonist has been found to prevent morphine tolerance and hyperalgesia, while chronic administration of morphine downregulates the expression of melanocortin-receptor 4 in the periaqueductal gray/striatum, and of pro-opiomelanocortin in the hypothalamus (Kalange et al., 2007).

ACTH also stimulates the brown adipose tissue (van den Beukel et al., 2014), and the expression of the thermogenin gene in cultured brown adipocytes (Biswas, 2017). The hormone promotes lipolysis and down- or upregulation of the melanocortin-receptor 2 accessory protein, respectively, decreased or increased ACTH-induced lipolysis (Zhang et al., 2018). Thus, this protein plays a role in the body's energy balance and in ACTH-induced adipose lipolysis.

An increase in feed efficiency has been reported in mice lacking the melanocortin-receptor 3 (Mul & Adan, 2013). The increased adipose mass observed in these animals occurs due to an increase in feed efficacy, and not due to changes in their ingestive behavior (Mul & Adan, 2013). Additionally, a deficiency of pro-opiomelanocortin-derived peptides has been found to produce obesity by hyperphagia (Yaswen, Diehl, Brennan, & Hochgeschwender, 1999), while a central administration of non-specific melanocortin-receptor 3 and 4 antagonists or agonists, respectively, increases or decreases food intake (Mul & Adan, 2013). It has been also suggested that the melanocortin system (via the melanocortin-receptor 4 located in the paraventricular hypothalamic nucleus) is involved in meal termination rather than in meal initiation (Adan et al., 2006). Paraventricular hypothalamic neurons expressing the melanocortin-receptor 4 gene also contain corticotropin-releasing hormone (Adan et al., 2006), and it has been suggested that the hormone appears to function downstream of melanocortin-receptor 4 signaling, contributing to processes by which the melanocortin system mediates food intake behavior (Adan et al., 2006). Melanocortin-receptors 4 located in the dorsomedial hypothalamic nucleus also mediate ingestive behavior (e.g., decrease or increase food intake) (Adan et al., 2006). The melanocortin-receptor 4 has been found in brain areas that control autonomic functions (nucleus of the solitary tract, dorsal motor nucleus of the vagus) (Cone, 2005), and administration of melanocortin-receptor 4 antagonists or agonists into the dorsal motor nucleus of the vagus or into the fourth ventricle has been demonstrated to affect food intake (Adan et al., 2006). It has been also reported that administration into the fourth ventricle of a non-specific melanocortin-receptor 3 and 4 agonist induces a reduction in meal size, but not in meal frequency (Adan et al., 2006), and that the melanocortin-receptors 4 located in the brainstem play an important role in the inhibition of food intake mediated by cholecystokinin (Tao, 2010). The melanocortin-receptor 4 has also been found in the central amygdala; and although a melanocortin-receptor 4 agonist administered into this brain region did not produce any change in food intake, the administration of an antagonist increased it (Adan et al., 2006). Finally, the administration of a melanocortin-receptor 4 agonist into the nucleus accumbens, arcuate nucleus, lateral hypothalamus or ventromedial hypothalamus did not significantly modify food intake (Adan et al., 2006).

8. Pathophysiology

Table 2 shows the involvement of enkephalins and ACTH in several pathologies.

8.1 Enkephalins

In an animal model of multiple sclerosis, met-enkephalin has been reported to delay and attenuate the behavioral signs of the disease, and to reduce neuronal damage, demyelination, and the activation of astrocytes (Rahn et al., 2011).

Table 2 Involvement of enkephalins/ACTH in pathologies.

Pathology	Enkephalins	References
Addictions	Crucial role in the behavioral and molecular expression of cocaine sensitization	Mongi-Bragato et al. (2016)
Alcoholism	Administration of μ or δ agonists into the paraventricular hypothalamic nucleus increases ethanol intake. The administration of δ agonists into the ventral tegmental area decreases alcohol consumption	Barson et al. (2010) and Margolis, Fields, Hjelmstad, and Mitchell (2008)
	The expression of pro-enkephalin is downregulated in the caudate nucleus of alcoholics	Sarkisyan et al. (2015)
	μ and δ receptors are involved in the expression and reinstatement of ethanol conditioned seeking behavior and enkephalin analogues inhibited alcohol withdrawal-induced anxiety behavior	Gibula-Bruzda, Marszalek-Grabska, Gawel, et al. (2015) and Gibula-Bruzda, Marszalek-Grabska, Witkowska, Izdebski, and Kotlinska (2015)
Cancer	Low fasting plasma concentration of pro-enkephalin is associated with an increased risk of future breast cancer. Met-enkephalin restores the damage of the immune system induced by chemotherapy	Melander et al. (2015) and Wang et al. (2014)
Depression	δ receptor agonists exert an anti-depressant effect	Dripps and Jutkiewicz (2018)
Eating disorders	Involved in feeding and diet-induced obesity	Mendez, Ostlund, Maidment, and Murphy (2015)
Gastrointestinal disorders	After a pelvic or an abdominal surgery, μ receptors are involved in the worsening of post-operative ileus	Holzer (2010) and Panchal, Muller-Schwefe, and Wurzelmann (2007)
	Opioid peptides are involved in intestinal inflammatory processes. μ receptor agonists exert a protective action during the intestinal inflammation	Philippe et al. (2006, 2003) and Saccani et al. (2012)

Table 2 Involvement of enkephalins/ACTH in pathologies.—cont'd

Pathology	Enkephalins	References
	In inflammatory bowel disease and in intestinal inflammation, an upregulation of μ-receptor mRNA has been reported	Philippe et al. (2006)
Glioma	Met-enkephalin promotes apoptotic mechanisms	Lu et al. (2018)
Hepatic damage	Met-enkephalin exerts an hepatoprotective effect	Martinic et al. (2014)
Huntington disease	Immunoreactivity decreases in basal ganglia and substantia nigra. δ receptor agonists protect glial/neuronal cells from ischemia, facilitate the survival of neurons and mitigate apoptotic pathways	Lee et al. (2018)
Hypertension	After electro acupuncture, enkephalins decrease hypertension	Li, Tjen-Looi, Guo, and Longhurst (2016)
Hypoxia	Analogues of leu-enkephalin normalize cerebral consequences after hypoxia	Simankova, Sazonova, and Lebed'ko (2017)
Influenza A virus	Inhibition of virus replication and decrease in cytokine levels	Tian et al. (2018)
Lower limb amputation	Immunoreactivity increase in the dorsal horn of the spinal cord	Pioro et al. (1990)
Melanoma	Met-enkephalin exerts an anti-tumor action	Wang et al. (2017)
Moyamoya disease	Increase of the level of pro-enkephalin$_{143-183}$ in the cerebrospinal fluid	Yokoyama et al. (2018)
Multiple sclerosis	Delay and attenuate the behavioral signs of the disease and reduce the damage, demyelination and activation of astrocytes	Rahn, McLaughlin, and Zagon (2011)
Stress	Upregulation or downregulation of enkephalins in the hippocampus	Li et al. (2018)
Pathology	**ACTH**	**References**
Addison's/ Graves' diseases	In both diseases, the administration of tetracosactide ($ACTH_{1-24}$) generates anti-$ACTH_{1-24}$ antibodies and a spontaneous anti-ACTH reactivity	Gan et al. (2016)

Continued

Table 2 Involvement of enkephalins/ACTH in pathologies.—cont'd

Pathology	ACTH	References
Alcoholism	A high level of ACTH has been reported in the blood of alcohol-dependent patients	Geisel, Schlemm, Hellweg, Wiedemann, and Müller (2019)
Atopic dermatitis	ACTH promotes the deterioration of atopic dermatitis symptoms	Hiramoto, Yamate, and Yokoyama (2018)
Autism	ACTH exerts beneficial effects on stuttering in attention deficit and hyperactivity	Altunel, Sever, and Altunel (2017)
Brain injury	Semax, an analogue of $ACTH_{4-7}$, regulates the expression of the immune response genes. Semax affects learning and exerts a protective action against metal induce cell toxicity	Medvedeva, Dmitrieva, Limborska, Myasoedov, and Dergunova (2017) and Tabbì et al. (2015)
Bone diseases	ACTH is protective for lumbar bone mineral density in patients with Cushing's disease	Guo et al. (2018)
	ACTH increases the expression of bone structural/regulating mRNAs of biglycan, vitamin D-receptor, TGF-β and collagen I	Tourkova et al. (2017)
	ACTH plays an important role in bone maintenance and increases alkaline phosphatase activity, bone mineralization and the levels of osterix and calcium-sensing receptor (RNA and protein)	Tourkova et al. (2017)
Cancer	High levels of stress hormones (e.g., ACTH) in the normal microenvironment are predictive for chemically induced cancer	Valente et al. (2018)
	The selective Jak2 inhibitor (SD-1029) or the inhibition of the heat shock protein 90 blocks ACTH synthesis and the proliferation of corticotroph tumor cells	Asari et al. (2017) and Sugiyama et al. (2015)
Diabetic neuropathy	High ACTH level and an increase in the number of ACTH binding sites in the muscle	Jégou et al. (2013)

Table 2 Involvement of enkephalins/ACTH in pathologies.—cont'd

Pathology	ACTH	References
Encephalitis	ACTH immunotherapy has been suggested for the treatment of anti-N-methyl-D-aspartate receptor encephalitis	Hatanaka et al. (2018)
Epilepsy	ACTH has been administered to children suffering from intractable seizures	Nasiri, Sarajan, Salari, and Sedghi (2017)
	ACTH is effective against infantile spasms	Yin et al. (2017)
	ACTH protects learning/memory function in epileptic experimental animals	Scantlebury, Chun, Ma, Rho, and Kim (2017)
	ACTH has been suggested for the treatment of intractable epilepsy in hemiconvulsion-hemiplegia-epilepsy syndrome	Shimakawa et al. (2015)
Erectile dysfunction	Melanocortin receptor agonists are effective for the treatment of erectile dysfunction and also increase the sexual desire and arousal	Safarinejad (2008) and Wessels et al. (2000)
Guillain-Barré syndrome	ORG-2766, an analogue of $ACTH_{4-9}$, protects against experimental allergic neuritis	Duckers, Verhaagen, and Gispen (1993)
Multiple sclerosis	Adjunctive treatment of ACTH to interferon β results in better examiner-assessed cumulative rates of relapses per patient than intravenous methylprednisolone adjunctive therapy	Berkovich et al. (2017)
	In female patients, yoga training increased the level of ACTH	Najafi and Moghadasi (2017)
Nerve injury	Increase of ACTH immunoreactivity in the spinal cord motoneurons	Jégou et al. (2013)
Nystagmus	$ACTH_{4-10}$ accelerates the ocular motor symptom spontaneous nystagmus compensation	Gilchrist, Darlington, and Smith (1993)

Continued

Table 2 Involvement of enkephalins/ACTH in pathologies.—cont'd

Pathology	ACTH	References
Podocytopathies and adrenal myelolipoma	ACTH analogues/ACTH have been used for the treatment of podocytopathies/membranous nephropathy. The ACTH analogue tetracosactide has been suggested for the treatment of nephrotic patients. ACTH is involved in the development of adrenal myelolipoma	Filippone et al. (2016), Kittanamongkolchai, Cheungpasitporn, and Zand (2016), Lorusso et al. (2015), and Park et al. (2017)
Sleep	Administration of ACTH over a long period of time causes sleep disturbances	Tsutsui, Shinomiya, Sendo, Kitamura, and Kamei (2015)
Stress	The serum level of ACTH increased after an acute heat stress exposure	Wang, Liu, Luo, Zhu, and Li (2015)
	Melanocortin-receptor 4 is involved in the control of emotional states and in stress responses	Chaki and Okubo (2007)
Weight disorder	A weight loss occurs when the secretion of ACTH by the pituitary gland is altered	McLaughlin and Zagon (2013)
	A deficiency of pro-opiomelanocortin increases appetite and weight gain	McLaughlin and Zagon (2013)
	Loss-of-function mutations in melanocortin-receptor 4 resulted in increased overall and lean body mass and in hyperphagia	Farooqi et al. (2003)
	The melanocortin-receptor 4 limits ingestive behavior and stimulates the energy expenditure	Adan et al. (2006), Cone (2005), Garfield, Lam, Marston, Przydzial, and Heisler (2009), and Tao (2010)
West syndrome	ACTH has been suggested for the treatment of patients suffering the West syndrome with severe hemophilia A	Kubota et al. (2016)

Met-enkephalin also decreases the number of stimulated T and B lymphocytes (Zagon, Donahue, Bonneau, & MacLaughlin, 2011), and a low dosage of naltrexone (a non-selective opioid antagonist with a preference for μ sites) upregulates the synthesis of both enkephalins and opioid receptors (McLaughlin, 2013). The use of naltrexone is a promising treatment against autoimmune diseases (e.g., multiple sclerosis, lupus, Crohn's disease). During multiple sclerosis relapses, enkephalin levels decrease in both humans and in experimental animals; and it has been reported that met-enkephalin or naltrexone reverses the course of the disease (Ludwig, Zagon, & McLaughlin, 2017a, 2017b). In humans, treatment with naltrexone also decreases patients' fatigue and improves their quality of life.

In patients suffering from Huntington's disease, and in those with a striatopallidal infarction, a decrease in enkephalin immunoreactivity was observed in the basal ganglia and substantia nigra, respectively. The activation of δ receptors by agonists is known to protect glial/neuronal cells from ischemia (stroke), facilitate the survival of neurons and mitigate apoptotic pathways (Lee et al., 2018). The δ receptor agonist DADLE ([D-Ala^2, D-Leu^5]-enkephalin) has been found to improve permanent focal cerebral ischemia and hepatic ischemia/reperfusion injury (Fu, Liu, Zhu, Li, & Yao, 2016; Zhou et al., 2017).

In persons with a lower limb amputation, the immunoreactivity for enkephalin in the dorsal horn of the spinal cord increases, and it appears that a reorganization of the peptide circuitry occurs (Pioro et al., 1990). It is also known that colitis and its associated pain are decreased by the enkephalins released from T-lymphocytes (Basso et al., 2018).

It has been also reported the therapeutic/prophylactic effects of met-enkephalin against the influenza A virus infection. For example, the pentapeptide inhibited the replication of the virus and decreased the level of several cytokines (e.g., IL-1β, IL-6) (Tian et al., 2018). Additionally, it has been reported that pro-enkephalin is a risk predictor for poor outcome in heart failure (Siong Chan, Cao, & Ng, 2018); that low fasting plasma concentration of pro-enkephalin is associated with an increased risk of future breast cancer (Melander et al., 2015); that stress upregulated or downregulated, respectively, met-enkephalin and leu-enkephalin in the hippocampus (Li et al., 2018); that the administration of a non-opioid analogue of leu-enkephalin partially normalizes cerebral consequences after hypoxia (Simankova et al., 2017); and that met-enkephalin exerts an anti-tumor action against melanoma (Wang et al., 2017) and promotes apoptotic mechanisms in glioma cells (Lu et al., 2018). Met-enkephalin has been proposed as a possible treatment for cancer patients whose immune systems have been damaged by chemotherapy, as the pentapeptide is able to restore such system (Wang et al., 2014).

δ receptors agonists generating antidepressant effects without convulsions are also being tested in humans (Dripps & Jutkiewicz, 2018). In Moyamoya disease, an increase in the level of pro-enkephalin$_{143-183}$, has been noted in the cerebrospinal fluid (Yokoyama et al., 2018) and it has been reported that enkephalins decreased cold-induced hypertension after repetitive electro acupuncture (Li et al., 2016); that leu-enkephalin promotes wound repair (Yang et al., 2016); and that, in parkinsonian animals treated with L-DOPA, high levels of pre-pro-enkephalin are needed in the striatum for a locomotor sensitization to L-DOPA (Sgroi, Capper-Loup, Paganetti, & Kaelin-Lang, 2016). Enkephalins have been found to be involved in feeding and diet-induced obesity (Mendez et al., 2015), and enkephalin is crucial for the behavioral and molecular expression of cocaine sensitization (Mongi-Bragato et al., 2016). A hepatoprotective action of met-enkephalin has also been reported (Martinic et al., 2014). The actions exerted by nicotine, ethanol or fat are known to be mediated by the same enkephalinergic neurons located in the limbic system and hypothalamus (Chang, Karatayev, Barson, Liang, & Leibowitz, 2014).

Opioid peptides are also involved in alcoholism (Rodríguez & Coveñas, 2011). The administration of μ or δ agonists into the paraventricular hypothalamic nucleus caused an increase in ethanol intake, whereas the administration of δ agonists or δ antagonists into the ventral tegmental area decreased or increased, respectively, alcohol consumption (Barson et al., 2010; Margolis et al., 2008). Moreover, it is known that pro-enkephalin is downregulated in the caudate nucleus of alcoholics (Sarkisyan et al., 2015); that μ and δ receptors are involved in the expression and reinstatement of ethanol conditioned seeking behavior (Gibula-Bruzda, Marszalek-Grabska, Gawel, et al., 2015); and that an enkephalin analogue inhibited alcohol withdrawal-induced anxiety behavior (Gibula-Bruzda, Marszalek-Grabska, Witkowska, et al., 2015). Additionally, prenatal ethanol exposure alters the expression of met-enkephalin in regions of the brain (mesocorticolimbic and nigrostriatal systems, hippocampus, hypothalamus) involved in reinforcement (Abate, Hernández-Fonseca, Reyes-Guzmán, Barbosa-Luna, & Méndez, 2014). The authors suggest the involvement of mesocorticolimbic enkephalins in ethanol reinforcement in offspring.

After a chronic administration of opiates, serious side-effects can appear in the gastrointestinal tract (e.g., opioid-bowel dysfunction) (Panchal et al., 2007). This dysfunction, affecting patients (60–90%) that received a chronic opioid treatment, is mediated by the μ receptor and is characterized by constipation, abdominal bloating and cramping, decreased gastric emptying, and delayed gastrointestinal transit. After a pelvic or abdominal surgery, when opiates are administered as post-surgery analgesics, μ receptors can be also involved in the worsening of post-operative ileus (vomiting, bloating, delayed gastrointestinal transit with bowel distension, nausea) (Holzer, 2010). The best choice for the post-operative ileus is the administration of peripheral and selective acting μ receptor antagonists (e.g., alvimopam). In the intestine, opioid peptides are also involved in inflammatory processes (Philippe et al., 2006, 2003). Immune cells express opioid peptides (which are released from these cells) and μ, κ and δ receptors and opioid peptides control the release of pro-inflammatory cytokines (e.g., tumor necrosis factor). Opioid antagonists decreased inflammation in a moderate experimental colitis model and improved inflammation in patients suffering from Crohn's disease (Matters et al., 2008; Smith et al., 2007). In an experimental animal model of acute intestinal inflammation, μ-receptor agonists ameliorated inflammatory processes, and animals lacking the μ-receptor were found to be more susceptible to develop colon inflammation than control animals. Upregulation of μ-receptor mRNA has been reported in patients suffering from inflammatory bowel disease and in intestinal inflammation (Philippe et al., 2006). In those patients, μ-receptors have been discovered in immunocytes and enteric neurons and, in patients with active inflammatory bowel disease, these receptors were found to be overexpressed in inflammatory cells. Cytokines increase the expression of μ receptors, and agonists of this receptor reduce the expression of tumor necrosis factor-α in the mucosa of patients with active inflammatory bowel disease. Moreover, μ receptors play a protective role during the intestinal inflammation induced by ischemia/reperfusion, since DAMGO (a selective μ receptor agonist) reduces epithelial damage, inflammatory processes, neutrophil infiltration, and pro-inflammatory cytokines (Saccani et al., 2012). In intestinal inflammation, μ-receptors appear to exert a protective effect; thus, μ-receptor agonists could be used as new therapeutic compounds for the treatment of inflammatory bowel disease.

8.2 ACTH

ACTH stimulates cell proliferation in the adrenal cortex, and an excess of ACTH in plasma promotes adrenal hyperplasia. In Cushing's disease, the appearance of hyperpigmentation occurs due to an excess of ACTH production. Additionally, it has been reported that ORG-2766, an analogue of $ACTH_{4-9}$, protected against experimental allergic neuritis (an animal model of the Guillain-Barré syndrome) (Duckers et al., 1993), and that another ACTH analogue has been used for the treatment of podocytopathies (Filippone et al., 2016). ACTH is involved in the development of adrenal myelolipoma (a mixture of hematopoietic and adipose tissues) (Park et al., 2017) and, in blood, a high level of ACTH has been reported in alcohol-dependent patients (Geisel et al., 2019).

Moreover, it is known that the basal serum ACTH level is normal in atopic dermatitis patients (Tehranchinia, Rahimi, & Lotfi, 2017) and that in experimental animals, ultraviolet B irradiation increased the level of ACTH in plasma and promoted the deterioration of atopic dermatitis symptoms; the latter being mediated by ACTH (Hiramoto et al., 2018). It has been suggested that high levels of stress hormones (e.g., ACTH) in the normal microenvironment are predictive for chemically induced cancer in rats (Valente et al., 2018). ACTH immunotherapy has been proposed for the treatment of anti-N-methyl-D-aspartate receptor encephalitis (Hatanaka et al., 2018) and it has been also reported that the selective Jak2 inhibitor (SD-1029) or the inhibition of the heat shock protein 90 blocked ACTH synthesis and the proliferation of corticotroph tumor cells (Asari et al., 2017; Sugiyama et al., 2015). It appears that GADD45β (growth arrest and DNA damage-inducible protein) and PTTG1 (pituitary tumor-transforming 1) are involved in these mechanisms. The authors conclude that therapies targeting EGFR-dependent Jak2/STAT3 may be useful for the treatment of Cushing's disease (Asari et al., 2017). Moreover, inhibition of the activity of ubiquitin-specific peptidase 8 blocked both growth and ACTH synthesis in corticotroph cells (Jian et al., 2016).

The administration of ACTH has been suggested for the treatment of children suffering from intractable seizures who are resistant to conventional antiepileptic drugs (Nasiri et al., 2017). ACTH is known to be effective against infantile spasms (Yin et al., 2017), and protects learning/memory function in epileptic experimental animals (Scantlebury et al., 2017). ACTH has also been suggested for the treatment of intractable epilepsy in hemiconvulsion-hemiplegia-epilepsy syndrome (Shimakawa et al., 2015), and improves attention deficit hyperactivity disorder-like symptoms in patients suffering from encephalopathy with electrical status epilepticus in sleep (Altunel, Altunel, & Sever, 2017). Genetic variants of the *SCN1A* gene (encoding the α1 subunit of the neuronal voltage-gated sodium channel) are associated with infantile spasms and ACTH responsiveness (Zhang, Zou, Ding, He, & Yang, 2014). Another study found that ACTH yields beneficial effects on stuttering in attention deficit and hyperactivity, and in autism spectrum disorder patients with electrical status epilepticus in sleep (Altunel, Sever, & Altunel, 2017).

It appears that ACTH is protective for lumbar bone mineral density in patients with Cushing's disease (Guo et al., 2018), and it is known that ACTH increased the expression of bone structural/regulating mRNAs of biglycan, vitamin D-receptor, TGF-β, and collagen I (Tourkova et al., 2017). ACTH also increased alkaline phosphatase activity and bone mineralization, and levels of osterix (a transcription factor for osteoblast differentiation) and calcium-sensing receptor (RNA and protein). The data indicate that ACTH plays an important role in bone maintenance (Tourkova et al., 2017).

In patients with multiple sclerosis, adjunctive treatment of ACTH to interferon β results in better examiner-assessed cumulative rates of relapses per patient than the intravenous methylprednisolone adjunctive therapy (Berkovich et al., 2017). Yoga training increases the level of ACTH in female patients with multiple sclerosis, but decreases that of cortisol (Najafi & Moghadasi, 2017).

It has been reported that normalized early post-operative values for ACTH predict non-remission after trans-sphenoidal surgery for Cushing's disease (Asuzu et al., 2017), and that an analogue of $ACTH_{4-7}$ (semax) regulates the expression of the immune response genes during ischemic brain injury in experimental animals (Medvedeva et al., 2017). Semax affects learning processes and also exerts a protective action against metal-induced cell toxicity (Tabbì et al., 2015). Additionally, ACTH appears to play an important role in the pathogenesis of post-traumatic acute diffuse brain swelling (Chen et al., 2017). In the adrenal cortex, long-term neonatal stressors lead to lower ACTH receptor sensitivity, which may be a sign of stress adaptation in adulthood (Holubová, Stofková, Jurcovicová, & Slamberová, 2016). Serum ACTH levels increased after an acute heat stress exposure (Wang et al., 2015). The melanocortin-receptor 4 has been involved in the control of emotional states and in stress responses, suggesting that melanocortin-receptor 4 antagonists could be used for the treatment of depression and anxiety (Chaki & Okubo, 2007). These antagonists may also be of therapeutic value against morphine tolerance and nociceptive mechanisms (Kalange et al., 2007).

In patients suffering from aldosterone-producing adenomas, it has been shown that blood pressure is affected by ACTH (Kobayashi et al., 2017). ACTH therapy has been suggested for the treatment of patients suffering the West syndrome with severe hemophilia A (Kubota et al., 2016), and ACTH was a well-tolerated and efficient therapy for the treatment of patients with membranous nephropathy (Kittanamongkolchai et al., 2016). Moreover, a low-dose administration of the synthetic ACTH-analogue tetracosactide has been suggested for the treatment of nephrotic patients (Lorusso et al., 2015). It has been also reported a relationship between basal serum ACTH and non-alcoholic fatty liver disease in male patients with idiopathic hypogonadotropic hypogonadism, whose basal serum ACTH levels were higher than those of healthy controls (Wang et al., 2016). In experimental animals, it has been shown that the administration of ACTH over a long period of time caused sleep disturbances (Tsutsui et al., 2015).

In humans, the administration of tetracosactide ($ACTH_{1-24}$) generated anti-$ACTH_{1-24}$ antibodies, and spontaneous anti-ACTH reactivity has been also reported in patients with autoimmune Addison's or with Graves' diseases (Gan et al., 2016). Compared to controls, term infants born with meconium-stained amniotic fluid and respiratory distress showed a lower level of blood ACTH (Prasanth, Kamat, Khilfeh, & Davis, 2014). It is also known that $ACTH_{4-10}$ accelerates the ocular motor symptom spontaneous nystagmus compensation (Gilchrist et al., 1993). In spinal cord motoneurons, an increase in ACTH immunoreactivity has been observed after nerve injury (Jégou et al., 2013). In diabetic neuropathy, the level of ACTH is also high, and an increase in the number of ACTH binding sites has been reported in muscle (Jégou et al., 2013). This observation indicates that the melanocortin-receptor 5 may be involved, since this receptor is exclusively expressed in muscle. It has been also reported that melanocortin receptor agonists are effective for the treatment of erectile dysfunction and also increase sexual desire and arousal (Safarinejad, 2008; Wessels et al., 2000).

Weight loss occurs when the secretion of ACTH by the pituitary gland is altered (McLaughlin & Zagon, 2013), and a deficiency of pro-opiomelanocortin (the precursor of ACTH) increased appetite and weight gain (McLaughlin & Zagon, 2013). Genetic variances in terms of single-nucleotide polymorphism in melanocortin-receptor 4 have been related to increased body mass index, early-onset obesity, adipose mass and risk to develop obesity (Speliotes et al., 2010). Loss-of-function mutations in melanocortin-receptor 4 resulted in increased mass and lean body mass and in hyperphagia, and it has been suggested that energy balance regulation is sensitive to variation in functional melanocortin-receptor 4 levels (Farooqi et al., 2003). It has been reported that the overexpression of agouti promotes obesity, because agouti blocks melanocortin-receptor 4 functions in the brain (Fan, Boston, Kesterson, Hruby, & Cone, 1997). In mice lacking this receptor, excess weight gain occurs due to increased ingestive behavior, reduced locomotor activity, and changes in energy expenditure. This finding implies that melanocortin-receptor 4 limits ingestive behavior and stimulates energy expenditure (Adan et al., 2006; Cone, 2005; Garfield et al., 2009; Tao, 2010).

9. Conclusion and future perspectives

The widespread anatomical distribution on met-enkephalin, leu-enkephalin, ACTH and their receptors suggests that the three peptides are involved in many physiological mechanisms. Although the physiological actions of enkephalins and ACTH (e.g., analgesia, stress) are well known, an in-depth understanding of the involvement of these peptides in other physiological actions remains an important goal. The enkephalin/opioid and ACTH/melanocortin receptor systems are involved in the molecular bases of many human pathologies, and thus a thorough analysis of these systems is crucial for a better clinical management of these pathologies. In the coming years, additional experiments (e.g., by using δ/melanocortin receptor agonists or antagonists) should be performed to fully explore the involvement of enkephalins in multiple sclerosis, Huntington's disease, alcoholism, virus infection, cancer, stroke, depression, heart failure, hypoxia, hepatoprotective action, and in the protection of the immune system after the use of chemotherapy, as well as the involvement of ACTH in Guillain-Barré syndrome, dermatitis, alcoholism, bone maintenance, intractable seizures, podocytopathies, multiple sclerosis, West syndrome with severe hemophilia A, anti-N-methyl-D-aspartate receptor encephalitis, and in the pathogenesis of port-traumatic acute diffuse brain swelling. Profound knowledge of the involvement of enkephalins/ACTH in the aforementioned pathologies would be of tremendous value in the development of beneficial therapeutic strategies for eventual testing in human clinical trials.

Acknowledgments

This work was supported by the School of Medicine of the Universidad de Santander, UDES and the "Programa XI: Programa de Financiación de Unidades de Excelencia de la Universidad de Salamanca," Salamanca, Spain. The authors wish to thank Mr. Emile Blanche for editing the English text.

References

Abate, P., Hernández-Fonseca, K., Reyes-Guzmán, A. C., Barbosa-Luna, I. G., & Méndez, M. (2014). Prenatal ethanol exposure alters met-enkephalin expression in brain regions related with reinforcement: Possible mechanism for ethanol consumption in offspring. *Behavioural Brain Research, 274*, 194–204.

Abrams, G. M., Nilaver, G., Hoffman, D., Zimmerman, E. A., Ferin, M., Krieger, D. T., et al. (1980). Immunocytochemical distribution of corticotropin (ACTH) in monkey brain. *Neurology, 30*, 1106–1110.

Adan, R. A., Tiesjema, B., Hillebrand, J. J., la Fleur, S. E., Kas, M. J., & de Krom, M. (2006). The MC4 receptor and control of appetite. *British Journal of Pharmacology, 149*, 815–827.

Akil, H., Watson, S. J., Young, E., Lewis, M. E., Khachaturian, H., & Walker, J. M. (1984). Endogenous opioids: Biology and function. *Annual Review of Neuroscience, 7*, 223–255.

Aleknaviciute, J., Tulen, J. H., Timmermans, M., de Rijke, Y. B., van Rossum, E. F., de Jong, F. H., et al. (2016). Adrenocorticotropic hormone elicits gonadotropin secretion in premenopausal women. *Human Reproduction, 31*, 2360–2368.

Altunel, A., Altunel, E. Ö., & Sever, A. (2017). Response to adrenocorticotropic in attention deficit hyperactivity disorder-like symptoms in electrical status epilepticus in sleep syndrome is released to electroencephalographic improvement: A retrospective study. *Epilepsy & Behavior, 74*, 161–166.

Altunel, A., Sever, A., & Altunel, E. Ö. (2017). ACTH has beneficial effects on stuttering in ADHD and ASD patients with ESES: A retrospective study. *Brain & Development, 39*, 130–137.

Argiolas, A., Melis, M. R., Murgia, S., & Schioth, H. B. (2000). ACTH- and alpha-MSH-induced grooming, stretching, yawning and penile erection in male rats: Site of action in the brain and role of melanocortin receptors. *Brain Research Bulletin, 51*, 425–431.

Asari, Y., Kageyama, K., Nakada, Y., Tasso, M., Takayasu, S., Niioka, K., et al. (2017). Inhibitory effects of a selective Jak2 inhibitor on adrenocorticotropic hormone production and proliferation of corticotroph tumor AtT20 cells. *OncoTargets and Therapy, 10*, 4329–4338.

Asuzu, D., Chatain, G. P., Hayes, C., Benzo, S., McGlotten, R., Keil, M., et al. (2017). Normalized early postoperative cortisol and ACTH values predict nonremission after surgery for Cushing disease. *Journal of Clinical Endocrinology and Metabolism, 102*, 2179–2187.

Babischkin, J. S., Aberdeen, G. W., Pepe, G. J., & Albrecht, E. D. (2016). Estrogen suppresses interaction of melanocortin 2 receptor and its accessory protein in the primate fetal adrenal cortex. *Endocrinology, 157*, 4588–4601.

Bagnol, D., Mansour, A., Akil, H., & Watson, S. J. (1997). Cellular localization and distribution of the cloned mu and kappa receptors in rat gastrointestinal tract. *Neuroscience, 81*, 579–591.

Barnes, M. J., Primeaux, S. D., & Bray, G. A. (2008). Food deprivation increases the mRNA expression of μ-opioid receptors in the ventral medial hypothalamus and the arcuate nucleus. *American Journal of Physiology. Regulatory, Integrative and Comparative Physiology, 295*, 1385–1390.

Barson, J. R., Carr, A. J., Soun, J. E., Sobhani, N. C., Rada, P., Leibowitz, S. F., et al. (2010). Opioids in the hypothalamic paraventricular nucleus stimulate ethanol intake. *Alcoholism, Clinical and Experimental Research, 34*, 214–222.

Basso, L., Garnier, L., Bessac, A., Boué, J., Blanpied, C., Cenac, N., et al. (2018). T-lymphocyte-derived enkephalins reduce Th1/Th17 colitis and associated pain in mice. *Gastroenterology, 53*, 215–226.

Berkovich, R., Bakshi, R., Amezcua, L., Axtell, R. C., Cen, S. Y., Tauhid, S., et al. (2017). Adrenocorticotropic hormone versus methylprednisolone added to interferon β in

patients with multiple sclerosis experiencing breakthrough disease: A randomized, rater-blinded trial. *Therapeutic Advances in Neurological Disorders*, *10*, 3–17.
Bertolini, A., Tacchi, R., & Vergoni, A. N. (2009). Brain effects of melanocortins. *Pharmacological Reviews*, *59*, 13–47.
Biswas, H. M. (2017). Effect of adrenocorticotropic hormone on UCP1 gene expression in brown adipocytes. *Journal of Basic and Clinical Physiology and Pharmacology*, *28*, 267–274.
Bohm, S., Grady, E. F., & Bunnett, N. W. (1997). Mechanisms attenuating signaling by G protein-coupled receptors. *Biochemical Journal*, *322*, 1–8.
Büch, T. R., Heling, D., Damm, E., Gudermann, T., & Breit, A. (2009). Pertussis toxin-sensitive signaling of melanocortin-4 receptors in hypothalamic GT1-7 cells defines agouti-related protein as a biased agonist. *Journal of Biological Chemistry*, *284*, 26411–26420.
Chai, B., Li, J. Y., Zhang, W., Wang, H., & Mulholland, M. W. (2009). Melanocortin-4 receptor activation inhibits c-Jun N-terminal kinase activity and promotes insulin signaling. *Peptides*, *30*, 1098–1104.
Chaki, S., & Okubo, T. (2007). Melanocortin-4 receptor antagonists for the treatment of depression and anxiety disorders. *Current Topics in Medicinal Chemistry*, *7*, 1145–1151.
Chang, G. Q., Karatayev, O., Barson, J. R., Liang, S. C., & Leibowitz, S. F. (2014). Common effects of fat, ethanol, and nicotine on enkephalin in discrete areas of the brain. *Neuroscience*, *277*, 665–678.
Chen, W., Sheng, J., Peng, G., Yang, J., Wang, S., & Li, K. (2017). Early stage alterations of catecholamine and adrenocorticotropic hormone levels in post-traumatic acute diffuse brain swelling. *Brain Research Bulletin*, *130*, 47–52.
Collip, J. B., Anderson, E., & Thomson, D. L. (1933). The adrenotropic hormone of the anterior pituitary lobe. *Lancet*, *222*, 347–348.
Cone, R. D. (2005). Anatomy and regulation of the central melanocortin system. *Nature Neuroscience*, *8*, 571–578.
Coveñas, R., de León, M., Narváez, J. A., Tramu, G., Aguirre, J. A., & González-Barón, S. (1996). An immunocytochemical mapping of ACTH/CLIP in the cat diencephalon. *Journal of Chemical Neuroanatomy*, *11*, 191–197.
Coveñas, R., Mangas, A., & Narváez, J. A. (2007). Introduction to neuropeptides. In R. - Coveñas, A. Mangas, & J. A. Narváez (Eds.), *Focus on neuropeptide research* (pp. 1–26). Trivandrum: Transworld Research Network.
Davis, C., Zai, C., Levitan, R. D., Kaplan, A. S., Carter, J. C., Reid-Westoby, C., et al. (2011). Opiates, overeating and obesity: A psychogenetic analysis. *International Journal of Obesity*, *35*, 1347–1354.
del Pozo, E., Martín-Pérez, J., Stadelmann, A., Girard, J., & Brownell, J. (1980). Inhibiting action of met-enkephalin on ACTH release in man. *Journal of Clinical Investigation*, *65*, 1531–1534.
Dietis, N., Rowbotham, D. J., & Lambert, D. G. (2011). Opioid receptor subtypes: Fact or artifact? *British Journal of Anesthesia*, *107*, 8–18.
Douglass, J., McKinzie, A. A., & Pollock, K. M. (1994). Identification of multiple DNA elements regulating basal and protein kinase A-induced transcriptional expression of the rat prodynorphin gene. *Molecular Endocrinology*, *8*, 333–344.
Dripps, I. J., & Jutkiewicz, E. M. (2018). Delta opioid receptors and modulation of mood and emotion. *Handbook of Experimental Pharmacology*, *47*, 179–197. https://doi.org/10.1007/164_2017_42.
Duckers, H. J., Verhaagen, J., & Gispen, W. H. (1993). A neurotrophic analogue of $ACTH_{4-9}$ protects against experimental allergic neuritis. *Annals of the New York Academy of Sciences*, *680*, 493–495.

Duque, E., Mangas, A., Díaz-Cabiale, Z., Narváez, J. A., & Coveñas, R. (2011). Neuropeptides in the monkey brain. In R. M. Williams (Ed.), *Monkeys: Biology, behavior and disorders* (pp. 151–166). New York: Nova Science Publishers.
Duque-Díaz, E., Díaz-Cabiale, Z., Narváez, J. A., & Coveñas, R. (2017). Mapping of enkephalins and adrenocorticotropic hormone in the squirrel monkey brainstem. *Anatomical Science International, 92*, 275–292.
Eberle, A. N. (1988). *The melanotropins: Chemistry, physiology and mechanisms of action*. Basel: Karger.
Endo, N., Yamane, H., Rahayu, L. P., & Tanaka, T. (2018). Effect of repeated adrenocorticotropic hormone administration on reproductive function and hair cortisol concentration during the estrous cycle in goats. *General and Comparative Endocrinology, 259*, 207–212.
Fan, W., Boston, B. A., Kesterson, R. A., Hruby, V. J., & Cone, R. D. (1997). Role of melanocortinergic neurons in feeding and the agouti obesity syndrome. *Nature, 385*, 165–168.
Farooqi, I. S., Keogh, J. M., Yeo, G. S., Lank, E. J., Cheetham, T., & O'Rahilly, S. (2003). Clinical spectrum of obesity and mutations in the melanocortin 4 receptor gene. *New England Journal of Medicine, 348*, 1085–1095.
Filippone, E. J., Dopson, S. J., Rivers, D. M., Monk, R. D., Udani, S. M., Jafari, G., et al. (2016). Adrenocorticotropic hormone analog use for podocytopathies. *International Medical Case Reports, 9*, 125–133.
Fridmanis, D., Petrovska, R., Pjanova, D., Schiöth, H. B., & Klovins, J. (2014). Replacement of short segments within transmembrane domains of MC2R disrupts retention signal. *Journal of Molecular Endocrinology, 53*, 201–215.
Fridmanis, D., Roga, A., & Klovins, J. (2017). ACTH receptor (MC2R) specificity: What do we know about underlying molecular mechanisms? *Frontiers in Endocrinology, 8*, 13.
Fu, D., Liu, H., Zhu, H., Li, S., & Yao, J. (2016). Protective effect of delta opioid receptor agonist (D-Ala2, D-Leu5) enkephalin on permanent focal cerebral ischemia in rats. *Neuroreport, 27*, 749–754.
Furness, J. B., Costa, M., & Miller, R. J. (1983). Distribution and projections of nerves with enkephalin-like immunoreactivity in the guinea-pig small intestine. *Neuroscience, 8*, 644–653.
Gan, E. H., MacArthur, K., Mitchell, A. L., Joshi, A., Crock, P., & Pearce, S. H. (2016). Spontaneous and tetracosactide-induced anti-ACTH antibodies in man. *Clinical Endocrinology, 84*, 489–495.
Garfield, A. S., Lam, D. D., Marston, O. J., Przydzial, M. J., & Heisler, L. K. (2009). Role of central melanocortin pathways in energy homeostasis. *Trends in Endocrinology and Metabolism, 20*, 203–215.
Geisel, O., Schlemm, L., Hellweg, R., Wiedemann, K., & Müller, C. A. (2019). Hypothalamic-pituitary-adrenocortical axis activity in alcohol-dependent patients during treatment with high-dose baclofen. *Pharmacopsychiatry, 52*, 32–37. https://doi.org/10.1055/s-0043-124189.
Gentilin, E., di Pasquale, C., Gagliano, T., Tagliati, F., Benfini, K., Ambrosio, M. R., et al. (2016). Protein kinase C delta restrains growth in ACTH-secreting pituitary adenoma cells. *Molecular and Cellular Endocrinology, 419*, 252–258.
Ghaddhab, C., Vuissoz, J. M., & Deladoëy, J. (2017). From bioinactive ACTH to ACTH antagonist: The clinical perspective. *Frontiers in Endocrinology, 8*, 17.
Gibula-Bruzda, E., Marszalek-Grabska, M., Gawel, K., Witkowska, E., Izdebski, J., & Kotlinska, J. H. (2015). The influence of the new enkephalin derivative, cyclo [N (ε), N (β)-carbonyl-d-Lys (2), Dap (5)] enkephalinamide (cUENK6), on reinstatement of ethanol-induced conditioned place preference in rats. *Physiology & Behavior, 145*, 50–56.

Gibula-Bruzda, E., Marszalek-Grabska, M., Witkowska, E., Izdebski, J., & Kotlinska, J. H. (2015). Enkephalin analog, cyclo [N (ε), N (β)-carbonyl-d-Lys (2), Dap (5)] enkephalinamide (cUENK6), inhibits the ethanol withdrawal-induced anxiety-like behavior in rats. *Alcohol, 49*, 229–236.

Gilchrist, D. P. D., Darlington, C. L., & Smith, P. F. (1993). Comparison of the effects of adrenocorticotropic hormone fragments on the compensation of spontaneous nystagmus and posture in guinea pig. *Annals of the New York Academy of Sciences, 680*, 524–526.

Gorostizaga, A. B., Mori Sequeiros García, M. M., Acquier, A. B., López-Costa, J. J., Méndez, C. F., Maloberti, P. M., et al. (2016). ACTH modulates PTP-PEST activity and promotes its interaction with paxillin. *Journal of Cellular Biochemistry, 117*, 2170–2181.

Gosnell, B. A., Kotz, C. M., Billington, C. J., & Levine, A. S. (2013). Opioids. In A. J. Kastin (Ed.), *Handbook of biologically active peptides* (pp. 1149–1153). Amsterdam: Elsevier.

Gray, A. C., Coupar, I. M., & White, P. J. (2006). Comparison of opioid receptor distributions in the rat ileum. *Life Sciences, 78*, 1610–1616.

Guo, W., Li, F., Zhu, C., Wang, B., Wang, K., Dai, C., et al. (2018). Effect of hypercortisolism on bone mineral density and bone metabolism: A potential protective effect of adrenocorticotropic hormone in patients with Cushing's disease. *Journal of International Medical Research, 46*, 492–503.

Guyon, A., Morselli, L. L., Balbo, M. L., Tasali, E., Leproult, R., L'Hermite-Balériaux, M., et al. (2017). Effects of insufficient sleep on pituitary-adrenocortical response to CRH stimulation in healthy men. *Sleep, 40*, 1–10.

Hatanaka, M., Shimakawa, S., Okumura, A., Natsume, J., Fukui, M., Nomura, S., et al. (2018). The efficacy of adrenocorticotropin hormone in a girl with anti-N-Methyl-D-aspartate receptor encephalitis. *Brain & Development, 40*, 247–250.

Hayward, M. D., Pintar, J. E., & Low, M. J. (2002). Selective reward deficit in mice lacking beta-endorphin and enkephalin. *Journal of Neuroscience, 22*, 8251–8258.

Hiramoto, K., Yamate, Y., & Yokoyama, S. (2018). Ultraviolet B eye irradiation aggravates atopic dermatitis via adrenocorticotropic hormone and NLRP3 inflammasome in NC/Nga mice. *Photodermatology, Photoimmunology & Photomedicine, 34*, 200–210.

Ho, A., Lievore, A., Patierno, S., Kohlmeier, S. E., Tonini, M., & Sternini, C. (2003). Neurochemically distinct classes of myenteric neurons express the μ-opioid receptor in the guinea pig ileum. *Journal of Comparative Neurology, 458*, 404–411.

Holubová, A., Stofková, A., Jurcovicová, J., & Slamberová, R. (2016). The effect of neonatal maternal stress on plasma levels of adrenocorticotropic hormone, corticosterone, leptin, and ghrelin in adult male rats exposed to acute heterotypic stressor. *Physiological Reviews, 65*(Suppl. 5), S557–S566.

Holzer, P. (2010). Opioid antagonists for prevention and treatment of opioid-induced gastrointestinal effects. *Current Opinion in Anaesthesiology, 23*, 616–622.

Hu, D., Liu, T., & Wang, X. (2017). Preliminary studies on the differential expression of adrenocorticotropic hormone receptor in adrenal adenomas. *Endokrynologia Polska, 68*, 512–518.

Hughes, J., Smith, T. W., Kosterlitz, H. W., Fothergill, L. A., Morgan, B. A., & Morris, H. R. (1975). Identification of two related pentapeptides from the brain with potent opiate agonist activity. *Nature, 258*, 577–580.

Humphreys, M. H., Ni, X. P., & Pearce, D. (2011). Cardiovascular effects of melanocortins. *European Journal of Pharmacology, 660*, 43–52.

Jégou, S., Cone, R. D., Eberle, A. N., & Vaudry, H. (2013). Melanocortins. In A. J. Kastin (Ed.), *Handbook of biologically active peptides* (pp. 838–844). Amsterdam: Elsevier.

Jian, F. F., Li, Y. F., Chen, Y. F., Jiang, H., Chen, X., Zheng, L. L., et al. (2016). Inhibition of ubiquitin-specific peptidase 8 suppresses adrenocorticotropic hormone production and tumorous corticotroph cell growth in AtT20 cells. *Chinese Medical Journal*, *129*, 2102–2108.

Joseph, S. A., Plicher, W. H., & Bennett-Clarke, C. (1983). Immunocytochemical localization of ACTH perikarya in nucleus tractus solitarius: Evidence for a second opiocortin neuronal system. *Neuroscience Letters*, *38*, 221–225.

Kalange, A. S., Kokare, D. M., Singru, P. S., Upadhya, M. A., Chopde, C. T., & Subhedar, N. K. (2007). Central administration of selective melanocortin 4 receptor antagonist HS014 prevents morphine tolerance and withdrawal hyperalgesia. *Brain Research*, *1181*, 10–20.

Kas, M. J., van den Bos, R., Baars, A. M., Lubbers, M., Lesscher, H. M., Hillebrand, J. J., et al. (2004). Mu-opioid receptor knockout mice show diminished food-anticipatory activity. *European Journal of Neuroscience*, *20*, 1624–1632.

Kieffer, B. L., & Evans, C. J. (2009). Opioid receptors: From binding sites to visible molecules *in vivo*. *Neuropharmacology*, *56*(Suppl. 1), 205–212.

Kim, Y., McGee, S., Czeczor, J. K., Walker, A. J., Kale, R. P., Kouzani, A. Z., et al. (2016). Nucleus accumbens deep-brain stimulation efficacy in ACTH-pretreated rats: Alterations in mitochondrial function relate to antidepressant-like effects. *Translational Psychiatry*, *6*, e842.

Kitahama, K., Sakai, K., & Jouvet, M. (1984). Mise en evidence de la projection directe des neurons immunoreactifs à l'ACTH sur le complexe du locus coeruleus chez le chat. *Comptes Rendus de l'Académie des Sciences Paris*, *298*, 163–167.

Kitahama, K., Sallanon, M., Buda, C., Janin, M., Dubois, M. P., & Jouvet, M. (1986). ACTH-immunoreactive neurons and their projections in the cat forebrain. *Peptides*, *7*, 801–807.

Kittanamongkolchai, W., Cheungpasitporn, W., & Zand, L. (2016). Efficacy and safety of adrenocorticotropic hormone treatment in glomerular diseases: A systematic review and meta-analysis. *Clinical Kidney Journal*, *9*, 387–396.

Kobayashi, H., Kaketa, A., Takahiro, U., Otsuka, H., Tanaka, S., Hatanaka, Y., et al. (2017). Plasma adrenocorticotropic hormone but not aldosterone is correlated with blood pressure in patients with aldosterone-producing adenomas. *Journal of Clinical Hypertension*, *19*, 280–286.

Kramer, P. R., Umorin, M., & Bellinger, L. L. (2015). Attenuation of myogenic orofacial nociception and mechanical hypersensitivity by viral mediated enkephalin overproduction in male and female rats. *BMC Neurology*, *15*, 34.

Kren, N. P., Zagon, I. S., & McLaughlin, P. J. (2015). Mutations in the opioid growth factor receptor in human cancers alter receptor function. *International Journal of Molecular Medicine*, *36*, 289–293.

Kromer, W. (1988). Endogenous and exogenous opioids in the control of gastrointestinal motility and secretion. *Pharmacological Reviews*, *40*, 121–162.

Kubota, K., Kinomura, Y., Yamamoto, T., Ozeki, M., Kawamoto, M., Kawamoto, N., et al. (2016). ACTH therapy for West syndrome with severe hemophilia A. *Epilepsy & Behavior Case Reports*, *6*, 1–2.

Larhammar, D., Sundström, G., & Dores, R. M. (2013). Evolution of the opioid system. In A. J. Kastin (Ed.), *Handbook of biologically active peptides* (pp. 1562–1569). Amsterdam: Elsevier.

Lee, J. Y., Liska, M. G., Crowley, M., Xu, K., Acosta, S. A., Borlongan, C. V., et al. (2018). Multifaceted effects of delta opioid receptors and DADLE in diseases of the nervous system. *Current Drug Discovery Technologies*, *15*, 94–108.

Li, C. H., Geschwind, H., Dixon, J. S., Levy, A. L., & Harris, J. E. (1955). Corticotropins (ACTH). I. Isolation of alpha-corticotropin from sheep pituitary glands. *Journal of Biological Chemistry*, *213*, 171–185.

Li, W., Papilloud, A., Lozano-Montes, L., Zhao, N., Ye, X., Zhang, X., et al. (2018). Stress impacts the regulation neuropeptides in the rat hippocampus and prefrontal cortex. *Proteomics*, *18*, e1700408.

Li, M., Tjen-Looi, S. C., Guo, Z. L., & Longhurst, J. C. (2016). Repetitive electroacupuncture attenuates cold-induced hypertension through enkephalin in the rostral ventral lateral medulla. *Scientific Reports*, *6*, 35791.

Lindner, J. M., Suhr, A. C., Grimm, S. H., Möhnle, P., Vogeser, M., & Briegel, J. (2017). The dynamics of a serum steroid profile after stimulation with intravenous ACTH. *Journal of Pharmaceutical and Biomedical Analysis*, *151*, 159–163.

Lisak, R. P., Nedelkoska, L., Bealmear, B., & Benjamins, J. A. (2015). Melanocortin receptor agonist ACTH 1-39 protects rat forebrain neurons from apoptotic, excitotoxic and inflammation-related damage. *Experimental Neurology*, *273*, 161–167.

Lisak, R. P., Nedelkoska, L., & Benjamins, J. A. (2016). The melanocortin ACTH 1-39 promotes protection of oligodendrocytes by astroglia. *Journal of the Neurological Sciences*, *362*, 21–26.

Lord, J. A. H., Waterfield, A. A., Hughes, J., & Kosterlitz, H. W. (1977). Endogenous opioid peptides: Multiple agonists and receptors. *Nature*, *267*, 495–499.

Lorusso, P., Bottai, A., Mangione, E., Innocenti, M., Cupisti, A., & Egigi, M. F. (2015). Low-dose synthetic adrenocorticotropic hormone-analog therapy for nephrotic patients: Results from a single-center pilot study. *International Journal of Nephrology and Renovascular Disease*, *8*, 7–12.

Lu, W. C., Xie, H., Tie, X. X., Wang, R., Wu, A. H., & Shan, F. P. (2018). NFAT-1 hyper-activation by methionine enkephalin (MENK) significantly induces cell apoptosis of rats C6 glioma *in vivo* and *in vitro*. *International Immunopharmacology*, *56*, 1–8.

Ludwig, M. D., Zagon, I. S., & McLaughlin, P. J. (2017a). Serum [Met5]-enkephalin levels are reduced in multiple sclerosis and restored by low-dose naltrexone. *Experimental Biology and Medicine*, *242*, 1524–1533.

Ludwig, M. D., Zagon, I. S., & McLaughlin, P. J. (2017b). Elevated serum [Met5]-enkephalin levels correlate with improved clinical and behavioral outcomes in experimental autoimmune encephalomyelitis. *Brain Research Bulletin*, *134*, 1–9.

Lukiw, W. J. (2006). Endogenous signaling complexity in neuropeptidase-leucine- and methionine-enkephalin. *Cellular and Molecular Neurobiology*, *26*, 1003–1010.

Machado, I. D., Santin, J. R., Drewes, C. C., Gil, C. D., Oliani, S. M., Perretti, M., et al. (2014). Alterations in the profile of blood neutrophil membrane receptor caused by *in vivo* adrenocorticotrophic hormone actions. *American Journal of Physiology. Endocrinology and Metabolism*, *307*, E754–E-763.

Manso, B., Sánchez, M. L., Medina, L. E., Aguilar, L. A., Díaz-Cabiale, Z., Narváez, J. A., et al. (2014). Immunohistochemical mapping of pro-opiomelanocortin- and prodynorphin-derived peptides in the alpaca (*Lama pacos*) diencephalon. *Journal of Chemical Neuroanatomy*, *59*, 36–50.

Margolis, E. B., Fields, H. L., Hjelmstad, G. O., & Mitchell, J. M. (2008). Delta-opioid receptor expression in the ventral tegmental area protects against elevated alcohol consumption. *Journal of Neuroscience*, *28*, 12672–12681.

Martin, W. R., Eades, C. C., Thompson, J. A., Gilbert, P. E., & Huppler, R. E. (1976). The effect of morphine and nalorphine-like drugs in the nondependent and morphine dependent chronic spinal dog. *Journal of Pharmacology and Experimental Therapeutics*, *197*, 517–532.

Martinic, R., Sosic, H., Turcic, P., Konjevoda, P., Fucic, A., Stojkovic, R., et al. (2014). Hepatoprotective effects of met-enkephalin on acetaminophen-induced liver lesions in male CBA mice. *Molecules*, *19*, 11833–11845.

Matters, G. L., Harms, J. F., McGovern, C., Fitzpatrick, L., Parikh, A., Nilo, N., et al. (2008). The opioid antagonist naltrexone improves murine inflammatory bowel disease. *Journal of Immunotoxicology*, *5*, 179–187.

McLaughlin, P. J. (2013). Proenkephalin-derived peptides. In A. J. Kastin (Ed.), *Handbook of biologically active peptides* (pp. 1602–1609). Amsterdam: Elsevier.

McLaughlin, P. J., & Zagon, I. S. (2013). POMC-derived opioid peptides. In A. J. Kastin (Ed.), *Handbook of biologically active peptides* (pp. 1592–1595). Amsterdam: Elsevier.

Medvedeva, E. V., Dmitrieva, V. G., Limborska, S. A., Myasoedov, N. F., & Dergunova, L. V. (2017). Semax, an analog of ACTH (4-7), regulates expression of immune response genes during ischemic brain injury in rats. *Molecular Genetics and Genomics*, *292*, 635–653.

Melander, O., Orho-Melander, M., Manjer, J., Svensson, T., Almgren, P., Nilsson, P. M., et al. (2015). Stable peptide of the endogenous opioid enkephalin precursor and breast cancer risk. *Journal of Clinical Oncology*, *33*, 2632–2638.

Mena, J. D., & Sadeghian Baldo, B. A. (2011). Induction of hyperphagia and carbohydrate intake by mu-opioid receptor stimulation in circumscribed regions of frontal cortex. *Journal of Neuroscience*, *31*, 3249–3260.

Mendez, L. A., Ostlund, S. B., Maidment, N. T., & Murphy, N. P. (2015). Involvement of endogenous enkephalins and β-endorphin in feeding and diet-induced obesity. *Neuropsychopharmacology*, *40*, 2103–2112.

Minnis, J., Patierno, S., Kohlmeier, S. E., Brecha, N., Tonini, M., & Sternini, C. (2003). Ligand-induced μ opioid receptor endocytosis and recycling in enteric neurons. *Neuroscience*, *119*, 33–42.

Mongi-Bragato, B., Zamponi, E., García-Keller, C., Assis, M. A., Virgolini, M. B., Mascó, D. H., et al. (2016). Enkephalin is essential for the molecular and behavioral expression of cocaine sensitization. *Addiction Biology*, *21*, 326–338.

Mountjoy, K. G., Robbins, L. S., Mortrud, M. T., & Cone, R. D. (1992). The cloning of a family of genes that encode the melanocortin receptors. *Science*, *257*, 1248–1251.

Mudge, A. W., Leeman, S. E., & Fischbach, G. D. (1979). Enkephalin inhibits release of substance P from sensory neurons in culture and decreases action potential duration. *Proceedings of the National Academy of Sciences of the United States of America*, 77, 526–530.

Mul, J. D., & Adan, R. A. H. (2013). Melanocortins. In A. J. Kastin (Ed.), *Handbook of biologically active peptides* (pp. 1135–1142). Amsterdam: Elsevier.

Mul, J. D., van Boxtel, R., Bergen, D. J., Brans, M. A., Brakkee, J. H., Toonen, P. W., et al. (2011). Melanocortin receptor 4 deficiency affects body weight regulation, grooming, behavior, and substrate preference in the rat. *Obesity*, *20*, 612–621.

Najafi, P., & Moghadasi, M. (2017). The effect of yoga training on enhancement of adrenocorticotropic hormone (ACTH) and cortisol levels in female patients with multiple sclerosis. *Complementary Therapies in Clinical Practice*, *26*, 21–25.

Nasiri, J., Sarajan, A., Salari, M., & Sedghi, M. (2017). Therapeutic effects of adrenocorticotropic hormone ACTH in children with severely intractable seizure. *Iran Journal of Child Neurology*, *11*, 19–26.

Nøhr, A. C., Shehata, M. A., Hauser, A. S., Isberg, V., Mokrosinski, J., Andersen, K. B., et al. (2017). The orphan G protein-coupled receptor GPR139 is activated by the peptides: Adrenocorticotropic hormone (ACTH), α-, and β-melanocyte stimulating hormone (α-MSH, and β-MSH), and the conserved core motif HFRW. *Neurochemistry International*, *102*, 105–113.

Palkovits, M. (1988). Neuropeptides in the brain. In L. Martini & F. Ganong (Eds.), *Frontiers in neuroendocrinology* (pp. 1–44). New York: Raven Press.

Pállinger, É., Kiss, G. A., & Csaba, G. (2016). Hormone (ACTH, T3) content of immunophenotyped lymphocyte subpopulations. *Acta Microbiologica et Immunologica Hungarica*, *63*, 373–385.

Panchal, S. J., Muller-Schwefe, P., & Wurzelmann, J. I. (2007). Opioid-induced bowel dysfunction: Prevalence, pathophysiology and burden. *International Journal of Clinical Practice*, *61*, 1181–1187.

Park, S. Y., Kwak, M. K., Kim, H. J., Park, H. K., Suh, K. I., Yoo, M. H., et al. (2017). Case report of a bilateral adrenal myelolipoma associated with Cushing disease. *Medicine (Baltimore)*, *96*, e9455.

Patierno, S., Anselmi, L., Jaramillo, I., Scott, D., García, R., & Sternini, C. (2011). Morphine induces mu opioid receptor endocytosis in guinea pig enteric neurons following prolonged receptor activation. *Gastroenterology*, *140*, 18–26.

Patierno, S., Raybould, H. E., & Sternini, C. (2004). Abdominal surgery induces μ opioid receptor endocytosis in enteric neurons of the guinea pig ileum. *Neuroscience*, *123*, 101–109.

Penhoat, A., Jaillard, C., & Sáez, J. M. (1989). Corticotropin positively regulates its own receptors and CAMP response in cultured bovine adrenal cells. *Proceedings of the National Academy of Sciences of the United States of America*, *86*, 4978–4981.

Pert, C. B., & Snyder, S. H. (1973). Opiate receptor demonstration in nervous tissue. *Science*, *179*, 1011–1014.

Pesini, P., Pego-Reigosa, R., Tramu, G., & Coveñas, R. (2004). Distribution of ACTH immunoreactivity in the diencephalon and the brainstem of the dog. *Journal of Chemical Neuroanatomy*, *27*, 275–282.

Philippe, D., Chakass, D., Thuru, X., Zerbid, P., Tsicopoulos, A., Geboes, A., et al. (2006). Mu opioid receptor expression in increased inflammatory bowel diseases: Implications for homeostatic intestinal inflammation. *Gut*, *55*, 815–823.

Philippe, D., Dubuquoy, L., Groux, H., Brun, V., Chuoi-Mariot, M. T., Gaveriaux-Ruff,-C., et al. (2003). Anti-inflammatory properties of the mu opioid receptor support its use in the treatment of colon inflammation. *Journal of Clinical Investigation*, *111*, 1329–1338.

Pioro, E. P., Mai, J. K., & Claudio Cuello, A. (1990). Distribution of substance P- and enkephalin-immunoreactive neurons and fibers. In G. Paxinos (Ed.), *The human nervous system* (pp. 1051–1094). San Diego: Academic Press.

Poole, D. P., Pelayo, J. C., Scherrer, G., Evans, C. J., Kieffer, B. L., & Bunnett, N. W. (2011). Localization and regulation of fluorescently labeled delta opioid receptor, expressed in enteric neurons of mice. *Gastroenterology*, *141*, 982–991.

Poonyachoti, S., Kulkarni-Narla, A., & Brown, D. R. (2002). Chemical coding of neurons expressing σ- and κ-opioid receptor and type I vanilloid receptor immunoreactivities in the porcine ileum. *Cell and Tissue Research*, *307*, 23–33.

Prasanth, K., Kamat, M., Khilfeh, M., & Davis, V. (2014). Adrenocorticotropic hormone and cortisol levels in term infants born with meconium-stained amniotic fluid. *Journal of Perinatal Medicine*, *42*, 699–703.

Rahn, K. A., McLaughlin, P. J., & Zagon, I. S. (2011). Prevention and diminished expression of experimental autoimmune encephalomyelitis by low dose naltrexone (LDN) or opioid growth factor (OGF) for an extended period: Therapeutic implications for multiple sclerosis. *Brain Research*, *1381*, 243–253.

Reece, A. S., & Hulse, G. K. (2016). Elevation of the ACTH/cortisol ratio in female opioid dependent patients: A biomarker of aging and correlate of metabolic and immune activation. *Neuro Endocrinology Letters*, *37*, 325–336.

Renquist, B. J., Lippert, R. N., Sebag, J. A., Ellacott, K. L., & Cone, R. D. (2011). Physiological roles of the melanocortin MC3 receptor. *European Journal of Pharmacology*, *660*, 13–20.

Rodríguez, D., & Coveñas, R. (2011). Targeting opioid and neurokinin-1 receptors to treat alcoholism. *Current Medicinal Chemistry*, *18*, 4321–4334.

Roelfsema, F., Yang, R. J., Olson, T. P., Joyner, M. J., Takahashi, P. Y., & Veldhuis, J. D. (2017). Enhanced coupling within gonadotropic and adrenocorticotropic axes by moderate exercise in healthy men. *Journal of Clinical Endocrinology and Metabolism*, *102*, 2482–2490.

Rose, A. K., Shaw, S. G., Prendergast, M. A., & Little, H. J. (2010). The importance of glucocorticoids in alcohol dependence and neurotoxicity. *Alcoholism, Clinical and Experimental Research*, *34*, 2011–2018.

Roy, S., Roy, S. J., Pinard, S., Taillefer, L. D., Rached, M., Parent, J. L., et al. (2011). Mechanisms of melanocortin-2 receptor (MC2R) internalization and recycling in human embryonic kidney (hek) cells: Identification of key Ser/Thr (S/T) amino acids. *Molecular Endocrinology*, *25*, 1961–1977.

Ruggiero, C., & Lalli, E. (2016). Impact of ACTH signaling on transcriptional regulation of steroidogenic genes. *Frontiers in Endocrinology*, *7*, 24.

Saccani, F., Anselmi, L., Jaramillo, I., Bertoni, S., Barocelli, E., & Sternini, C. (2012). Protective role of μ opioid receptor activation in intestinal inflammation induced by mesenteric ischemia/reperfusion in mice. *Journal of Neuroscience Research*, *90*, 2146–2153.

Safarinejad, M. R. (2008). Evaluation of the safety and efficacy of bremelanotide, a melanocortin receptor agonist, in female subjects with arousal disorder: A double-blind placebo-controlled, fixed dose, randomized study. *The Journal of Sexual Medicine*, *5*, 887–897.

Sarkisyan, D., Hussain, M. Z., Watanabe, H., Kononenko, O., Bazov, I., Zhou, X., et al. (2015). Downregulation of the endogenous opioid peptides in the dorsal striatum of human alcoholics. *Frontiers in Cellular Neuroscience*, *9*, 187.

Scantlebury, M. H., Chun, K. C., Ma, S. C., Rho, J. M., & Kim, D. Y. (2017). Adrenocorticotropic hormone protects learning and memory function in epileptic Kcna1-null mice. *Neuroscience Letters*, *645*, 14–18.

Sgroi, S., Capper-Loup, C., Paganetti, P., & Kaelin-Lang, A. (2016). Enkephalin and dynorphin neuropeptides are differently correlated with locomotor hypersensitivity and levodopa-induced dyskinesia in parkinsonian rats. *Experimental Neurology*, *280*, 80–88.

Shah, A. J., Kriska, T., Gauthier, K. M., Falck, J. R., & Campbell, W. B. (2018). Effect of angiotensin II and ACTH on adrenal blood flow in the male rat adrenal gland *in vivo*. *Endocrinology*, *159*, 217–226.

Shimakawa, S., Nomura, S., Ogino, M., Fukui, M., Kashiwagi, M., Tanabe, T., et al. (2015). ACTH therapy on intractable epilepsy in hemiconvulsion-hemiplegia-epilepsy syndrome. *Brain & Development*, *37*, 733–737.

Simankova, A. A., Sazonova, E. N., & Lebed'ko, O. A. (2017). Delayed effects of neonatal administration of non-opioid analog of leu-enkephalin on cerebral consequences of antenal hypoxia. *Bulletin of Experimental Biology and Medicine*, *163*, 594–598.

Simon, E. J. (1991). Opioid receptors and endogenous opioid peptides. *Medical Research Reviews*, *11*, 357–374.

Simon, E. J., Hiller, J. M., & Edelman, I. (1973). Stereospecific binding of the potent narcotic analgesic [3H]-etorphine to rat-brain homogenate. *Proceedings of the National Academy of Sciences of the United States of America*, *70*, 1947–1949.

Siong Chan, D. C., Cao, T. H., & Ng, L. L. (2018). Proenkephalin in heart failure. *Heart Failure Clinics*, *14*, 1–11.

Smith, J. P., Stock, H., Bingaman, S., Mauger, D., Rogosnitzky, M., & Zagon, I. S. (2007). Low-dose naltrexone therapy improves active Crohn's disease. *American Journal of Gastroenterology*, *102*, 820–828.

Snyder, S. H. (2004). Opiate receptors and beyond: 30 years of neural signaling research. *Neuropharmacology*, *47*, 274–285.

Spampinato, S., Bedini, A., & Baiula, M. (2013). Prodynorphin-derived peptides. In A. J. Kastin (Ed.), *Handbook of biologically active peptides* (pp. 1596–1601). Amsterdam: Elsevier.
Speliotes, E. K., Willer, C. J., Berndt, S. I., Monda, K. L., Thorleifsson, G., Jackson, A. A., et al. (2010). Association analyses of 249,796 individuals reveal 18 new loci associated with body mass index. *Nature Genetics*, *42*, 937–948.
Steele, P. A., & Costa, M. (1990). Opioid-like immunoreactive neurons in secretomotor pathways of the guinea-pig ileum. *Neuroscience*, *38*, 771–786.
Sternini, C., Duraffourd, C., & Anselmi, L. (2013). Opioids. In A. J. Kastin (Ed.), *Handbook of biologically active peptides* (pp. 1283–1288). Amsterdam: Elsevier.
Sternini, C., Patierno, S., Selmer, I. S., & Kirchgessner, A. (2004). The opioid system in the gastrointestinal tract. *Neurogastroenterology and Motility*, *16*(Suppl. 2), 3–16.
Sugiyama, A., Kageyama, K., Murasawa, S., Ishigame, N., Niioka, K., & Daimon, M. (2015). Inhibition of heat shock protein 90 decreases ACTH production and cell proliferation in AtT-20 cells. *Pituitary*, *18*, 542–553.
Tabbì, G., Magrì, A., Giuffrida, A., Lanza, V., Pappalardo, G., Naletova, I., et al. (2015). Semax, an ACTH4-10 peptide analog with high affinity for copper (II) ion and protective ability against metal induced cell toxicity. *Journal of Inorganic Biochemistry*, *142*, 39–46.
Taha, S. A. (2010). Preference or fat? Revisiting opioid effects on food intake. *Physiology & Behavior*, *100*, 429–437.
Talaber, G., Tuckermann, J. P., & Okret, S. (2015). ACTH controls thymocytes homeostasis independent of glucocorticoids. *The FASEB Journal*, *29*, 2526–2534.
Tao, Y. X. (2010). The melanocortin-4 receptor: Physiology, pharmacology and pathophysiology. *Endocrine Reviews*, *31*, 506–543.
Tatro, J. B. (2000). Melanocortins receptor expression and function in the nervous system. In R. D. Cone (Ed.), *The melanocortin receptors* (pp. 173–207). Totowa: Humana Press Inc.
Tehranchinia, Z., Rahimi, H., & Lotfi, S. (2017). Basal serum cortisol and adrenocorticotropic hormone levels in patients with atopic dermatitis. *Dermatology Practical & Conceptual*, *7*, 25–29.
Terenius, L. (1973). Characteristics of the "receptor" for narcotic analgesics in synaptic plasma membrane from rat brain. *Acta Pharmacologica et Toxicologica*, *33*, 377–384.
Thakur, N. A., DeBoyace, S. D., & Margulies, B. S. (2016). Antagonism of the Met5-enkephalin-opioid growth factor receptor-signaling axis promotes MSC to differentiate into osteoblasts. *Journal of Orthopaedic Research*, *34*, 1195–1205.
Tian, J., Jiao, X., Wang, X., Geng, J., Wang, R., Liu, N., et al. (2018). Novel effect of methionine enkephalin against influenza A virus infection through inhibiting TLR7-MyD88-TRAF6-NF-κB p65 signaling pathway. *International Immunopharmacology*, *55*, 38–48.
Tourkova, I. L., Liu, L., Sutjarit, N., Larrouture, Q. C., Luo, J., Robinson, L. J., et al. (2017). Adrenocorticotropic hormone and 1, 25-dihydroxyvitamin D3 enhance human osteogenesis *in vitro* by synergistically accelerating the expression of bone-specific genes. *Laboratory Investigation*, *97*, 1072–1083.
Tsutsui, R., Shinomiya, K., Sendo, T., Kitamura, Y., & Kamei, C. (2015). Effects of the 5-HT(1A) receptor agonist tandospirone on ACTH-induced sleep disturbance in rats. *Biological and Pharmaceutical Bulletin*, *38*, 884–888.
Valente, V. B., Verza, F. A., Lopes, F. Y. K., Ferreira, J. Z., Dos Santos, P. S. P., Sundefeld, M. L. M. M., et al. (2018). Stress hormones concentrations in the normal microenvironment predict risk for chemically induced cancer in rats. *Psychoneuroendocrinology*, *89*, 229–238.
van den Beukel, J. C., Grefhorst, A., Quarta, C., Steenbergen, J., Mastroberardino, P. G., Lombès, M., et al. (2014). Direct activating effects of adrenocorticotropic hormone (ACTH) on brown adipose tissue are attenuated by corticosterone. *The FASEB Journal*, *28*, 4857–4867.

van der Ploeg, L. H., Martin, W. J., Howard, A. D., Nargund, R. P., Austin, C. P., Guan, X., et al. (2002). A role for the melanocortin 4 receptor in sexual function. *Proceedings of the National Academy of Sciences of the United States of America*, *99*, 11381–11386.

Vaudry, H., & Eberle, A. N. (1993). The melanotropic peptides. *Annals of the New York Academy of Sciences*, *680*, 1–687.

Vucetic, Z., Kimmel, J., & Reyes, T. M. (2011). Chronic high-fat diet drives postnatal epigenetic regulation of mu-opioid receptor in the brain. *Neuropsychopharmacology*, *36*, 1199–1206.

Wang, Q., Gao, X., Yuan, Z., Meng, Y., Cao, Y., Plotnikoff, N. P., et al. (2014). Methionine enkephalin (MENK) improves lymphocyte subpopulations in human peripheral blood of 50 cancer patients by inhibiting regulatory T cells (Tregs). *Human Vaccines & Immunotherapeutics*, *10*, 1836–1840.

Wang, D. M., Jiao, X., Plotnikoff, N. P., Griffin, N., Qi, R. Q., Gao, X. H., et al. (2017). Killing effect of methionine enkephalin on melanoma *in vivo* and *in vitro*. *Oncology Reports*, *38*, 2132–2140.

Wang, L. I., Liu, F., Luo, Y., Zhu, L., & Li, G. (2015). Effect of acute heat stress on adrenocorticotropic hormone, cortisol, interleukin-2, interleukin-12 and apoptosis gene expression in rats. *Biomedical Reports*, *3*, 425–429.

Wang, W. B., She, F., Xie, L. F., Yan, W. H., Ouyang, J. Z., Wang, B. A., et al. (2016). Evaluation of basal serum adrenocorticotropic hormone and cortisol levels and their relationship with nonalcoholic fatty liver disease in male patients with idiopathic hypogonadotropic hypogonadism. *Chinese Medical Journal*, *129*, 1147–1153.

Wessels, H., Gralnek, D., Dorr, R., Hruby, V. J., Hadley, M. E., & Levin, N. (2000). Effect on an alpha-melanocyte stimulating hormone analog on penile erection and sexual desire in men with organic erectile dysfunction. *Urology*, *56*, 641–646.

Xu, X., Gao, Y., Wen, L., Zhai, Z., Zhang, S., Shan, F., et al. (2016). Methionine enkephalin regulates microglia polarization and function. *International Immunopharmacology*, *40*, 90–97.

Yang, Y. (2011). Structure, function and regulation of the melanocortin receptors. *European Journal of Pharmacology*, *660*, 125–130.

Yang, D. J., Lee, K. S., Ko, C. M., Moh, S. H., Song, J., Hur, L. C., et al. (2016). Leucine-enkephalin promotes wound repair through the regulation of hemidesmosome dynamics and matrix metalloprotease. *Peptides*, *76*, 57–64.

Yang, Y., Mishra, V., Crasto, C. J., Chen, M., Dimmitt, R., & Harmon, C. M. (2015). Third transmembrane domain of the adrenocorticotropic receptor is critical for ligand selectivity and potency. *Journal of Biological Chemistry*, *290*, 7685–7692.

Yaswen, L., Diehl, N., Brennan, M. B., & Hochgeschwender, U. (1999). Obesity in the mouse model of pro-opiomelanocortin deficiency responds to peripheral melanocortin. *Nature Medicine*, *5*, 1066–1070. See comment in PubMed Commons below.

Yin, J., Lu, Q., Yin, F., Wang, Y., He, F., Wu, L., et al. (2017). Effectiveness and safety of different once-daily doses of adrenocorticotropic hormone for infantile spasms. *Paediatric Drugs*, *19*, 357–365.

Yokoyama, K., Maruwaka, M., Yoshikawa, K., Araki, Y., Okamoto, S., Sumitomo, M., et al. (2018). Elevation of proenkephalin 143-183 in cerebrospinal fluid in Moyamoya disease. *World Neurosurgery*, *109*, e446–e459.

Zadina, J. E., Hackler, L., Ge, L.-J., & Kastin, A. (1997). A potent and selective endogenous agonist for the μ-opiate receptor. *Nature*, *386*, 499–502.

Zagon, I. S., Donahue, R. N., Bonneau, R. H., & MacLaughlin, P. J. (2011). T lymphocyte proliferation is suppressed by the opioid growth factor ([Met^5]-enkephalin)-opioid growth factor receptor axis: Implication for the treatment of autoimmune diseases. *Immunobiology*, *216*, 579–590.

Zhang, M., & Kelley, A. E. (2000). Enhanced intake of high-fat food following striatal mu-opioid stimulation: Microinjection mapping and Fos expression. *Neuroscience*, *99*, 267–277.

Zhang, X., Saarinen, A. M., Campbell, L. E., de Filippis, E. A., & Liu, J. (2018). Regulation of lipolytic response and energy balance by melanocortin 2 receptor accessory protein (MRAP) in adipocytes. *Diabetes*, *67*, 222–234.

Zhang, Y., Zou, L. P., Ding, Y. X., He, B., & Yang, G. (2014). Association of SCN1A gene polymorphisms with infantile spasms and adrenocorticotropic hormone responsiveness. *European Review for Medical and Pharmacological Sciences*, *18*, 2500–2506.

Zhao, C., Du, H., Xu, L., Wang, J., Tang, L., Cao, Y., et al. (2015). Metabolomic analysis revealed glycylglycine accumulation in astrocytes after methionine enkephalin administration exhibiting neuron protective effects. *Journal of Pharmaceutical and Biomedical Analysis*, *115*, 48–54.

Zhou, Y., Zhang, J., Lei, B., Liang, W., Gong, J., Zhao, C., et al. (2017). DADLE improves hepatic ischemia/reperfusion injury in mice via activation of the Nrf2/HO-1 pathway. *Molecular Medicine Reports*, *16*, 6214–6221.

CHAPTER EIGHT

Opioids and the hormone oxytocin

Leda Kovatsi[a,*], Kakia Nikolaou[b]
[a]Laboratory of Forensic Medicine and Toxicology, School of Medicine, Aristotle University of Thessaloniki, Thessaloniki, Greece
[b]Addiction Department IANOS, General Hospital of Thessaloniki 'G. Papanikolaou'—Psychiatric Hospital of Thessaloniki, Thessaloniki, Greece
*Corresponding author: e-mail address: kovatsi@hotmail.com

Contents

Abstract

The neuropeptide Oxytocin (OT) is involved as a neurohormone, a neurotransmitter, or a neuromodulator in an extensive range of central and peripheral effects, complex emotional and social human behaviors, memory and learning processes. It is implicated in homeostatic, neuroadaptive processes associated with stress responses and substance use via interactions with the hypothalamic–pituitary–adrenal (HPA) axis and the dopamine mesolimbic reward stress system. This chapter reviews the preclinical and clinical literature on the complicated relationships between endogenous and exogenous opioids and OT systems and attempts to highlight key findings to date on the effectiveness of intranasal OT administration to treat opioid use disorders. OT seems to attenuate, even inhibit, the development of opioid use disorders in preclinical models but is still under clinical research as a promising pharmacological agent in the treatment of opioid use related behaviors. Evidence suggests a role for OT as an adjunctive or stand-alone treatment of behavioral, cognitive and emotional deficits associated with substance use,

Vitamins and Hormones, Volume 111
ISSN 0083-6729
https://doi.org/10.1016/bs.vh.2019.05.003

which may be responsible for seeking behavior and relapse. The mechanisms by which oxytocin acts to reverse the neural substrates of these deficits, partially due to substance induced alterations of the endogenous OT system, and thus modify the behavioral response to substance use are discussed. Other clinically relevant issues are also discussed.

1. Introduction

Oxytocin (OT) is a fascinating neuropeptide, which has captured the attention of the scientific community ever since Sir Henry Hallett Dale described the action of neurohypophysis hormones on the human uterus (Dale, 1906) and since Blair-Bell (1909) introduced its clinical application in parturition and lactation. OT was the first nonapeptide hormone to be sequenced and synthesized by du Vigneaud (1953). OT is synthesized in the magnocellular cells of the neurosecretory neurons of the supraoptic (SON) and paraventricular (PVN) nuclei and in the parvocellular neurons of the PVN of the hypothalamus. It is secreted by the axonal terminals of the neurohypophysis into the bloodstream. In addition there is a somatodendritic release from magnocellular neurons and a direct release from parvocellular neurons, projecting to several brain structures and the spinal cord. OT is also produced in peripheral tissues (Gimpl & Fahrenholz, 2001).

Besides its classical actions OT seems to exert an anxiolytic effect (Heinrichs & Domes, 2008). It interferes with the hypothalamus–pituitary–adrenal axis (HPA) in a very complex manner regulating the hormonal responsiveness of the axis to stress. In rodents various emotional and physical stressors cause the release of OT within the hypothalamic SON, (threefold), and PVN nuclei, (fivefold with an increase tendency upon repeated stress exposure) (Wotjak et al., 1998), accompanied by a secretion into the systemic circulation (Ondrejcakova et al., 2010; Yayou, Ito, Yamamoto, Kitagawa, & Okamura, 2010). Studies in humans agree with the preclinical findings (Uvnäs-Moberg & Petersson, 2005). Stress induced cortisol release was shown to have a central effect on OT release causing an increase in salivary (Jong et al., 2015) and plasma OT levels and attenuation of anxiety (Taylor et al., 2006; Tops, van Peer, Korf, Wijers, & Tucker, 2007), but adrenocorticotropic hormone (ACTH) had no effect on OT levels (Tops et al., 2007). OT administration on the other hand seems to be able to stimulate as well as to inhibit the activity in the HPA-axis within a short- and long-term perspective, respectively: acute subcutaneous (SC) OT administration stimulates plasma levels of ACTH and corticosterone (Petersson, Hulting, & Uvnäs-Moberg, 1999), whereas repeated

OT administration, intracerebroventricularly (ICV) or subcutaneously (SC), inhibits activation of the HPA-axis, of Corticotropin Releasing Factor (CRF) and of corticosterone release (Dabrowska et al., 2011; Windle et al., 2004), decreasing anxiety levels (Slattery & Neumann, 2010) and causing sedation in high doses in rats (Uvnäs-Moberg, Ahlenius, Hillegaart, & Alster, 1994). Similar effects were observed after OT administration in humans as well: decrease of corticosol release and anxiety relief (Ditzen et al., 2009; Heinrichs, Baumgartner, Kirschbaum, & Ehlert, 2003; Kubzansky, Mendes, Appleton, Block, & Adler, 2012; Quirin, Kuhl, & Dusing, 2011) (also see Sections 2.2.2.2, 3.1.2, and 3.2.5).

This anxiolytic effect is possibly responsible for facilitating the development of complex emotional and social behavior in many species, necessary for a gradual formation of relationships: social recognition, social approach behaviour, trust, attachment, affiliation, bonding behavior (Baumgartner, Heinrichs, Vonlanthen, Fischbacher, & Fehr, 2008; MacDonald & MacDonald, 2010; Ross & Young, 2009).

There seem to be multiple and complex interactions and overlaps, between OT and the development of substance use disorders (SUD), supporting the OT involvement in a wide range of SUD, at various stages of the addiction cycle, for numerous substances (Baskerville & Douglas, 2010; Burkett & Young, 2012; Kovács, Sarnyai, & Szabó, 1998; Lee, Rohn, Tanda, & Leggio, 2016; Lee & Weerts, 2016; McGregor & Bowen, 2012; McGregor, Callaghan, & Hunt, 2008; Sarnyai, 2011; Sarnyai & Kovács, 1994), eventhough this role of OT is still largely unexplained and may vary according to the individual psychological and environmental context (Baskerville & Douglas, 2010). This chapter attempts a systematic investigation of the available evidence concerning the particularly complicated interactions between opioids and OT, focusing on the potential role OT may play in the treatment of opioid use disorders (OUD). Further emphasis is placed on the progressive emergence of a number of clinical, unresolved issues, concerning the OT administration, which create controversies and methodological problems.

2. Effects of endogenous and exogenously administered opioids on brain OT systems

A great amount of preclinical research shows that brain OT systems undergo major neuroadaptations following acute and long-term substance exposure due to their neuroplasticity. The endogenous OT system is downregulated as a result of, mainly, long-term abuse of most substances. Acute exposure to stimulants may initially lead to increase in OT brain levels (especially 3,4-methylenedioxy-methamphetamine, MDMA, and gamma-hydroxybutyrate, GHB, may preferentially activate brain OT systems to produce their characteristic prosocial and prosexual effects) (Kamilar-Britt & Bedi, 2015), but long-term use also results in "lesions" in the endogenous oxytocin system (McGregor et al., 2008).

Such neuroadaptations may contribute or be responsible for enduring deficits in OT-dependent behaviors, such as sociability, trust and anxiety, reflecting the dysfunctional behaviors and effects of psychoactive substances (negative mood, low stress threshold, aggression, irritation) and leading to craving, drug seeking behavior, loss of interest in social rewards and relapse (McGregor et al., 2008). Lasting alterations have been observed during acute and chronic substance administration to rodents according to dose and brain region: irreversible loss of OT neurons from alcohol (Silva, Madeira, Ruela, & Paula-Barbosa, 2002), OT-receptor upregulation from alcohol (Hansson et al., 2018) and psychostimulants (Georgiou et al., 2015; Zanos et al., 2014), as well as changes in plasma and brain OT levels, OT production and release from stimulants (Baracz et al., 2016; Carson, Guastella, Taylor, & McGregor, 2013). Alterations in plasma OT levels have also been recorded in humans from stimulants (Dumont et al., 2009; Kirkpatrick, Francis, Lee, de Wit, & Jacob, 2014; Light et al., 2004; Wolff et al., 2006) and alcohol use (Marchesi, Chiodera, Brusamonti, Volpi, & Coiro, 1997). A series of post mortem studies in human brains of alcoholics offer a further validation of the results in rodents (Hansson et al., 2018; Lee et al., 2017; Sivukhina, Dolzhikov, Morozov, Jirikowski, & Grinevich, 2006).

2.1 Effects of endogenous opioids on brain OT systems

OT neurons in the hypothalamus synthesize and process opioid peptides (Bicknell, 1985). The terminals of the nucleus arcuatus and of the nucleus tractus solitarius, which contain opioid peptides, result in the neurohypophysis (Bicknell & Leng, 1982). The SON of rats contains both κ-and μ- but, according to most authors, no δ-opioid receptors (Sumner, Coombes, Pumford, & Russell, 1990; Zhao, Chapman, & Bicknell, 1988). About half of SON OT neurons seem to possess μ-opioid receptors (Li, You, Chen, Song, & Lu, 2001). The neurohypophysis possesses κ-opioid receptors (Watson, Akil, Chazarossian, et al., 1981).

Endogenous opioid peptide systems from the neurohypophysis regulate OT secretion restraining the basal secretion of OT at several anatomical and functional levels (Brown, Russell, & Leng, 2000):

- Presynaptically by inhibiting noradrenergic afferent inputs from the Cholecystokinin-Noradrenaline-Oxytocin Circuit to the supraoptic and paraventricular nuclei and by inhibiting noradrenaline release at the cell bodies of afferent input cells (Zhao, Chapman, & Bicknell, 1988; Zhao, Chapman, Brown, & Bicknell, 1988), to a much lesser extent than the other ways.
- Directly at OT magnocellular cell bodies in the hypothalamus (Russell, Leng, & Bicknell, 1995)
- At neurosecretory axonal terminals in the neurohypophysis (Bicknell, 1985).

These effects are mediated primarily via κ-receptors. κ-receptor agonists, dynorphins, were found to mediate opioid inhibition of OT secretion in SON neurons, whereas the μ-receptor agonist proenkephalin (also released by OT terminals) was found to cause only a minor inhibition of OT secretion on OT cell bodies (Inenaga, Imura, Yanaihara, & Yamashita, 1990; Russell et al., 1995). δ-opioid receptors were found to exert only some inhibition on OT neurons (Inenaga et al., 1990).

Several OT actions are mediated through endogenous opioids:

- *Positive OT effect on the HPA-axis: it caused* increase of plasma corticosterone levels in male rats. Naloxone alone produced the same effect, but less strongly as OT (Montilla, Ríos, & Muñoz, 1989).
- *Stress induced OT secretion,* mainly mediated by μ-opioid receptors and secondary by κ-receptors (Russell et al., 1995). This endogenous opioid regulation of the stress induced OT secretion is partly determined by circulating gonadal hormones (Carter, Williams, & Lightman, 1986).
- *The OT antinociceptive effect* is mediated via both opioid and cannabinoid systems (Russo et al., 2012) and is reversed by the administration of opioid antagonists (Miranda-Cardenas et al., 2006; Yang et al., 2011). Leucine-enkephalin, methionine-enkephalin and β-endorphine, but not Dynorphin A (1 − 13), have been identified as the opioid peptides involved in OT mediated pain modulation in the periaquaductal gray area. Exogenous OT administration increased b-endorphin and L-encephalin contents in the spinal cord (Yang et al., 2011).
- *Regulation of the OT involvement in in the onset of labour and parturition by modifying the inhibitory control on OT secretion in late pregnancy.* During pregnancy OT neurons are inhibited by κ-receptor opioids

(Leng, Dyes, & Bicknell, 1997). In mid-pregnancy, the effectiveness of opioid mechanisms in the neurohypophysis increases, leading to OT accumulation in view of parturition (Russell et al., 1995). By the end of pregnancy a separate endogenous opioid system, acting via μ-receptors, with κ-receptors having no effect (Kutlu et al., 2004), restrains the electrical activity of oxytocin neurones and retains the ability to control OT neurones during parturition, and can interrupt established parturition by inhibiting OT neuron firing rate in negative environmental conditions (Russell, Leng, & Douglas, 2003). The μ-opioid receptor antagonist naloxone caused an increased OT secretion (Brown, Brunton, & Russell, 2008).

- *Regulation of lactation through inhibition of OT release.* κ-receptor opioids are responsible for restraining oxytocin secretion lactation in rats (Russell et al., 2003).

2.2 Effects of exogenous opioids on brain OT systems

2.2.1 Effects of acute opioid administration on brain OT systems

Acute administration of exogenous opioids also inhibits OT secretion in animals (Vuong, Van Uum, O'Dell, Lutfy, & Frieman, 2010). The sites of action are, similarly to the ones of the endogenous opioid system (also see Section 2.1), at the level of the hypothalamus (suppression of the firing rate of magnocellular oxytocinergic cells by action on their perikarya) and at the level of the axon terminals at the neurohypophysis (Vuong et al., 2010). The hypothalamus inhibition involves the supraoptical (and not the paraventricular) oxytocin neurons, the μ-antagonist naloxone was effective only when given into the SON (Ludwig, Brown, Russell, & Leng, 1997), via G(i/o)-protein-linked μ-opioid receptors (Brown, Johnstone, Murphy, Leng, & Russell, 2000) and is mediated presynaptically by inhibition of afferent noradrenergic endings and noradrenaline release in the SON with the activity of noradrenergic cells in the nucleus tractus solitarius remaining unaffected (Leng, Brown, Murphy, Onaka, & Russell, 1995).

The mechanism explaining the inhibition of OT neurons after acute morphine administration may be a small but consistent dose-dependent membrane hyperpolarization in all SON neurons. The unexpectedly recorded, enhanced after-depolarisation amplitude was mainly without effect and there was no alteration in the spike broadening (Ruan, Russell, & Brown, 2011) (also see Section 2.2.2.2).

2.2.2 *Effects of chronic opioid administration on brain OT systems*

Brain OT systems are altererd by chronic opioid exposure. OT content, mRNA expression and OT synthesis are decreased after chronic morphine administration in SON, median eminence, hippocampus and nucleus accumbens (Li et al., 2001; You, Li, Song, Wang, & Lu, 2000) and increased in the ventral tegmental area and locus coeruleus in rats (Kovács et al., 1987). OT receptor binding was also increased in the amygdala, piriform cortex, medial septum, and anterior olfactory nucleus and decreased in the HT (Zanos et al., 2014).

2.2.2.1 Development of tolerance and dependence of OT neurons under chronic opioid exposure

OT neurons in the brain develop tolerance and dependence under chronic opioid exposure conditions (Ludwig et al., 1997). Tolerance occurs both at the cell bodies of OT neurons in the SON and in their brainstem noradrenergic afferent input (Brown, Munro, Murphy, Leng, & Russell, 1996), whereas dependence is apparent predominantly at the magnocellular OT neurons (Brown et al., 1996; Ludwig et al., 1997). It seems that OT neurons projecting to the mediolateral septum selectively develop morphine dependence (Russell, Neumann, & Landgraf, 1992). Tolerance involves reduction in μ-opioid receptor density (Russell et al., 1995) and a decrease in the oxytocin content in the median eminence and in the SON (Laorden, Milanés, Chapleur-Château, & Burlet, 1997), while dependence may involve compensatory upregulation of mechanisms regulating Ca2+ influx (Russell et al., 1995) and requires the presence of functional G(i/o)-proteins (Brown, Johnstone, et al., 2000). During chronic exposure to morphine the cholecystokinin-noradrenergic-induced OT cell activation recovers from initial inhibition and cholecystokinin receptor mechanisms involved in the control of OT neuron activation are upregulated contributing to the development of tolerance and perhaps dependence (Brown et al., 1996; Munro, Pumford, & Russell, 1998).

2.2.2.2 OT neuron withdrawal from opioids

Hypothalamic, SON OT neurons in rats, dependent on morphine following chronic exposure, exhibit, upon withdrawal or antagonism by the μ-opioid antagonist naloxone, a withdrawal hyperexcitation, i.e., increase of their firing rate, which results in a massive and prolonged secretion of OT into the circulation and the brain (Brown et al., 2005; Bull, Russell, Scott, & Brown, 2011) (also see Sections 3.1.2 and 3.2.5). This excitation is described as an increased action potential (spike) firing rate associated with an increased afterdepolarisation (post-spike) excitability, i.e., morphine withdrawal enhances the after-depolarisation amplitude in SON OT neurons, resulting in a rebound depolarisation, a similar effect to the one acute morphine exposure exerts on the OT neurons (Bull et al., 2011; Ruan et al., 2011) (also see Section 2.2.1).

This enhanced afterdepolarisation potential alone is probably not cabable of inducing a rebound depolarisation of SON neurons. A series of other underlying mechanisms may also contribute to the increased, post-spike, withdrawal excitation.

Alterations in the intrinsic membrane properties of the SON OT neurons may be involved (Brown et al., 2005; Bull et al., 2011). Transient outward rectification and after-hyperpolarization were reduced by 20% and 60% respectively, with multiple K+ channel inhibition increasing post-spike excitability for short interspike intervals (Brown et al., 2005). A functional increase of synaptic input, i.e., of the after-depolarisation potential amplitude was observed (Ruan et al., 2011). An increased Ca^{2+} influx was found at higher firing rates from pre and post- synaptic L-type Ca^{2+} channels (administration of Ca^{2+} blockers leads to an increased electrical activity of SON neurons and a subsequent release of OT within the SON during withdrawal) (Blackburn-Munro, Brown, Neumann, Landgraf, & Russell, 2000).

Afferent inputs to the SON neurons may be additionally used to enhance the magnitude of OT neuron withdrawal excitation (Ludwig et al., 1997). Noradrenergic afferent inputs seem not to be essential, but to have only a partial function (Brown et al., 1998; Murphy, Onaka, Brown, & Leng, 1997), chronic destruction of the noradrenergic inputs to the hypothalamus did not affect the magnitude of withdrawal-induced OT secretion (Brown et al., 1998), whereas acute pharmacological blockade of the noradrenergic input was able to suppress withdrawal (Brown et al., 1996; Leng et al., 1995).

Identical basal amplitudes of afterdepolarisation potential of morphine tolerant rats during withdrawal and of morphine naïve rats indicate tolerance development to the effects of morphine on the afterdepolarisation potential (Ruan et al., 2011). These effects are probably mediated by μ-receptors (Brown, Johnstone, et al., 2000; Li et al., 2001) and require the functional integrity of G(i/o)-proteins within the SON (Brown, Johnstone, et al., 2000).

The effects of endogenous and exogenously administered opioids on brain OT systems are summarized in Table 1.

Table 1 Human studies investigating the association between OT plasma levels and opioid use disorders.

Behavior	Sample	Addiction Phase	Setting	Instrument for behavior measurement	Results: Severity of behavior	Results: Plasma OT	Reference
Novelty-seeking	77 heroin users	Methadone maintenance	MMC	TPQ	↑	↓	Lin et al. (2015)
Heroin craving	77 heroin users	Methadone maintenance	MMC	Chinese Craving Scale	↑	↓	Lin et al. (2018)
Heroin withdrawal stress	57 male heroin users	Heroin withdrawal	RD	COWS, VAS-Craving, HAM-A	↑	↑	Nikolaou, Kapoukranidou, Ndungu, Floros, and Kovatsi (2017)
Heroin withdrawal symtoms				COWS	↑	↑	
Craving				VAS-Craving	↑	↑	
Anxiety				HAM-A	↑	↑	
Treatment completion					↓	↑	
Heroin use duration	57 male heroin users	Heroin withdrawal	RD		↔	↔	Nikolaou (2017)
Heroin dosage (last month)					↔	↔	

Continued

Table 1 Human studies investigating the association between OT plasma levels and opioid use disorders.—cont'd

Behavior	Sample	Addiction Phase	Setting	Instrument for behavior measurement	Results: Severity of behavior	Results: Plasma OT	Reference
Former heroin use No heroin use	18 ex-heroin users 18 non-users	Abstinence				↑ ↓	Gerra et al. (2017)
Psychiatric symptoms				Symptoms Check List 90	↑ ↑	↑ ↑	
Aggressiveness				BDHI	↑	↑	
Mother neglect				CECA			
Heroin use duration					↔	↔	
Heroin Dosage							

↓ = Decreased; ↑ = Increased; ↔ = no significant correlation; OT = oxytocin; MMC = Methadone Maintenance Clinic; RD = Residential Detoxification; COWS = Clinical Opioid Withdrawal Scale; VAS-Craving = Visual-Analogue Craving Scale; HAM-A = Hamilton-Anxiety-Rating Scale; TPQ = novelty seeking subscale of the Tridimensional Personality Questionnaire; LDL = low-density lipoprotein cholesterol; BDHI = Buss Durkee Hostility Inventory; CECA-Q = Child Experience of Care and Abuse-Questionnaire

3. Effects of endogenous and exogenously administered oxytocin on opioid use disorders

OT is one of the neurotransmitters involved in the brain rewarding systems and in modulating the reinforcing effects of substances on these processes (Gardner, 2011; Koob, 2000, 2008).

In their pioneer work Sarnyai and Kovács have demonstrated repeatedly that OT may act as a *"neuro-modulator of dopaminergic neurotransmission in the mesolimbic-mesocortical dopaminergic pathway, the reward circuitry, regulating neuroadaptation during the development of addiction phenomena, in order to maintain homeostasis under substance exposure"* (Sarnyai, 2011; Sarnyai & Kovács, 1994).

A number of shared anatomical and neurochemical substrates enhance this hypothesis (Baskerville & Douglas, 2010; Kovács et al., 1998; McGregor et al., 2008; Sarnyai & Kovács, 1994): OT receptors are located in limbic and basal forebrain structures (Sarnyai et al., 1988). The human amygdala function and that of other limbic structures seem to be strongly modulated by OT, via alterations on dopaminergic transmission (Kovács, Sarnyai, Babarczi, Szabó, & Telegdy, 1990; McGregor et al., 2008; Sarnyai & Kovács, 1994) and administration of OT into the SON of rats reduces dopamine-mediated rewarding effects (Baracz & Cornish, 2013).

OT-dependent behaviors (affiliative behavior, attachment, social bonding, trust and anxiolysis), on the other hand, seem to exert a protective effect of their own on the development of substance use disorders (Burkett & Young, 2012; Young, Gobrogge, & Wang, 2011).

3.1 Endogenous OT systems

May modulate acute and long-term psychoactive substance use disorders by altering associated substance-induced changes in dopamine, glutamate and Fos expression in cortical and basal ganglia sites (McGregor et al., 2008; McGregor & Bowen, 2012). The baseline activity of the endogenous OT system is probably not sufficient to cause these effects. OT secretion may be enhanced by exogenous OT administration, which has a boosting effect on the oxytocinergic system (Burkett & Young, 2012) (also see Section 3.2).

3.1.1 Inhibition of the development of tolerance to opioids

Has been demonstrated by the fact that OT-receptor antagonists injected ICV (Kovács & Telegdy, 1988), or locally into the basal forebrain and the limbic structures, known to contain OT receptors, facilitated the development of morphine tolerance (Sarnyai et al., 1988). OT injection into the nucleus caudatus (with no OT receptors) or the lateral cerebral ventricle was without effect (Sarnyai et al., 1988).

3.1.2 Increase in OT release during opioids withdrawal syndrom

It was already described how naloxone-precipitated morphine withdrawal syndrome in tolerant/dependent rats results in an abrupt increase in the OT secretion in the magnocellular SON neurons and their axons, which represent the major origin of central OT secretion during morphine withdrawal (Bull et al., 2011; Ruan et al., 2011) (also see Section 2.2.2.2.). An increase in OT levels was recorded in the hippocampus, amygdala and in the basal forebrain structures (Coombes, Robinson, Antoni, & Russell, 1991), as well as in the PVN and median eminence (Laorden, Milanés, Chapleur-Château, & Burlet, 1998; You et al., 2000) and in the lateral septum (Russell et al., 1992). The OT secretion resulted in increased OT plasma levels, which lasted for over 90 min, as well as increased CSF concentrations. According to the author OT plasma levels were up to 150% (Bicknell, Zhao, Chapman, Heavens, & Sirinathsinghji, 1988), 80 (Coombes et al., 1991) or 36 times more (Russell et al., 1992) and were maintained for 60 min at least (Bicknell et al., 1988), both in portal as well as in peripheral blood (Sheward, Coombes, Bicknell, Fink, & Russell, 1990), and were still 10 times the initial concentration 90 min after naloxone (Coombes et al., 1991). CSF oxytocin was increased fivefold 40 min after naloxone (Coombes et al., 1991). A decrease of hypothalamic OT levels and a concomitant increase of OT receptor binding were observed after 7 days abstinence in morphine-tolerant male mice suggestive of compensatory mechanisms to the reduced OT-ergic tone in the lateral septum and amygdala (Zanos, Georgiou, et al., 2014).

3.2 The positive effects of exogenous OT administration on opioid use disorders

A great number of studies in rodents, on the other hand, show that administration of exogenous OT attenuates many of the acute and long-term effects of many psychoactive substances: cocaine (Leong, Zhou, Ghee, See, & Reichel, 2016; Sarnyai, 1998), methamphetamine (Baracz, Everett, McGregor, & Cornish, 2016; Cox et al., 2017; Ferland, Reichel, & McGinty, 2016), MDMA (Thompson, Callaghan, Hunt, Cornish, & McGregor, 2007), alcohol (King et al., 2017; MacFadyen et al., 2016), NMDA antagonist PCP (Lee, Brady, Shapiro, Dorsa, & Koenig, 2005). Some studies show that OT treatment may even have a preventive effect on later development of alcohol (Bowen, Carson, Spiro, Arnold, & McGregor, 2011) and methamphetamine (Hicks, Cornish, Baracz, Suraev, & McGregor, 2016) use disorders.

This is another example of the ability of brain OT systems to adaptive neuroplasticity and to a feed-forward effect, suggestive of a long-lasting upregulation of endogenous oxytocinergic systems and increased plasma levels after exogenous OT administration (Bowen et al., 2011), an effect Carson et al. (2010) had earlier described as "a self-stimulation effect of exogenous OT on its own hypothalamic oxytonergic neurons". According to earlier research, a positive feedback circuit is established between both in vivo and in vitro stimulation of hypothalamic OXT receptors and dendritic and peripheral release of OT resulting in hypertrophy of the oxytocinergic system, which causes an enduring (lasting for months, depending on the magnitude and length of the stimulation) induction of endogenous OT release (Theodosis, 2002).

The effects of OT administration specifically on OUD document the important inhibitory role of the neuropeptide on the adaptive responses to opioids in animals and seem to be centrally mediated (Lee et al., 2016).

3.2.1 *Inhibition of the development of tolerance to and dependence on opioids*

ICV and local injection of OT and synthetic OT C-terminal derivatives into various brain structures, as well as systemic (SC) injections of OT inhibited the development, in naïve as well as in tolerant rats, of tolerance to and dependence on morphine, heroin, D-Pro2-Met5-enkephalinamide and to the endogenous opioid beta-endorphin (Kovács & Telegdy, 1988). OT was more effective at an earlier stage, before tolerance is established (Lee et al., 2016): in naïve rats only a single, average OT dose was required, whereas in tolerant rats a higher amount was required to reverse opioid tolerance developement (Kovács, Horvath, Sarnyai, Faludi, & Telegdy, 1985). The central role of the nucleus accumbens is documented by the complete or partial prevention of the OT and OT analogues' effects on morphine tolerance/dependence following a lesion (Kovács, Izbéki, Horváth, & Telegdy, 1984). Chronic OT treatment decreased the enhanced dopamine utilization in the mesencephalon induced by morphine tolerance and dependence, probably by decreasing the number of oxytocinergic binding sites of dopamine in the basal forebrain area (Kovács et al., 1998; Sarnyai & Kovács, 1994) thus attenuating morphine tolerance.

3.2.2 Inhibition of cross-tolerance between opioids

Peripheral treatment with and ICV administration of OT blocked the development of cross-tolerance between the μ-agonist heroin and the δ-agonist enkephalin but the development of heroin-fentanyl cross-tolerance in mice was inhibited only by ICV administration of OT (Kriván, Szabó, Sarnyai, Kovács, & Telegdy, 1992, 1995).

3.2.3 Enhancement of the analgesic effect of opioids

OT attenuated the development of tolerance to the analgesic action of morphine, heroin, beta-endorphin and D-Pro2-Met5-enkephalinamide in morphine-tolerant rodents resulting in an increase of the antinociceptive effect of opioids (Kovács & Telegdy, 1988) and of its duration (Arletti, Benelli, & Bertolini, 1993).

3.2.4 Self-administration of opioids

Was decreased by SC OT administration (Kovács & Telegdy, 1988) and by local injection of OT into the nucleus accumbens and into the ventral hippocampus in heroin-tolerant rats (Ibragimov, Kovacs, Szabo, & Telegdy, 1987).

3.2.5 Amelioration or even suppression of opioid-withdrawal symptoms by raising brain and plasma OT levels in rodents

OT attenuated various symptoms of morphine naloxone-induced withdrawal in rats (e.g., stereotyped jumping, hypothermia, defecation, changes in body mass) in a dose-dependent manner (Kovács et al., 1984; Kovács & Telegdy, 1988). Apart from these acute physical withdrawal symptoms, which disappear after the first week, negative emotional symptoms (anxiety, depression and impaired sociability) persisting after 7 days of morphine abstinence in morphine tolerant male mice, were also attenuated by the OT analog carbetocin (Zanos, Georgiou, et al., 2014) (also see Sections 2.2.2.2. and 3.1.2.).

3.2.6 Prevention of reinstatement to opioids-seeking behavior

The OT analog Carbetocin was able to prevent stress-induced (i.e., due to persisting withdrawal symptoms a week after opioid withdrawal) and priming-induced reinstatement to morphine-seeking behavior, thus impeding relapse in rodents. The authors discuss the involvement of striatal noradrenergic and dopaminergic systems as well as interactions between

OT- and μ-opioid receptors in the modulation of relapse (Georgiou et al., 2015; Zanos, Georgiou, et al., 2014).

The effects of endogenous and exogenously administered oxytocin on opioid use disorders are summarized in Table 2.

4. Discussion

The phenomenon of dependence is a complex one, reflecting the biopsychosocial model for intrerpreting human behavior like no other. The social behavioral deficits, cognitive and emotional impairment (anxiety, depression, anger, frustration, low self-esteem, low threshold for stress tolerance, inadequate and dysfunctional stress coping and adaptive mechanisms, avoidance behavior of stressful events, aggressive assertiveness) observed in high risk personalities represent a certain vulnerability for a future SUD. These features are also present in active and in abstinent users and are associated with drug-seeking behavior and relapse. There is accumulating evidence by now that pre-existing or substance-induced, or both, alterations of the brain OT systems may be, at least partially, responsible for, or contribute to this clinical picture, together with environmental factors such as stress and traumatic social experiences which also affect the endogenous OT system (Buisman-Pijlman et al., 2014).

There is still disagreement concerning the role OT may play as a therapeutic agent in SUD, or if it has one in the first place. The substance itself, severity of dependence and individual differences may have an impact on the effect of OT on SUD. Lee and Weerts (2016) conclude that "it is still unknown if administration of exogenous OT will reverse these changes in OT synthesis created by exposure to drugs of abuse and whether this would translate to reversal of addictive behaviors or if administration of OT can prevent the development of addiction after exposure to environmental risk factors."

This long-standing interest in the clinical promise of oxytocin for a number of neuro-psychiatric disorders, among them SUD, raises numerous controversial issues that should be addressed and resolved before its therapeutic potential is validated in more clinical trials.

Most methodological issues that are not easy to overcome are due to the still largely unknown pharmacodynamics and pharmacokinetics of OT. Oxytocin is a very unstable molecule with a short halflife (1–6 min) (Weisman, Zagoory-Sharon, & Feldman, 2012), with substantial difficulties for clinical utility (Lee & Weerts, 2016), such as clinical relevance of dosing

Table 2 Human studies investigating the effect of intranasal OT on opioid use disorders.

Behavior	Sample: size, gender	Addiction Phase	Setting	OT intranasal dose (IU)	Instrument for behavior measurement	Results	Reference
Tolerability of a single OT dose	36 male heroin users	Methadone or Buprenorphine replacement therapy	Medical Centre	40	OAWS	↔	Woolley et al. (2016)
					VAS	↑	
Opioid craving					Cue-induced craving task	↔	
Implicit associations with opioids					IAT	↔	
Social perception					RMET	↓	
					TASIT	↓	
Social perception	34 healthy controls				TASIT	↓	

↓ = Decreased; ↑ = Increased; ↔ = no significant effect; OT = oxytocin; OAWS = opioid agonist and withdrawal scale; VAS = Visual analog scale for perceived "bad" and "high" OT effects; IAT = Implicit Association Task; RMET = Reading the Mind in the Eyes Task; TASIT = The Awareness of Social Inference Test.

(no confirmation of a biologically active dose, no clinically meaningful dose range, no known dose-response curve, unknown duration of dosing and unknown therapeutic plasma levels) and uncertainty in what are the effective standard operative procedures for dosing (storage, no standardized intranasal administration).

Lastly the relationship between OT plasma and central levels and the brain penetrance of intranasal route are mainly responsible for the ongoing skepticism and debate (Carson, Yuan, & Labuschagne, 2016; Leng & Ludwig, 2016; Quintana & Woolley, 2016). Standardization of OT plasma levels is particularly important for the establishement and ongoing evaluation of sufficient, therapeutic levels and of interpreting the activity of both endogenous and administered OT (Guastella et al., 2013). Plasma levels do not necessarily reflect the secretory activity of central OT systems (Carson et al., 2013). Central release of OT may be followed by a plasma level increase (Wotjak et al., 1998), but central and peripheral OT can be independently regulated, meaning that peripheral OT levels may not be correlated with brain OT levels and may provide an unreliable indicator for central OT levels (Landgraf & Neumann, 2004). To make things even more complicated central OT substrates (extracellular fluid of SON and PVN nuclei, cerebrospinal fluid) may not be an accurate representation of central OT activity (Landgraf & Neumann, 2004) in the first place, and it is still unclear to what extent peripheral sources contribute to plasma OT levels (Leckman, 2011).

Intranasal administration provides the most effective delivery method (Quintana, Smerud, Andreassen, & Djupesland, 2018), since the OT molecule size does not pass the blood-brain-barrier in significant quantities. Central, endogenous OT activity is stimulated approximately 40 min later (Heinrichs et al., 2003), CSF levels are still elevated 20–80 min post-administration in rodents (Neumann, Maloumby, Beiderbeck, Lukas, & Landgraf, 2013) and in humans (Born et al., 2002; Striepens et al., 2013) and peak plasma levels are still detected 30–40 min after intranasal administration in men (Gossen et al., 2012). OT begins to exert its effects within 30 min after administration and they last for at least 90 min (Norman et al., 2011). In a meta-analysis of 17 studies in animals and humans, Valstad, Alvares, Egknud, et al. (2017) record a positive association between central and peripheral OT concentrations after intranasal OT administration and after experimentally induced stress, indicating a coordination of central and peripheral OT release, but no association under basal conditions. Lee et al. (2018) though didn't find peripheral OT administration to have an effect on central release of endogenous OT at all, nor did they find intranasal OT peripheral route superior to the intravenous one in attaining higher CSF OT concentrations in rhesus macaques. The degree and the way the intranasal route of OT administration reaches the receptors in the brain (Churchland & Winkielman, 2012) and if it is the optimal choice for increasing central OT levels remain to be clarified. Furthermore it is still not clear if it reaches the brain as an intact peptide at all (Lee & Weerts, 2016) and to what extent the "self-stimulation effect of exogenous oxytocin on its own hypothalamic oxytonergic neurons" may also contribute to the increased central levels (Carson et al., 2010) (also see Section 3.2). Further, it also seems to do so only after increasing peripheral levels (Guastella et al., 2013).

Quintana suggests that 26.1% of 32 non-significant intranasal oxytocin results and 73.5% of 34 unpublished non-significant effects studied in a very recent meta-analysis (Quintana, 2017) may not necessarily mean absence of an effect, but reflect methodological problems instead (data insensitivity), which may be reversed if equivalence tests are included in the study design.

Inspite of all the drawbacks and limitations a significant amount of preclinical evidence indicate that OT does have a therapeutic potential and may play an, undoubtedly, intriguing part in the therapeutic process of OUD, as well as of other SUD, taking in account the fact that individual differences may differentially have an impact on this effect.

This may be a result of the oxytocinergic effect on reducing stress, which facilitates improvement of dysfunctional emotional states, social skills and affiliative processes. The association between stress, dysregulation of HPA-axis, dependence development and relapse is considered to be the main mechanism sustaining substance use (Lee & Weerts, 2016), with the biochemical equivalent being a stress induced activation of reward processing by increasing mesolimbic dopamine transmission (Koob et al., 2014). Withdrawal symptoms and stress increase craving, generate a powerful motivation to reinstate the drug-seeking behavior and lead to relapse. The anxiety present in the withdrawal syndrome is associated with the production (in the amygdala) and the consequent release of CRF into the ventral tegmental area and nucleus accumbens (Koob, 2008). Chronic OT administration may affect the opioids withdrawal syndrome by inhibiting the HPA-axis and thus causing a decrease in CRF release (McGregor et al., 2008, also see p. 3). Zanos, Georgiou, et al. (2014) and Georgiou, Zanos, Garcia-Carmona, et al. (2015) showed that the withdrawal associated hypercortisolism in rats returned to levels similar to those of controls after the 7-day opiates withdrawal period, demonstrating a normalization of the HPA axis.

McGregor et al. (2008) suggest that OT might inoculate against stress-induced relapse to substance seeking, since OT exerts an antagonistic action on AVP involvement in the amygdala in the autonomic expression of fear (Huber, Veinante, & Stoop, 2005), effective in the aversive emotional states of opioids withdrawal, as well as in the consequences of these negative emotional states on the reinstatement of substance-seeking behavior (Zhou, Leri, Cummins, Hoeschele, & Kreek, 2008). In humans, normalization of HPA axis and hypercortisolism were correlated with withdrawal symptoms, craving and heroin use indicating their important part as the main goal of the pharmacological treatment of OUD (Nava, Caldiroli, Premi, & Lucchini, 2006).

In addition to its anti-stress effects OT may be able to restore social deficits, caused by the effects of substance use on brain OT systems, and help social stimuli prevail over substance use (Lee et al., 2016; Lee & Weerts, 2016), especially when given within a therapeutic context (Heinrichs et al., 2003; Meyer-Lindenberg, Domes, Kirsch, & Heinrichs, 2011). Pedersen (2017) also suggests diminishing established tolerance as another possible mechanism of the OT effect on SUD. OT may be uniquely effective in reducing negative reinforcement that maintains chronic substance use.

The existing literature in humans shows OT to be quite safe, with minimal adverse, systemic side-effects when delivered in doses of 18–40 IU, for short term use, in controlled research settings (MacDonald et al., 2011). Trials have been largely exploratory so far, with small, mainly male samples in different stages of treatment and different OT dosing (Lee & Weerts, 2016; MacDonald et al., 2011).

It is not clear whether OT itself is rewarding. According to current opinion OT has low risk of abuse potential and intranasal OT administration to humans is not euphorogenic. In most human studies, participants weren't able to differentiate OT from placebo (MacDonald et al., 2011). More potent, orally bioavailable, non-peptide OT agonists though (Ring et al., 2010) may have abuse potential. There is evidence of positive reinforcing properties of OT in an extremely high dose (much higher than that usually used in humans) or in a dose-dependent manner on reward mechanisms (László et al., 2016; Liberzon, Trujillo, Akil, & Young, 1997). Studies in rodents using the conditioned place preference (CPP) model demonstrate contradicting results in the OT role in modulating reward-related behaviors. Rae et al. (2018) showed that an oxytocinergic-dependent mechanism underlies the stimulatory effect of environmental enrichment on alcohol reward in mice, an effect replicated with Carbetocin administration.

In contrast to these results, the acquisition of the conditioned behavior for alcohol (Bahi, 2015) and psychostimulants (Qi et al., 2009) was decreased by OT administration. In regard to opioids Zanos, Georgiou, et al. (2014) also observed a diminishing impact by the OT analog carbetocin: given peripherally, it prevented stress induced reinstatement of CPP for morphine. In a more recent study though, OT administered ICV resulted in an increase in expression of morphine-induced CPP, with no effect on its acquisition, at least in part, via activation of oxytocin receptors within the nucleus accumbens shell (Moaddab, Hyland, & Brown, 2015).

There are very few studies investigating the role that OT administration may play in the treatment of SUD disorders in humans and even less focusing on the relationship between opioids use disorder and OT: Woolley et al. (2016) found that the intranasal administration of a single OT dose to opioid users under opioid replacement therapy was well-tolerated; it had no effect on craving though and no significant, implicit association with heroin cues. OT worsened performance on social perceptual ability. There was a direct association between OT plasma levels aggressive behavior and co-occurring psychiatric symptoms. In contrast to these findings Lin et al. (2018) did record a negative association between the plasma oxytocin and craving levels in heroin-dependent patients on methadone maintenance, the more so in subjects with lower scores of novelty-seeking behavior. The same team recorded earlier a negative association between plasma oxytocin levels and novelty seeking in a similar population (Lin et al., 2015). Nikolaou et al. (2017) showed, among heroin users in a controlled setting, that the severity of opioid withdrawal stress was correlated with alterations in plasma OT levels and treatment outcome: participants who didn't complete a detoxification program had higher withdrawal stress at the peak of the withdrawal syndrome, reflected in higher scores in the Clinical Opioid Withdrawal Scale (COWS), the Visual-Analogue Craving Scale (VAS-Craving) and the Hamilton-Anxiety-Rating Scale compared to the ones who did. This higher withdrawal stress induced higher plasma oxytocin levels. There was also a positive correlation between plasma oxytocin values and scores achieved in the COWS, VAS-Craving and the Hamilton-Anxiety scales in all participants. Gerra et al. (2017) found higher OT plasma levels among abstinent heroin users, comparing to non-users, which correlated positively with psychiatric symptoms, aggressiveness and mother neglect scores. Similarly to the results by Nikolaou (2017) there was no correlation between oxytocin plasma levels and heroin exposure duration or dosage.

5. Conclusions

Clinically, OT is an attractive candidate for the treatment of OUD. Bowen and Neumann (2018), describe its therapeutic effect as a "potential, multidimensional, therapeutic utility at each stage of the addiction cycle for numerous drugs of abuse, probably due to its interactions with key biological systems that underlie the development and maintenance of addiction and are able to reverse the neuroadaptations occurring with substance use."

OT has proved to be effective in attenuating the development of tolerance and dependence to opioids, in reducing opioid self-administration, in alleviating withdrawal and in preventing against stress-induced relapse in animals. The involvement of OT in the homeostatic regulation of stress responses, by normalizing the HPA-axis, decreasing CRF release and raising brain OT levels, may be the most significant contribution to the reduction of relapse risk.

Opioid withdrawal seems to be the most promising field at the moment (Lee et al., 2016). OT may have an adjunctive role to play in opioid detoxification in controlled treatment settings, since severity of withdrawal symptoms possibly plays a significant role in early discharge from treatment (Specka, Buchholz, Kuhlmann, Rist, & Scherbaum, 2011). Detoxification represents an important initial phase in a long-term recovery process by relieving withdrawal discomfort, thus increasing retention and completion rates, by motivating patients, facilitating transition and engaging them in treatment continuation, and by decreasing the risk for relapse (Blodgett, Maisel, Fuh, Wilbourne, & Finney, 2014). Moreover, oxytocin administration may alleviate early withdrawal symptoms and increase adherence when initiating buprenorphine treatment to heroin users (due to the partial agonist-antagonist profile of buprenorphine an abstinence period may be indispensable before buprenorphine initiation).

The currently available scientific data do warrant the rationale to further investigate OT's potential and limitations as a novel treatment approach for OUD in clinical trials, notwithstanding any drawbacks. Further research on humans is imperative, focusing on the specific role of OT and its therapeutic effectiveness. Answers to clinically relevant issues are of paramount priority: treatment objectives and target groups ought to be defined, treatment protocol (adjustment process of oxytocin dosage, dosage, optimal dosing interval and duration of treatment), safety issues of longterm administration, gender differences, interactions with other neurotransmitters (especially AVP) should also be studied. Lastly and most importantly the brain penetrance and the relationship between central and plasma levels have to be clarified so that plasma levels may be standardized according to current state, otherwise the measurement of plasma OT levels as a surrogate for central OT function is of limited value (Lee & Weerts, 2016). Finally the development of non-peptide OT receptor agonists with longer half-life and less frequent administrations would maximize patients' compliance to the treatment (Lee et al., 2016; Lee & Weerts, 2016).

Quintana and Woolley (2016) argue that "poor understanding of the mechanisms underlying intranasal OT administration shouldn't overshadow the goal to define the how (mechanism), the where (site of action) and for whom (target group)."

References

Bicknell, R. J. (1985). Endogenous opioid peptides and hypothalamic neuroendocrine neurones. *Journal of Endocrinology*, *107*(3), 437–446.

Arletti, R., Benelli, A., & Bertolini, A. (1993). Influence of oxytocin on nociception and morphine antinociception. *Neuropeptides*, *24*, 125–129. 1993.

Bahi, A. (2015). The oxytocin receptor impairs ethanol reward in mice. *Physiology & Behavior*, *139*, 321–327.

Baracz, S. J., & Cornish, J. L. (2013). Oxytocin modulates dopamine-mediated reward in the rat subthalamic nucleus. *Hormones and Behavior*, *63*(2), 370–375. Erratum in Horm Behav. 2013 Jun;64(1):172.

Baracz, S. J., Everett, N. A., McGregor, I. S., & Cornish, J. L. (2016). Oxytocin in the nucleus accumbens core reduces reinstatement of methamphetamine-seeking behaviour in rats. *Addiction Biology*, *21*(2), 316–325.

Baracz, S. J., Parker, L. M., Suraev, A. S., Everett, N. A., Goodchild, A. K., McGregor, I. S., et al. (2016). Chronic methamphetamine self-administration dysregulates oxytocin plasma levels and oxytocin receptor fibre density in the nucleus accumbens core and subthalamic nucleus of the rat. *Journal of Neuroendocrinology*, *28*. https://doi.org/10.1111/jne.12337.

Baskerville, T. A., & Douglas, A. J. (2010). Dopamine and oxytocin interactions underlying behaviors: Potential contributions to behavioral disorders. *CNS Neuroscience & Therapeutics*, *16*(3), 92–123.

Baumgartner, T., Heinrichs, M., Vonlanthen, A., Fischbacher, U., & Fehr, E. (2008). Oxytocin shapes the neural circuitry of trust and trust adaptation in humans. *Neuron*, *58*(4), 639–650.

Bicknell, R. J., & Leng, G. (1982). Endogenous opiates regulate oxytocin but not vasopressin secretion from the neurohypophysis. *Nature*, *298*(5870), 161–162.

Bicknell, R. J., Zhao, B. G., Chapman, C., Heavens, R. P., & Sirinathsinghji, D. J. (1988). Opioid inhibition of secretion from oxytocin and vasopressin nerve terminals following selective depletion of neurohypophysial catecholamines. *Neuroscience Letters*, *93*(2–3), 281–286.

Blackburn-Munro, G., Brown, C. H., Neumann, I. D., Landgraf, R., & Russell, J. A. (2000). Verapamil prevents withdrawal excitation of oxytocin neurones in morphine-dependent rats. *Neuropharmacology*, *39*(9), 1596–1607.

Blair-Bell, W. (1909). The pituitary body and the therapeutic value of the infundibular extract in shock, uterine atony and intestinal paresis. *British Medical Journal, 2*, 1609.

Blodgett, J. C., Maisel, N. C., Fuh, I. L., Wilbourne, P. L., & Finney, J. W. (2014). How effective is continuing care for substance use disorders? A meta-analytic review. *Journal of Substance Abuse Treatment, 46*(2), 87–97. https://doi.org/10.1016/j. jsat.2013.08.022.

Born, J., Lange, T., Kern, W., McGregor, G. P., Bickel, U., & Fehm, H. L. (2002). Sniffing neuropeptides: A transnasal approach to the human brain. *Nature Neuroscience, 5*(6), 514–516 [PubMed: 11992114].

Bowen, M. T., Carson, D. S., Spiro, A., Arnold, J. C., & McGregor, I. S. (2011). Adolescent oxytocin exposure causes persistent reductions in anxiety and alcohol consumption and enhances sociability in rats. *PLoS One, 6*(11), e27237. Published online 2011 Nov 16. https://doi.org/10.1371/journal.pone.0027237.

Bowen, M. T., & Neumann, I. D. (2018). The multidimensional therapeutic potential of targeting the brain oxytocin system for the treatment of substance use disorders. *Current Topics in Behavioral Neurosciences, 35*, 269–287. https://doi.org/10.1007/7854_2017_17.

Brown, C. H., Brunton, P. J., & Russell, J. A. (2008). Rapid estradiol-17beta modulation of opioid actions on the electrical and secretory activity of rat oxytocin neurons in vivo. *Neurochemical Research, 33*(4), 614–623.

Brown, C. H., Johnstone, L. E., Murphy, N. P., Leng, G., & Russell, J. A. (2000). Local injection of pertussis toxin attenuates morphine withdrawal excitation of rat supraoptic nucleus neurones. *Brain Research Bullettin, 52*(2), 115–121.

Brown, C. H., Munro, G., Murphy, N. P., Leng, G., & Russell, J. A. (1996). Activation of oxytocin neurones by systemic cholecystokinin is unchanged by morphine dependence or withdrawal excitation in the rat. *Journal of Physiology, 496*(3), 787–794.

Brown, C. H., Murphy, N. P., Munro, G., Ludwig, M., Bull, P. M., Leng, G., et al. (1998). Interruption of central noradrenergic pathways and morphine withdrawal excitation of oxytocin neurones in the rat. *Journal of Physiology, 507*(3), 831–842.

Brown, C. H., Russell, J. A., & Leng, G. (2000). Opioid modulation of magnocellular neurosecretory cell activity. *Neuroscience Research, 36*(2), 97–120.

Brown, C. H., Stern, J. E., Jackson, K. L., Bull, P. M., Leng, G., & Russell, J. A. (2005). Morphine withdrawal increases intrinsic excitability of oxytocin neurons in morphine-dependent rats. *The European Journal of Neuroscience, 21*(2), 501–512.

Buisman-Pijlman, F. T., Sumracki, N. M., Gordon, J. J., Hull, P. R., Carter, C. S., & Tops, M. (2014). Individual differences underlying susceptibility to addiction: Role for the endogenous oxytocin system. *Pharmacology, Biochemistry, and Behavior, 119*, 22–38 [PubMed: 24056025].

Bull, P. M., Russell, J. A., Scott, V., & Brown, C. H. (2011). Apamin increases post-spike excitability of supraoptic nucleus neurons in anaesthetized morphine-naïve rats and morphine-dependent rats: Consequences for morphine withdrawal excitation. *Experimental Brain Research, 212*(4), 517–528.

Burkett, J. P., & Young, L. J. (2012). The behavioral, anatomical and pharmacological parallels between social attachment, love and addiction. *Psychopharmacology, 224*(1), 1–26.

Carson, D. S., Guastella, A. J., Taylor, E. R., & McGregor, I. S. (2013). A brief history of oxytocin and its role in modulating psychostimulant effects. *Journal of Psychopharmacology, 27*(3), 231–247.

Carson, D. S., Hunt, G. E., Guastella, A. J., Barber, L., Cornish, J. L., Arnold, J. C., et al. (2010). Systemically administered oxytocin decreases methamphetamine activation of the subthalamic nucleus and accumbens core and stimulates oxytocinergic neurons in the hypothalamus. *Addiction Biology, 15*(4), 448–463.

Carson, D. S., Yuan, H., & Labuschagne, I. (2016). Improving research standards to restore Trust in Intranasal Oxytocin. *Biological Psychiatry, 79*(8), e53–e54. https://doi.org/10.1016/j.biopsych.2015.08.031.

Carter, D. A., Williams, T. D., & Lightman, S. L. (1986). A sex difference in endogenous opioid regulation of the posterior pituitary response to stress in the rat. *Journal of Endocrinology, 111*(2), 239–244.

Churchland, P. S., & Winkielman, P. (2012). Modulating social behavior with oxytocin: How does it work? What does it mean? *Hormones & Behaviour, 61*, 392–399.

Coombes, J. E., Robinson, I. C., Antoni, F. A., & Russell, J. A. (1991). Release of oxytocin into blood and into cerebrospinal fluid induced by naloxone in anaesthetized morphine-dependent rats: The role of the paraventricular nucleus. *Journal of Neuroendocrinology, 3*(5), 551–561.

Cox, B. M., Bentzley, B. S., Regen-Tuero, H., See, R. E., Reichel, C. M., & Aston-Jones, G. (2017). Oxytocin acts in nucleus Accumbens to attenuate methamphetamine seeking and demand. *Biological Psychiatry, 81*(11), 949–958. https://doi.org/10.1016/j.biopsych.2016.11.011.

Dabrowska, J., Hazra, R., Ahern, T. H., Guo, J. D., McDonald, A. J., Mascagni, F., et al. (2011). Neuroanatomical evidence for reciprocal regulation of the corticotrophin-releasing factor and oxytocin systems in the hypothalamus and the bed nucleus of the stria terminalis of the rat: Implications for balancing stress and affect. *Psychoneuroendocrinology, 36*, 1312–1326.

Dale, H. H. (1906). On some physiological actions of ergot. *Journal of Physiology (London), 34*(3), 163–206.

Ditzen, B., Schaer, M., Gabriel, B., Bodenmann, G., Ehlert, U., & Heinrichs, M. (2009). Intranasal oxytocin increases positive communication and reduces cortisol levels during couple conflict. *Biological Psychiatry, 65*(9), 728–731 [PubMed: 19027101].

du Vigneaud, V. (1953). Hormones of the posterior pituitary gland: Oxytocin and vasopressin. *Harvey Lectures, 50*, 1.

Dumont, G. J., Sweep, F. C., van der Steen, R., Hermsen, R., Donders, A. R., Touw, D. J., et al. (2009). Increased oxytocin concentrations and prosocial feelings in humans after ecstasy (3,4-methylenedioxymethamphetamine) administration. *Society for Neuroscience, 4*, 359–366.

Ferland, C. L., Reichel, C. M., & McGinty, J. F. (2016). Effects of oxytocin on methamphetamine-seeking exacerbated by predator odor pre-exposure in rats. *Psychopharmacology, 233*(6), 1015–1024.

Gardner, E. L. (2011). Addiction and brain reward and antireward pathways. *Advanced Psychosomatic Medicine, 30*, 22–60.

Georgiou, P., Zanos, P., Ehteramyan, M., Hourani, S., Kitchen, I., Maldonado, R., et al. (2015). Differential regulation of mGlu5 R and MOPr by priming- and cue-induced reinstatement of cocaine-seeking behaviour in mice. *Addiction Biology, 20*(5), 902–912.

Georgiou, P., Zanos, P., Garcia-Carmona, J. A., Hourani, S., Kitchen, I., Kieffer, B. L., et al. (2015). The oxytocin analogue carbetocin prevents priming-induced reinstatement of morphine-seeking: Involvement of dopaminergic, noradrenergic and MOPr systems. *European Neuropsychopharmacology, 25*(12), 2459–2464.

Gerra, L. M., Gerra, G., Mercolini, L., Manfredini, M., Somaini, L., Pieri, C. M., et al. (2017). Increased oxytocin levels among abstinent heroin addicts: Association with aggressiveness, psychiatric symptoms and perceived childhood neglect. *Progress in Neuro-Psychopharmacology & Biological Psychiatry, 75*, 70–76. https://doi.org/10.1016/j.pnpbp.2017.01.005.

Gimpl, G., & Fahrenholz, F. (2001). The oxytocin receptor system: Structure, function, and regulation. *Physiological Reviews, 81*, 629–683.

Gossen, A., Hahn, A., Westphal, L., Prinz, S., Schultz, R. T., Grunder, G., et al. (2012). Oxytocin plasma concentrations after single intranasal oxytocin administration—A study in healthy men. *Neuropeptides, 46*(5), 211–215 [PubMed: 22884888].

Guastella, A. J., Hickie, J. B., McGuiness, M. M., Otis, M., Woods, E. A., Disinger, H. M., et al. (2013). Recommendations for the standardization of oxytocin nasal administration and guidelines for its reporting in human research. *Psychoneuroendocrinology*, *38*, 612–625.

Hansson, A. C., Koopmann, A., Uhrig, S., Bühler, S., Domi, E., Kiessling, E., et al. (2018). Oxytocin reduces alcohol Cue-reactivity in alcohol-dependent rats and humans. *Neuropsychopharmacology*, *43*(6), 1235–1246. https://doi.org/10.1038/npp.2017.257.

Heinrichs, M., Baumgartner, T., Kirschbaum, C., & Ehlert, U. (2003). Social support and oxytocin interact to suppress cortisol and subjective responses to psychosocial stress. *Biological Psychiatry*, *54*, 1389–1398.

Heinrichs, M., & Domes, G. (2008). Neuropeptides and social behaviour: Effects of oxytocin and vasopressin in humans. *Progress in Brain Research*, *170*, 337–350. https://doi.org/10.1016/S0079-6123(08)00428-7.

Hicks, C., Cornish, J. L., Baracz, S. J., Suraev, A., & McGregor, I. S. (2016). Adolescent pre-treatment with oxytocin protects against adult methamphetamine-seeking behavior in female rats. *Addiction Biology*, *21*(2), 304–315.

Huber, D., Veinante, P., & Stoop, R. (2005). Vasopressin and oxytocin excite distinct neuronal populations in the central amygdale. *Science*, *308*, 245–248.

Ibragimov, R. S., Kovacs, G. L., Szabo, G., & Telegdy, G. (1987). Microinjection of oxytocin into limbic–mesolimbic brain structures disrupts heroin self-administration behavior: A receptor-mediated event? *Life Sciences*, *41*, 1265–1271.

Inenaga, K., Imura, H., Yanaihara, N., & Yamashita, H. (1990). Kappa-selective opioid receptor agonists leumorphin and dynorphin inhibit supraoptic neurons in rat hypothalamic slice preparations. *Journal of Neuroendocrinology*, *2*(3), 389–395.

Jong, T. R., Menon, R., Bludau, A., Grund, T., Biermeier, V., Klampfl, S. M., et al. (2015). Salivary oxytocin concentrations in response to running, sexual self-stimulation, breastfeeding and the TSST: The Regensburg oxytocin challenge (ROC) study. *Psychoneuroendocrinology*, *62*, 381–388. https://doi.org/10.1016/j.psyneuen.2015.08.027.

Kamilar-Britt, P., & Bedi, G. (2015). The prosocial effects of 3,4-methylenedioxymethamphetamine (MDMA): Controlled studies in humans and laboratory animals. *Neuroscience and Biobehavioral Reviews*, *57*, 433–446.

King, C. E., Griffin, W. C., Luderman, L. N., Kates, M. M., McGinty, J. F., & Becker, H. C. (2017). Oxytocin reduces ethanol self-Administration in Mice. *Alcoholism, Clinical and Experimental Research*, *41*(5), 955–964. https://doi.org/10.1111/acer.13359.

Kirkpatrick, M. G., Francis, S. M., Lee, R., de Wit, H., & Jacob, S. (2014). Plasma oxytocin concentrations following MDMA or intranasal oxytocin in humans. *Psychoneuroendocrinology*, *46*, 23–31.

Koob, G. F. (2000). Neurobiology of addiction. Toward the development of new therapies. *Annals of the New York Academy of Sciences*, *909*, 170–185.

Koob, G. F. (2008). A role for brain stress systems in addiction. *Neuron*, *59*(1), 11–34.

Koob, G. F., Buck, C. L., Cohen, A., Edwards, S., Park, P. E., Schlosburg, J. E., et al. (2014). Addiction as a stress surfeit disorder. *Neuropharmacology*, *76*(Pt B), 370–382 [PubMed: 23747571].

Kovács, G. L., & Telegdy, G. (1988). Hypothalamo-neurohypophyseal neuropeptides and experimental drug addiction. *Brain Research Bullettin*, *20*(6), 893–895.

Kovács, G. L., Horvath, Z., Sarnyai, Z., Faludi, M., & Telegdy, G. (1985). Oxytocin and a C-terminal derivative (Z-prolyl-D-leucine) attenuate tolerance to and dependence on morphine and interact with dopaminergic neurotransmission in the mouse brain. *Neuropharmacology*, *24*(5), 413–419.

Kovács, G. L., Izbéki, F., Horváth, Z., & Telegdy, G. (1984). Effects of oxytocin and a derivative (Z-prolyl-D-leucine) on morphine tolerance/withdrawal are mediated by the limbic system. *Behavioural Brain Research*, *14*(1), 1–8.

Kovács, G. L., Laczi, F., Vecsernyés, M., Hódi, K., Telegdy, G., & László, F. A. (1987). Limbic oxytocin and arginine 8-vasopressin in morphine tolerance and dependence. *Experimental Brain Research*, *65*(2), 307–311.
Kovács, G. L., Sarnyai, Z., Babarczi, E., Szabó, G., & Telegdy, G. (1990). The role of oxytocin dopamine interactions in cocaine-induced locomotor hyperactivity. *Neuropharmacology*, *29*, 365–368.
Kovács, G. L., Sarnyai, Z., & Szabó, G. (1998). Oxytocin and addiction: A review. *Psychoneuroendocrinology*, *23*(8), 945–962.
Kriván, M., Szabó, G., Sarnyai, Z., Kovács, G. L., & Telegdy, G. (1992). Oxytocin blocks the development of heroin-enkephalin cross-tolerance in mice. *Pharmacology, Biochemistry, and Behavior*, *43*(1), 187–192.
Kriván, M., Szabó, G., Sarnyai, Z., Kovács, G. L., & Telegdy, G. (1995). Oxytocin blocks the development of heroin-fentanyl cross-tolerance in mice. *Pharmacology, Biochemistry, and Behavior*, *52*(3), 591–594.
Kubzansky, L. D., Mendes, W. B., Appleton, A. A., Block, J., & Adler, G. K. (2012). A heartfelt response: Oxytocin effects on response to social stress in men and women. *Biological Psychology*, *90*(1), 1–9 [PubMed: 22387929].
Kutlu, S., Yilmaz, B., Canpolat, S., Sandal, S., Ozcan, M., Kumru, S., et al. (2004). μ- opioid modulation of oxytocin secretion in late pregnant and parturient rats. Involvement of noradrenergic neurotransmission. *Neuroendocrinology*, *79*, 197–203.
Landgraf, R., & Neumann, I. D. (2004). Vasopressin and oxytocin release within the brain: A dynamic concept of multiple and variable modes of neuropeptide communication. *Frontiers in Neuroendocrinology*, *25*, 150–176.
Laorden, M. L., Milanés, M. V., Chapleur-Château, M., & Burlet, A. (1997). Changes in hypothalamic oxytocin levels during morphine tolerance. *Neuropeptides*, *31*(2), 143–146.
Laorden, M. L., Milanés, M. V., Chapleur-Château, M., & Burlet, A. (1998). Changes in oxytocin content in rat brain during morphine withdrawal. *Neuropeptides*, *32*(1), 67–71.
László, K., Kovács, A., Zagoracz, O., Ollmann, T., Péczely, L., Kertes, E., et al. (2016). Positive reinforcing effect of oxytocin microinjection in the rat central nucleus of amygdala. *Behavioural Brain Research*, *296*, 279–285.
Leckman, J. F. (2011). Variations in maternal behavior—Oxytocin and reward pathways—Peripheral measures matter?! *Neuropsychopharmacology*, *36*(13), 2587–2588.
Lee, M. R., & Weerts, E. M. (2016). Oxytocin for the treatment of drug and alcohol use disorders. *Behavioural Pharmacology*, *27*(8), 640–648. https://doi.org/10.1097/FBP.0000000000000258.
Lee, P. R., Brady, D. L., Shapiro, R. A., Dorsa, D. M., & Koenig, J. I. (2005). Social interaction deficits caused by chronic phencyclidine administration are reversed by oxytocin. *Neuropsychopharmacology*, *30*, 1883–1894.
Lee, M. R., Rohn, M. C. H., Tanda, G., & Leggio, L. (2016). Targeting the oxytocin system to treat addictive disorders: Rationale and Progress to date. *CNS Drugs*, *30*(2), 109–123. https://doi.org/10.1007/s40263-016-0313-z.
Lee, M. R., Scheidweiler, K. B., Diao, X. X., Akhlaghi, F., Cummins, A., Huestis, M. A., et al. (2018). Oxytocin by intranasal and intravenous routes reaches the cerebrospinal fluid in rhesus macaques: Determination using a novel oxytocin assay. *Molecular Psychiatry*, *23*(1), 115–122. https://doi.org/10.1038/mp.2017.27.
Lee, M. R., Schwandt, M. L., Sankar, V., Suchankova, P., Sun, H., & Leggio, L. (2017). Effect of alcohol use disorder on oxytocin peptide and receptor mRNA expression in human brain: A post-mortem case-control study. *Psychoneuroendocrinology*, *85*, 14–19. https://doi.org/10.1016/j.psyneuen.2017.07.481.
Leng, G., Brown, C. H., Murphy, N. P., Onaka, T., & Russell, J. A. (1995). Opioid-noradrenergic interactions in the control of oxytocin cells. *Adv Exp Med Biol*, *395*, 95–104.

Leng, G., Dyes, S., & Bicknell, R. J. (1997). Kappa-opioid restraint of oxytocin secretion: Plasticity through pregnancy. *Neuroendocrinology, 66*, 378–383.
Leng, G., & Ludwig, M. (2016). Intranasal oxytocin: Myths and delusions. *Biological Psychiatry, 79*(3), 243–250. https://doi.org/10.1016/j.biopsych.2015.05.003.
Leong, K. C., Zhou, L., Ghee, S. M., See, R. E., & Reichel, C. M. (2016). Oxytocin decreases cocaine taking, cocaine seeking, and locomotor activity in female rats. *Experimental and Clinical Psychopharmacology, 24*(1), 55–64.
Li, J., You, Z., Chen, Z., Song, C., & Lu, C. (2001). Chronic morphine treatment inhibits oxytocin release from the supraoptic nucleus slices of rats. *Neuroscience Letters, 300*(1), 54–58.
Liberzon, I., Trujillo, K. A., Akil, H., & Young, E. A. (1997). Motivational properties of oxytocin in the conditioned place preference paradigm. *Neuropsychopharmacology, 17*, 353–359.
Light, K. C., Grewen, K. M., Amico, J. A., Boccia, M., Brownley, K. A., & Johns, J. M. (2004). Deficits in plasma oxytocin responses and increased negative affect, stress, and blood pressure in mothers with cocaine exposure during pregnancy. *Addictive Behaviors, 29*(8), 1541–1564.
Lin, S. H., Chen, P. S., Lee, L. T., Lee, S. Y., Tsai, H. C., Chen, W. T., et al. (2018). The association between the level of plasma oxytocin and craving among former heroin users. *European Addiction Research, 24*(2), 71–78. https://doi.org/10.1159/000485563.
Lin, S. H., Lee, L. T., Tsai, H. C., Chen, K. C., Chen, W. T., Lee, I. H., et al. (2015). Association between blood level of plasma oxytocin and novelty seeking among methadone-maintained heroin users. *Neuropsychobiology, 71*(2), 65–69. 9.
Ludwig, M., Brown, C. H., Russell, J. A., & Leng, G. (1997). Local opioid inhibition and morphine dependence of supraoptic nucleus oxytocin neurones in the rat in vivo. *Journal of Physiology, 505*(1), 145–152.
MacDonald, E., Dadds, M. R., Brennan, J. L., Williams, K., Levy, F., & Cauchi, A. J. (2011). A review of safety, side-effects and subjective reactions to intranasal oxytocin in human research. *Psychoneuroendocrinology, 36*, 1114–1126.
MacDonald, K., & MacDonald, T. M. (2010). The peptide that binds: A systematic review of oxytocin and its prosocial effects in humans. *Harvard Review of Psychiatry, 18*, 1–21.
MacFadyen, K., Loveless, R., DeLucca, B., Wardley, K., Deogan, S., Thomas, C., et al. (2016). Peripheral oxytocin administration reduces ethanol consumption in rats. *Pharmacology Biochemistry and Behavior, 140*, 27–32.
Marchesi, C., Chiodera, P., Brusamonti, E., Volpi, R., & Coiro, V. (1997). Abnormal plasma oxytocin and β-endorphin levels in alcoholics after short and long term abstinence. *Progress in Neuro-Psychopharmacology & Biological Psychiatry, 21*, 797–807.
McGregor, I. S., & Bowen, M. T. (2012). Breaking the loop: Oxytocin as a potential treatment for drug addiction. *Hormones and Behavior, 61*(3), 331–339.
McGregor, I. S., Callaghan, P. D., & Hunt, G. E. (2008). From ultrasocial to antisocial: A role for oxytocin in the acute reinforcing effects and long-term adverse consequences of drug use? *British Journal of Pharmacology, 154*(2), 358–368.
Meyer-Lindenberg, A., Domes, G., Kirsch, P., & Heinrichs, M. (2011). Oxytocin and vasopressin in the human brain: Social neuropeptides for translational medicine. *Nature Reviews. Neuroscience, 12*(9), 524–538. https://doi.org/10.1038/nrn3044.
Miranda-Cardenas, Y., Rojas-Piloni, G., Martínez-Lorenzana, G., Rodríguez-Jiménez, J., López-Hidalgo, M., Freund-Mercier, M. J., et al. (2006). Oxytocin and electrical stimulation of the paraventricular hypothalamic nucleus produce antinociceptive effects that are reversed by an oxytocin antagonist. *Pain, 122*(1–2), 182–189.
Moaddab, M., Hyland, B. I., & Brown, C. H. (2015). Oxytocin enhances the expression of morphine-induced conditioned place preference in rats. *Psychoneuroendocrinology, 53*, 159–169.

Montilla, P., Muñoz, J. L., Ríos, J. E., & Muñoz, M. C. (1989). Effect of oxytocin and naloxone on the plasma levels of corticosterone in the rat. *Revista Española de Fisiología*, *45*(4), 319–322.

Munro, G., Pumford, K. M., & Russell, J. A. (1998). Altered cholecystokinin binding site density in the supraoptic nucleus of morphine-tolerant and -dependent rats. *Brain Research*, *780*(2), 190–198.

Murphy, N. P., Onaka, T., Brown, C. H., & Leng, G. (1997). The role of afferent inputs to supraoptic nucleus oxytocin neurons during naloxone-precipitated morphine withdrawal in the rat. *Neuroscience*, *80*(2), 567–577.

Nava, F., Caldiroli, E., Premi, S., & Lucchini, A. (2006). Relationship between plasma cortisol levels, withdrawal symptoms and craving in abstinent and treated heroin addicts. *Journal of Addictive Diseases*, *25*(2), 9–16.

Neumann, I. D., Maloumby, R., Beiderbeck, D. I., Lukas, M., & Landgraf, R. (2013). Increased brain and plasma oxytocin after nasal and peripheral administration in rats and mice. *Psychoneuroendocrinology*, *38*(10), 1985–1993 [PubMed: 23579082].

Nikolaou, K. (2017). *Severity of withdrawal symptoms, plasma oxytocin and treatment outcome in heroin users undergoing detoxification*. Greek National Archive of PhD Theses. https://www.didaktorika.gr/eadd/.

Nikolaou, K., Kapoukranidou, D., Ndungu, S., Floros, G., & Kovatsi, L. (2017). Severity of withdrawal symptoms, plasma oxytocin levels and treatment outcome in heroin users undergoing acute withdrawal. *Journal of Psychoactive Drugs*, *49*, 233–241. Published online 26.4.2017. https://doi.org/10.1080/02791072.2017.1312644.

Norman, G. J., Cacioppo, J. T., Morris, J. S., Karelina, K., Malarkey, W. B., DeVries, A. C., et al. (2011). Selective influences of oxytocin on the evaluative processing of social stimuli. *Journal of Psychopharmacology*, *25*(10), 1313–1319. https://doi.org/10.1177/0269881110367452.

Ondrejcakova, M., Bakos, J., Garafova, A., Kovacs, L., Kvetnansky, R., & Jezova, D. (2010). Neuroendocrine and cardiovascular parameters during simulation of stress-induced rise in circulating oxytocin in the rat. *Stress*, *13*(4), 314–322 [PubMed: 20536333].

Pedersen, C. A. (2017). Oxytocin, tolerance, and the dark side of addiction. *International Review of Neurobiology*, *136*, 239–274. https://doi.org/10.1016/bs.irn.2017.08.003.Epub [2017 Oct 12].

Petersson, M., Hulting, A. L., & Uvnäs-Moberg, K. (1999). Oxytocin causes a sustained decrease in plasma levels of corticosterone in rats. *Neuroscience Letters*, *264*, 41–44.

Qi, J., Yang, J. Y., Wang, F., Zhao, Y. N., Song, M., & Wu, C. F. (2009). Effects of oxytocin on methamphetamine-induced conditioned place preference and the possible role of glutamatergic neurotransmission in the medial prefrontal cortex of mice in reinstatement. *Neuropharmacology*, *56*(5), 856–865.

Quintana, D. S. (2017). Revisiting non-significant effects of intranasal oxytocin using equivalence testing. *Psychoneuroendocrinology*, *87*, 127–130. https://doi.org/10.1016/j.psyneuen.2017.10.010.

Quintana, D. S., Smerud, K. T., Andreassen, O. A., & Djupesland, P. G. (2018). Evidence for intranasal oxytocin delivery to the brain: Recent advances and future perspectives. *Therapeutic Delivery*, *9*(7), 515–525. https://doi.org/10.4155/tde-2018-0002. Special Report.

Quintana, D. S., & Woolley, J. D. (2016). Intranasal oxytocin mechanisms can be better understood, but its effects on social cognition and behavior are not to be sniffed at. *Biological Psychiatry*, *79*(8), e49–e50. https://doi.org/10.1016/j.biopsych.2015.06.021.

Quirin, M., Kuhl, J., & Dusing, R. (2011). Oxytocin buffers cortisol responses to stress in individuals with impaired emotion regulation abilities. *Psychoneuroendocrinology*, *36*(6), 898–904.

Rae, M., Zanos, P., Georgiou, P., Chivers, P., Bailey, A., & Camarini, R. (2018). Environmental enrichment enhances conditioned place preference to ethanol via an oxytocinergic-dependent mechanism in male mice. *Neuropharmacology*, *138*, 267–274. https://doi.org/10.1016/j.neuropharm.2018.06.013.

Ring, R. H., Schechter, L. E., Leonard, S. K., Dwyer, J. M., Platt, B. J., Graf, R., et al. (2010). Receptor and behavioral pharmacology of WAY-267464, a non-peptide oxytocin receptor agonist. *Neuropharmacology, 58*, 69–77.

Ross, H. E., & Young, L. J. (2009). Oxytocin and the neural mechanisms regulating social cognition and affiliative behavior. *Frontiers in Neuroendocrinology, 30*(4), 534–547.

Ruan, M., Russell, J. A., & Brown, C. H. (2011). Acute morphine administration and withdrawal from chronic morphine increase after depolarization amplitude in rat supraoptic nucleus neurons in hypothalamic explants. *Neuropharmacology, 61*(4), 789–797.

Russell, J. A., Leng, G., & Bicknell, R. J. (1995). Opioid tolerance and dependence in the magnocellular oxytocin system: A physiological mechanism? *Experimental Physiology, 80*(3), 307–340.

Russell, J. A., Leng, G., & Douglas, A. J. (2003). The magnocellular oxytocin system, the fount of maternity: Adaptations in pregnancy. *Frontiers in Neuroendocrinology, 24*, 27–61.

Russell, J. A., Neumann, I., & Landgraf, R. (1992). Oxytocin and vasopressin release in discrete brain areas after naloxone in morphine-tolerant and -dependent anesthetized rats: Push-pull perfusion study. *Journal of Neuroscience, 12*(3), 1024–1032.

Russo, R., D'Agostino, G., Mattace Raso, G., Avagliano, C., Cristiano, C., Meli, R., et al. (2012). Central administration of oxytocin reduces hyperalgesia in mice: Implication for cannabinoid and opioid systems. *Peptides, 38*(1), 81–88.

Sarnyai, Z. (1998). Oxytocin and neuroadaptation to cocaine. *Progress in Brain Research, 119*, 449–466.

Sarnyai, Z. (2011). Oxytocin as a potential mediator and modulator of drug addiction. *Addiction Biology, 16*(2), 199–201.

Sarnyai, Z., & Kovács, G. L. (1994). Role of oxytocin in the neuroadaptation to drugs of abuse. *Psychoneuroendocrinology, 19*(1), 85–117[Review].

Sarnyai, Z., Viski, S., Krivan, M., Szabo, G., Kovács, G. L., & Telegdy, G. (1988). Endogenous oxytocin inhibits morphine tolerance through limbic forebrain oxytocin receptors. *Brain Research, 463*(2), 284–288.

Sheward, W. J., Coombes, J. E., Bicknell, R. J., Fink, G., & Russell, J. A. (1990). Release of oxytocin but not corticotrophin-releasing factor-41 into rat hypophysial portal vessel blood can be made opiate dependent. *Journal of Endocrinology, 124*(1), 141–150.

Silva, S. M., Madeira, M. D., Ruela, C., & Paula-Barbosa, M. M. (2002). Prolonged alcohol intake leads to irreversible loss of vasopressin and oxytocin neurons in the paraventricular nucleus of the hypothalamus. *Brain Research, 925*, 76–88.

Sivukhina, E. V., Dolzhikov, A. A., Morozov, I. E., Jirikowski, G. F., & Grinevich, V. (2006). Effects of chronic alcoholic disease on magnocellular and parvocellular hypothalamic neurons in men. *Hormone and Metabolic Research, 38*(6), 382–390.

Slattery, D. A., & Neumann, I. D. (2010). Chronic icv oxytocin attenuates the pathological high anxiety state of selectively bred Wistar rats. *Neuropharmacology, 58*, 56–61.

Specka, M., Buchholz, A., Kuhlmann, T., Rist, F., & Scherbaum, N. (2011). Prediction of the outcome of inpatient opiate detoxification treatment: Results from a multicenter study. *European Addiction Research, 17*(4), 178–184.

Striepens, N., Kendrick, K. M., Hanking, V., Landgraf, R., Wullner, U., Maier, W., et al. (2013). Elevated cerebrospinal fluid and blood concentrations of oxytocin following its intranasal administration in humans. *Scientific Reports, 3*, 3440.

Sumner, B. E., Coombes, J. E., Pumford, K. M., & Russell, J. A. (1990). Opioid receptor subtypes in the supraoptic nucleus and posterior pituitary gland of morphine tolerant rats. *Neuroscience, 37*(3), 635–645.

Taylor, S. E., Gonzaga, G. C., Cousino Klein, L., Peifeng, H., Greendale, G. A., & Seeman, T. E. (2006). Relation of oxytocin to psychological stress responses and hypothalamic-pituitary-adrenocortical axis activity in older women. *Psychosomatic Medicine, 68*, 238–245.

Theodosis, D. T. (2002). Oxytocin-secreting neurons: A physiological model of morphological neuronal and glial plasticity in the adult hypothalamus. *Frontiers in Neuroendocrinology*, *23*, 101–135.

Thompson, M. R., Callaghan, P. D., Hunt, G. E., Cornish, J. L., & McGregor, I. S. (2007). A role for oxytocin and 5-Ht(1a) receptors in the prosocial effects of 3,4 Methylenedioxymethamphetamine ("ecstasy"). *Neuroscience*, *146*, 509–514.

Tops, M., van Peer, J. M., Korf, J., Wijers, A. A., & Tucker, D. M. (2007). Anxiety, cortisol, and attachment predict plasma oxytocin. *Psychophysiology*, *44*(3), 444–449.

Uvnäs-Moberg, K., Ahlenius, S., Hillegaart, V., & Alster, P. (1994). High doses of oxytocin cause sedation and low doses cause an anxiolytic-like effect in male rats. *Pharmacology, Biochemistry, and Behavior*, *49*, 101–106.

Uvnäs-Moberg, K., & Petersson, M. (2005). Oxytocin, a mediator of anti-stress, well-being, social interaction, growth and healing. *Zeitschrift für Psychosomatische Medizin und Psychotherapie*, *51*(1), 57–80.

Valstad, M., Alvares, G. A., Egknud, M., et al. (2017). The correlation between central and peripheral oxytocin concentrations: A systematic review and meta-analysis. *Neuroscience and Biobehavioral Reviews*, *78*, 117–124. https://doi.org/10.1016/j.neubiorev.2017.04.017. [Review].

Vuong, C., Van Uum, S. H., O'Dell, L. E., Lutfy, K., & Frieman, T. C. (2010). The effects of opioids and opioids analogs on animal and human endocrine systems. *Endocrine Reviews*, *31*(1), 98–132.

Watson, S. J., Akil, H., Chazarossian, V. E., et al. (1981). Dynorphin immunocytochemical localization in brain and peripheral nervous tissue: Preliminary studies. *Proceedings of the National Academy of Sciences of the United States of America*, *78*, 1260.

Weisman, O., Zagoory-Sharon, O., & Feldman, R. (2012). Intranasal oxytocin administration is reflected in human saliva. *Psychoneuroendocrinology*, *37*(9), 1582–1586. https://doi.org/10.1016/j.psyneuen.2012.02.014.

Windle, R. J., Kershaw, Y. M., Shanks, N., Wood, S. A., Lightman, S. L., & Ingram, C. D. (2004). Oxytocin attenuates stress-induced c-fos mRNA expression in specific forebrain regions associated with modulation of hypothalamo-pituitary-adrenal activity. *Journal of Neuroscience*, *24*, 2974–2982.

Wolff, K., Tsapakis, E. M., Winstock, A. R., Hartley, D., Holt, D., Forsling, M. L., et al. (2006). Vasopressin and oxytocin secretion in response to the consumption of ecstasy in a clubbing population. *Journal of Psychopharmacology*, *20*(3), 400–410.

Woolley, J. D., Arcuni, P. A., Stauffer, C. S., Fulford, D., Carson, D. S., Batki, S., et al. (2016). The effects of intranasal oxytocin in opioid-dependent individuals and healthy control subjects: A pilot study. *Psychopharmacology*, *233*(13), 2571–2580. https://doi.org/10.1007/s00213-016-4308-8.

Wotjak, C. T., Ganster, J., Kohl, G., Holsboer, F., Landgraf, R., & Engelmann, M. (1998). Dissociated central and peripheral release of vasopressin, but not oxytocin, in response to repeated swim stress: New insights into the secretory capacities of peptidergic neurons. *Neuroscience*, *85*, 1209–1222.

Yang, J., Liang, J. Y., Li, P., Pan, Y. J., Qiu, P. Y., Zhang, J., et al. (2011). Oxytocin in the periaqueductal gray participates in pain modulation in the rat by influencing endogenous opiate peptides. *Peptides*, *32*(6), 1255–1261.

Yayou, K., Ito, S., Yamamoto, N., Kitagawa, S., & Okamura, H. (2010). Relationships of stress responses with plasma oxytocin and prolactin in heifer calves. *Physiology & Behavior*, *99*(3), 362–369.

You, Z. D., Li, J. H., Song, C. Y., Wang, C. H., & Lu, C. L. (2000). Chronic morphine treatment inhibits oxytocin synthesis in rats. *Neuroreport*, *11*(14), 3113–3116.

Young, K. A., Gobrogge, K. L., & Wang, Z. (2011). The role of mesocorticolimbic dopamine in regulating interactions between drugs of abuse and social behavior. *Neuroscience and Biobehavioral Reviews*, *35*, 498–515.

Zanos, P., Georgiou, P., Wright, S. R., Hourani, S. M., Kitchen, I., Winsky-Sommerer, R., et al. (2014). The oxytocin analogue Carbetocin prevents emotional impairment and stress-induced reinstatement of opioid-seeking in morphine-abstinent mice. *Neuropsychopharmacology*, *39*(4), 855–865.

Zanos, P., Wright, S. R., Georgiou, P., Hoon Yoo, J., Ledent, C., Susanna, M., et al. (2014). Chronic methamphetamine treatment induces oxytocin receptor up-regulation in the amygdala and hypothalamus via an adenosine A2A receptor-independent mechanism. *Pharmacology, Biochemistry and Behavior*, *119*, 72–79.

Zhao, B. G., Chapman, C., & Bicknell, R. J. (1988). Opioid-noradrenergic interactions in the neurohypophysis. I. Differential opioid receptor regulation of oxytocin, vasopressin, and noradrenaline release. *Neuroendocrinology*, *48*(1), 16–24.

Zhao, B. G., Chapman, C., Brown, D., & Bicknell, R. J. (1988). Opioid-noradrenergic interactions in the neurohypophysis. II. Does noradrenaline mediate the actions of endogenous opioids on oxytocin and vasopressin release? *Neuroendocrinology*, *48*(1), 25–31.

Zhou, Y., Leri, F., Cummins, E., Hoeschele, M., & Kreek, M. J. (2008). Involvement of arginine vasopressin and V1b receptor in heroin withdrawal and heroin seeking precipitated by stress and by heroin. *Neuropsychopharmacology*, *33*(2), 226–236.

Further reading

Kirsch, P., Esslinger, C., Chen, Q., Mier, D., Lis, S., Siddhanti, S., et al. (2005). Oxytocin modulates neural circuitry for social cognition and fear in humans. *The Journal of Neuroscience*, *25*(49), 11489–11493.

Kovács, G. L., & Telegdy, G. (1987). Beta-endorphin tolerance is inhibited by oxytocin. *Pharmacology, Biochemistry, and Behavior*, *26*(1), 57–60.

Pedersen, C. A., Smedley, K. L., Leserman, J., Jarskog, L. F., Rau, S. W., Kampov-Polevoi, A., et al. (2013). Intranasal oxytocin blocks alcohol withdrawal in human subjects. *Alcoholism, Clinical and Experimental Research*, *37*(3), 484–489.

CHAPTER NINE

Estrogens as arbiters of sex-specific and reproductive cycle-dependent opioid analgesic mechanisms

Alan R. Gintzler*, Emiliya M. Storman, Nai-Jiang Liu
Department of Obstetrics and Gynecology, State University of New York, Downstate Medical Center, Brooklyn, NY, United States
*Corresponding author: e-mail address: alan.gintzler@downstate.edu

Contents

Abstract

The organization of estrogenic signaling in the CNS is exceedingly complex. It is comprised of peripherally and centrally synthesized estrogens, and a plethora of types of estrogen receptor that can localize to both the nucleus and the plasma membrane. Moreover, CNS estrogen receptors can exist independent of aromatase (aka estrogen synthase) as well as oligomerize with it, along with a host of other membrane signaling proteins. This ability of CNS estrogen receptors to either to physically pair or exist separately enables locally produced estrogens to act on multiple spatial levels, with a high degree of gradated regulation and plasticity, signaling either in-phase or out-of phase with circulating estrogens. This complexity explains the numerous contradictory findings regarding sex-dependent pain processing and sexually dimorphic opioid

Vitamins and Hormones, Volume 111
ISSN 0083-6729
https://doi.org/10.1016/bs.vh.2019.06.002

antinociception. This review highlights the increasing awareness that estrogens are major endogenous arbiters of both opioid analgesic actions and the mechanisms used to achieve them. This behooves us to understand, and possibly intercede at, the points of intersection of estrogenic signaling and opioid functionality. Factors that integrate estrogenic actions at subcellular, synaptic, and CNS regional levels are likely to be prime drug targets for novel pharmacotherapies designed to modulate CNS estrogen-dependent opioid functionalities and possibly circumvent the current opioid epidemic.

1. Introduction

Most hormones and neurotransmitters that are present in males and females utilize similar molecular mechanisms in both sexes. Interestingly, this is not the case with opioids, which can act via substantially different mechanisms in males vs. females to achieve the same end result. Assuming that sex-dependent responsiveness to exogenous opioids reflects analogous divergence in *endogenous* opioid systems, the lack of parallelism in mechanisms underlying opioid actions in males and females not only reveals an extraordinary degree of plasticity among systems mediating opioid actions, but also provides insight into the ways in which endogenous opioid systems might be manipulated for medicinal purposes. Although sexual dimorphism in pain and opioid antinociception has long been recognized (Craft, 2003; Fillingim & Gear, 2004; Fillingim, King, Ribeiro-Dasilva, Rahim-Williams, & Riley, 2009; Ibironke & Aji, 2011; Liu & Gintzler, 2013; Loyd, Morgan, & Murphy, 2007; Mogil, Chesler, Wilson, Juraska, & Sternberg, 2000; Teepker, Peters, Vedder, Schepelmann, & Lautenbacher, 2010), we are only now beginning to understand its biochemical and molecular underpinnings, and the ability of endogenous estrogens to act as key arbiters of mechanisms utilized by opioids and sexual dimorphism thereof.

Estrogens (predominantly comprised of estradiol and estriol) belong to the steroidal class of hormones, which are derivatives of cholesterol. Enzymes largely control the synthesis and release of steroidal hormones since they are exceedingly lipid soluble and cannot be stored in vesicles. The ovaries are the predominant source of estrogens. However, we now know that throughout the central nervous system (CNS) there is expression of aromatase (Balthazart, Baillien, & Ball, 2001, 2006; Balthazart, Cornil, et al., 2006; Cornil, Ball, & Balthazart, 2006; Evrard, 2006; Evrard et al., 2000; Evrard & Balthazart, 2003; Evrard, Willems, Harada, & Balthazart, 2003; Jakab, Harada, & Naftolin, 1994; Liu, Chakrabarti, Schnell,

Wessendorf, & Gintzler, 2011; Liu, Murugaiyan, Storman, Schnell, Wessendorf, et al., 2017b) (a key enzyme in the synthesis of estrogens), along with multiple types of estrogen receptor (ER) (Hazell et al., 2009; Merchenthaler, Lane, Numan, & Dellovade, 2004; Mitra et al., 2003; Perez, Chen, & Mufson, 2003; Shughrue & Merchenthaler, 2001). This resulted in a profound change in the functional categories used to describe estrogenic actions, i.e., not exclusively as hormones, but also as neuroactive agents that are intrinsic to the CNS. Additionally, the realization that the CNS is an estrogen-producing organ, and that some CNS-derived estrogens are poised to be secreted into the periphery (Storman, Liu, Wessendorf, & Gintzler, 2018), illustrates that the brain should be considered as an neuroendocrine organ.

Estrogens are known to influence a myriad of CNS functions. These include mood, memory, cognition, neuroprotection and, more recently, nociception (pain processing) and opioid antinociception (pain relief) (reviewed in Boulware & Mermelstein, 2005). This review will focus exclusively on the effect of estrogens on opioid functionality (exogenous as well as endogenous), the importance of estrogens to sex-dependent analgesic responsiveness, and related underlying mechanisms.

Estrogens and their receptors are now established to influence the physiological consequences and underlying analgesic mechanism(s) used by morphine (Liu, von Gizycki, & Gintzler, 2007), endomorphin 2 (EM2) (Liu & Gintzler, 2013) (a highly mu-opioid receptor (MOR)-selective opioid (Zadina, Hackler, Ge, & Kastin, 1997)), and dynorphin (kappa opioid receptor (KOR) agonist) (Gear, Gordon, et al., 1996; Gear, Miaskowski, et al., 1996; Miaskowski & Levine, 1999), as well as the heterodimerization of MOR with KOR (Chakrabarti, Liu, & Gintzler, 2010; Liu et al., 2011), the endogenous release of spinal opioids (e.g., EM2, dynorphin) (Kumar, Storman, Liu, & Gintzler, 2015; Liu, Murugaiyan, Storman, Schnell, Kumar, et al., 2017a), and the physiological collaboration among them (e.g., EM2 and dynorphin) (Liu, Murugaiyan, Storman, Schnell, Kumar, et al., 2017a). Complicating the influence of estrogens on opioid sequelae is the fact that at least two physiological distinguishable pools of estrogens exist: peripheral (ovarian) and estrogens produced within CNS. Furthermore, CNS-produced estrogens are not necessarily synchronized with the synthesis of ovarian estrogens, i.e., the production of CNS-derived estrogens can be either in-phase or out-of-phase with the ebb and flow of circulating estrogens (Storman et al., 2018). This complicates associating estrogen-dependent CNS functionalities with a particular stage of reproductive cycle. Additionally, nuclear ERs also traffic to the plasma membrane (Bjornstrom & Sjoberg, 2005; Boulware & Mermelstein, 2005; Dewing et al., 2007; Levin, 2005, 2008, 2009a, 2009b, 2009c; Razandi et al., 2003; Vasudevan & Pfaff, 2007), where they can either oligomerize with aromatase (estrogen synthase) or exist independent (or "free") of it (Storman et al., 2018). This effectively creates multiple subpopulations of ERs that differ in function and regulation.

2. Sexually dimorphic opioid mechanisms mediate comparable opioid analgesia

Estrogens can profoundly influence the mechanisms that underlie analgesic responsiveness to the spinal application of morphine and their sexually dimorphic presentation. The spinal application of morphine produces thermal antinociception that is similar in magnitude and time course in rats of both sexes. However, despite the similarity in analgesic responsiveness in males and females, the antinociception results from strikingly sex-specific differential recruitment of spinal opioid analgesic components (Liu et al., 2007). Activation of spinal MOR, the predominant opioid receptor that is targeted for medicinal pain management, is critical for spinal morphine antinociception in both sexes. However, in females, but not males, activation of spinal KOR is also a prerequisite for spinal morphine antinociception, as is the recruitment of spinal dynorphin (an endogenous KOR-selective agonist) (Liu et al., 2007).

Gonadal hormonal action can be either "activational" or "organizational" (Phoenix, Goy, Gerall, & Young, 1959). The sexually dimorphic dynorphin/KOR component of spinal morphine analgesia is strikingly dependent on organizational actions of ovarian steroids. Elimination of acute activational effects of gonadal steroids via ovariectomy or orchiectomy fails to eliminate in females or unmask in males the dynorphin/KOR component of spinal morphine analgesia (Liu et al., 2007). This indicates that activational actions of gonadal hormones are not a determinant of phenotypic responsiveness to intrathecal morphine. In striking contrast, the spinal KOR component of spinal morphine antinociception is not manifest in sexually mature female rats that had been androgenized during the neonatal period (Liu et al., 2007). Thus, organizational effects of ovarian sex steroids are critical to the ability of intrathecal morphine to recruit KOR antinociceptive mechanisms. Notably, neonatal castration does not unveil a KOR component of spinal antinociception in sexually mature males. Thus, it is the presence of ovarian sex steroids during development, not the absence of testicular steroids, that is the critical ingredient for the ability of spinal morphine to harness KOR antinociception in females.

3. Influence of estrogens on the physical relationship of MOR and KOR

The female-specific dependence of spinal morphine antinociception on the concomitant utilization of spinal MOR and KOR/dynorphin bespeaks of an organization between spinal MOR and KOR in females that is not manifest in males. Indeed, MOR/KOR heterodimers are vastly more prevalent in the spinal cord of females than males, and in the spinal cord of proestrous than diestrous females (Chakrabarti et al., 2010).

Female-specific and estrous cycle stage-dependent formation of spinal MOR/KOR heterodimers suggests estrogens to be critical determinants of MOR/KOR oligomerization. Furthermore, the parallel occurrence of MOR/KOR heterodimerization and dynorphin/KOR mediation of spinal morphine antinociception during proestrus (when circulating levels of estrogens are high), but not during diestrus (when circulating estrogens are low) (Chakrabarti et al., 2010), is consistent with the ability of spinal dynorphin to activate the KOR protomer within heterodimeric MOR/KOR (Chakrabarti et al., 2010), providing the required KOR component of spinal morphine antinociception during proestrus. These findings strongly indicate MOR/KOR to be an estrogen-dependent female-specific molecular transducer for spinal morphine antinociception.

Abundant formation of spinal MOR/KOR during proestrus but not diestrus (or in males) suggests, but does not definitively establish, that estrogens are the driving force. This was validated by demonstrating the effect on MOR/KOR of spinally administered ER antagonists and ovariectomy. These studies indicated that CNS-derived as well as ovarian-derived estrogens are essential determinants of the ebb and flow of the heterodimerization of MOR and KOR (Chakrabarti et al., 2010) and thus the spinal antinociceptive mechanisms that are available, i.e., KOR-dependent and KOR-independent. Additionally, this oligomerization also requires concomitant, but non-additive, activation of multiple spinal mERs (ERα, ERβ and ER1 {aka GPR30, a G protein coupled ER}) (Liu et al., 2011).

Importantly, spinal estrogenic signaling, or lack thereof, interconverts the mechanistic underpinnings of spinal morphine analgesia (without altering its magnitude or temporal characteristics), shifting it between KOR-dependent and KOR-independent. There are substantial physiological and medicinal implications of the parallel existence of readily exchangeable morphine-activated spinal analgesic systems, which oscillate in accordance with stage of cycle. For example, in proestrous females (or the corresponding menstrual cycle stage in women), when spinal morphine analgesia requires dynorphin activation of the KOR protomer of MOR/KOR, impaired spinal dynorphin release could result in lack of analgesic responsiveness to spinal morphine. However, spinal morphine analgesic responsiveness can be rescued by interfering with spinal estrogenic signaling (via either spinal aromatase inhibition, or blockade of spinal mERs). This realization suggests that the spinal application of fadrozole (an aromatase inhibitor) or selective antagonists of mERα, mERβ, or GPR30 are likely to be effective adjuvants with intrathecal morphine in these situations.

4. Estrogens can influence the balance between antinociception and pronociception induced by dynorphin-KOR signaling

Dynorphin and KOR have been proposed to be pronociceptive (pain promoting) (Dubner & Ruda, 1992; Hollt, Haarmann, Millan, & Herz, 1987; Lai, Ossipov, Vanderah, Malan, & Porreca, 2001; Ruda, Iadarola, Cohen, & Young III, 1988; Stiller, Grubb, & Schaible, 1993; Wang et al., 2001) as well as antinociceptive (pain reducing) (Guirimand, Strimbu-Gozariu, Willer, & Le Bars, 1994; Han & Xie, 1982; Herman & Goldstein, 1984; Menendez, Andres-Trelles, Hidalgo, & Baamonde, 1993; Pelissier, Paeile, Soto-Moyano, Saavedra, & Hernandez, 1990; Przewlocki et al., 1983; Schmauss, 1987; Schmauss & Yaksh, 1984; Watkins, Wiertelak, & Maier, 1992). Formation of spinal cord MOR/KOR heterodimers provides a way to adjust the balance between antinociceptive vs. pronociceptive functions of the spinal dynorphin/KOR opioid system. The estrogen-driven chemical partnering of KOR with MOR facilitates spinal KOR-mediated antinociception to be manifest without being compromised by the pain-promoting functions that have also been associated with spinal KOR. Thus, dimeric MOR/KOR represents an estrogen-dependent female-specific signaling molecule that could underlie reports of much greater KOR-mediated antinociception in women than men

(Gear et al., 2003; Gear, Gordon, et al., 1996; Gear et al., 1999; Gear, Miaskowski, et al., 1996; Holtman & Wala, 2006). The more robust, estrogen-driven heterodimerization of MOR and KOR in spinal cord of females than males suggests that in females dynorphin is capable of subserving antinociception, whereas in males dynorphin is much more likely to act as a pronociceptive agent.

5. Spinal ER-mGluR1 signaling suppresses EM2 analgesia

In addition to modulating the mechanistic underpinnings of analgesic responsiveness to spinal morphine (without altering the analgesia itself), estrogens are also endogenous modulators of the analgesia elicited by EM2 (the predominant endogenous MOR ligand in the spinal cord (Martin-Schild, Gerall, Kastin, & Zadina, 1999)). Antinociception elicited by spinal EM2 manifests not only a striking sexual dimorphism, but also dependence on stage of cycle. In females, the spinal EM2 antinociceptive system oscillates between analgesically active and inactive states. During diestrus, when circulating estrogens are low, spinal EM2 antinociception is minimal, whereas during proestrus spinal EM2 antinociception is robust and comparable in magnitude to that observed in males. Additionally, in proestrous females, spinal EM2 antinociception also requires spinal dynorphin and KOR activation, concomitant with MOR activation (Liu & Gintzler, 2013), as was observed for morphine.

However, instead of altering the mechanistic underpinnings of comparable analgesic responsiveness (as occurs for spinal morphine), estrogens actively suppress the spinal EM2 analgesia that is elicited by intrathecal EM2. This effect is mediated via mERα activation of mGluR$_1$ (Liu & Gintzler, 2013). Furthermore, illustrating the complexity of estrogenic modulation, estrogenic suppression of spinal EM2 analgesia occurs, paradoxically, during diestrus (when circulating estrogens are low).

This lack synchronicity with circulating estrogens of spinal estrogenic signaling that suppresses spinal EM2 analgesia results from the actions of mERα within a spinal membrane-bound oligomer that contains aromatase (aka estrogen synthase), metabotropic glutamate receptors (mGluR$_1$, mGluR$_{2/3}$), MOR and KOR (Liu, Murugaiyan, Storman, Schnell, Kumar, et al., 2017a; Liu, Murugaiyan, Storman, Schnell, Wessendorf, et al., 2017b). In the case of EM2, during diestrus, suppressive spinal estrogenic signaling is mediated by mERα that signals via mGluR$_1$

(Boulware et al., 2005; Liu, Murugaiyan, Storman, Schnell, Wessendorf, et al., 2017b). Estrogen-activated mERα-mGluR$_1$ signaling reduces spinal dynorphin release, lessening KOR activation (Liu, Murugaiyan, Storman, Schnell, Kumar, et al., 2017a). This impedes spinal EM2 analgesic responsiveness since it requires both dynorphin and KOR activity, as is the case during proestrus (Liu & Gintzler, 2013).

6. Estrogenic signaling suppresses *endogenous* spinal EM2 release

The ability of estrogens to inhibit release of spinal dynorphin is not restricted to this opioid. Spinal estrogenic signaling is also a critical regulator of the utilization of *endogenous* spinal EM2. However, whereas mERα signaling (when coupled to mGluR$_1$) is sufficient to suppress spinal dynorphin release, suppression of spinal EM2 release requires the concomitant activity of both mERα and GPR30 (Kumar et al., 2015). The need for cooperative signaling by mERα and GPR30 is consistent with earlier reports of their co-expression by neurons of the spinal dorsal horn (Liu et al., 2011), and their cooperative effects on gene transcription (Albanito et al., 2007) and formation of spinal MOR/KOR heterodimers (Liu et al., 2011).

Notably, the ability of mERα/GPR30 activity to inhibit EM2 release is sexually dimorphic; mERα/GPR30 blockade does not have any effect on spinal EM2 release in spinal cord of males (Kumar et al., 2015). Furthermore, the magnitude of estrogenic inhibition of spinal EM2 release is dependent on stage of estrous cycle (Kumar et al., 2015); blockade of spinal mERα/GPR30 during proestrus results in the largest increase in spinal EM2 release, an intermediate increase during diestrus, and no detectable increment during estrus (when circulating estrogens are at their lowest).

The negative correlation between the magnitude of basal EM2 release and circulating estrogens levels (i.e., basal EM2 release is highest during estrus, when circulating estrogens are lowest), suggests that spinal estrogenic activity acts as a physiological clamp to suppress EM2 release. Strikingly, the enhancement of spinal EM2 release by concomitant blockade of ERα and GPR30 is eliminated by either inhibition of spinal aromatase (via intrathecal fadrozole) or ovariectomy (Kumar et al., 2015). This indicates that both ovarian and spinally synthesized estrogens act collaboratively to regulate release of spinal EM2.

7. Estrogenic signaling enables coordination of spinal EM2 utilization with physiological demand

It is interesting to note that suppression of spinal EM2 release by estrogens directly parallels estrogen-dependent heterodimerization of MOR and KOR, which is also greater in proestrus than diestrus (Chakrabarti et al., 2010; Liu et al., 2011). Many studies have concluded that monomeric spinal KOR subserves pronociception (see Lai et al., 2001 for review) but not antinociception (Piercey & Einspahr, 1989; Przewlocka, Dziedzicka, Lason, & Przewlocki, 1992; Schmauss, 1987; Stevens & Yaksh, 1986). However, activation of KOR with MOR within MOR/KOR reveals antinociceptive attributes of KOR. Thus, during diestrus, when levels of MOR/KOR are low, the pronociceptive functions of spinal KOR prevail, while during proestrus, the pronociceptive functions of monomeric KOR are counterbalanced by the increased formation of MOR/KOR. So, during proestrus, when there is maximum estrogenic suppression of the release of EM2, this opioid would be less essential for counterbalancing endogenous dynorphin/KOR pronociception than during diestrus, when levels of pronociceptive monomeric KOR are high.

8. Collaboration between spinally- and ovarian-produced estrogens

Several studies have reported the relevance of either peripheral or central estrogens in nociception. As examples, ovariectomized rats demonstrate diminished sensitivity to formalin-induced pain (Gaumond, Arsenault, & Marchand, 2002); inhibition of spinal aromatase decreases pain sensitivity in both male and female quail (Evrard & Balthazart, 2004); in proestrous female rats, inhibition of spinal aromatase reduces MOR/KOR dimerization and eliminates the dynorphin/KOR dependency of morphine antinociception (Liu et al., 2011). There are much fewer examples of collaboration between peripheral and central estrogens in nociception and opioid antinociception. Nevertheless, their functional interrelationship remains subject of much speculation (Schlinger, Remage-Healey, & Rensel, 2014).

CNS-derived estrogens are synthesized by CNS aromatase, which is located near synaptic structures. CNS-derived estrogens are thereby poised to activate nearby mERs relevant to pain modulation (Evrard, 2006;

Evrard & Balthazart, 2004; Hojo et al., 2004; Naftolin et al., 1996; Peterson, Yarram, Schlinger, & Saldanha, 2005). Such estrogenic signaling would be expected to be highly spatially restricted. In contrast, peripherally synthesized estrogens reach the CNS by penetrating the blood-brain barrier and diffusing from cerebrospinal fluid into extracellular fluid (Guyton & Hall, 2010) to act on CNS ERs. Such signaling, at first blush, would be expected to produce uniform activation of all CNS ERs, which is inconsistent with segregated, partitioned CNS functionality and is incongruous with the graded (not all or none) suppression of spinal EM2 release over the estrous cycle (Kumar et al., 2015). Thus, it is more likely that peripheral and central estrogens cooperate to modulate CNS functionality (the relative contributions of each and specific points of intersection remaining largely unknown). The mechanism(s) underpinning hypothesized synergistic interactions between centrally and peripherally synthesized estrogens remain obscure. Given reports that diffusion of centrally synthesized estrogens, and, by extension, diffusion of peripherally synthesized estrogens into the CNS, is highly spatially restricted (Charlier et al., 2011; Fokidis, Prior, & Soma, 2013; Remage-Healey, Maidment, & Schlinger, 2008; Schlinger et al., 2014), it is possible that CNS- and ovarian-derived estrogens activate different populations of spinal ERs that are functionally convergent and act in a cooperative, possibly synergistic fashion.

9. Multiple subpopulations of CNS aromatase and mERα

The plethora and complexity of the actions of estrogens on the CNS is undergirded by a multiplicity of types of ER (e.g., ERα, ERβ, GPR30, Gq-mER, ER-X). The multi-faceted nature of estrogenic modulation of neuronal function is supported by the ability of ERs to localize to the nucleus to regulate transcription, as well as to the plasma membrane (Filardo, Quinn, Bland, & Frackelton, 2000; Razandi, Pedram, Greene, & Levin, 1999; Revankar, Cimino, Sklar, Arterburn, & Prossnitz, 2005), where they activate (within sec to min) signaling cascades identical to those initiated by G protein-coupled receptors (Levin, 2009c; Mermelstein, 2009; Micevych & Dominguez, 2009; Vasudevan & Pfaff, 2008).

Actions of estrogens fall into five main categories: endocrine, paracrine, autocrine, synaptocrine and oligocrine. The first three differ only with respect to the distance between the point of synthesis and the site of action (reviewed in Saldanha, Remage-Healey, & Schlinger, 2011). A fourth category, synaptocrine (Saldanha et al., 2011), involves specific modulation of estrogen concentrations at the synapse. Synaptocrine estrogenic signaling is enabled by the presence of aromatase and ERs on dendritic/somatic membranes and presynaptic terminals (Beyer, Pawlak, & Karolczak, 2003; Blaustein, 1992; Blaustein, Lehman, Turcotte, & Greene, 1992; McEwen et al., 2001; McEwen & Alves, 1999; Schlinger & Callard, 1989; Wu et al., 2009). The fifth category, oligocrine, was defined by our recent finding that aromatase and mERs are present in the same oligomer along with other plasma membrane-associated signaling molecules. Oligocrine signaling, the ability of estrogens to function as intracellular messengers whose synthesis and actions occur within the same macromolecular signaling complex, confers exquisite discrete spatial and temporal specificity to estrogenic signaling. Furthermore, the presence of aromatase activity within a signaling complex containing mERs permits the differential activation/deactivation of discrete subcellular ER-coupled signaling, independent of variations in circulating levels of estrogens—either out of phase (as occurs for estrogenic suppression of spinal EM2 antinociception) or in-phase with ovarian steroid production (as occurs for estrogenic regulation of MOR/KOR heterodimerization).

Oligocrine estrogenic signaling in the CNS complements endocrine (ovarian estrogens acting on CNS ERs) and neurotransmitter-like synaptocrine estrogenic signaling. Maximum physiological utilization of the multiple modalities of CNS estrogenic signaling suggests the importance of coordinating the actions of peripheral and central estrogens. Additionally, the ability of the CNS as well as the ovary to make estrogens means that there are at least two functional pools of aromatase. This duality is further complicated by the presence of aromatase in a CNS plasma membrane oligomer that contains mERs, as well as other signaling proteins. This organization creates two additional populations of mERs and aromatase, those that are oligomerized with each other and those that exist separately, i.e., "free." Thus, the multi-dimensional nature of estrogenic actions in the CNS, particularly as pertains to modulating nociception and opioid antinociception, likely requires not only coordination between the actions of central and peripheral estrogens, but also the activity of spatially and functionally segregated pools of aromatase and mERs in the CNS. Given this complexity and multidimensional nature of estrogenic signaling, it is not surprising that estrogens have been reported to be pronociceptive (Bradshaw, Miller, Ling, Malsnee, & Ruda, 2000; Ji, Murphy, & Traub, 2003; Ji, Tang, &

Traub, 2011; Li et al., 2009; Lu, Chen, Wang, & Wu, 2009; Sanoja & Cervero, 2005), as well as antinociceptive (Aloisi et al., 2010; Cao, Ji, Tang, & Traub, 2012; Fischer et al., 2008; Giamberardino, Affaitati, Valente, Iezzi, & Vecchiet, 1997; Kramer & Bellinger, 2009; Lawson, Nag, Thompson, & Mokha, 2010; Mannino, South, Quinones-Jenab, & Inturrisi, 2007; Riley, Robinson, Wise, & Price, 1999; Sarajari & Oblinger, 2010).

The existence of mERα/aromatase oligomerized and free populations holds out the promise that their various subpopulations can be independently targeted for medicinal purposes. The putative utility of doing so is underscored by our recent finding that although aromatase and mERα physically associate with each other in both spinal cord and hypothalamus, the distribution between mERα-associated and free is essentially opposite in these areas—in the spinal cord, essentially all of the aromatase is oligomerized with mERα, whereas in the hypothalamus, essentially the opposite pertains (Storman et al., 2018). Thus, a drug that is selective for the mERα/aromatase oligomer would likely affect estrogenic signaling in the spinal cord, but markedly less so in hypothalamus whereas targeting free mERα or aromatase would likely alter estrogenic signaling in hypothalamus but not spinal cord.

Perhaps the best example of the potency of sex steroids in influencing opioid analgesia is the increase in nociceptive thresholds that occurs during physiological gestation. During physiological pregnancy, there is a naturally occurring elevation of pain tolerance, peaking just prior to parturition (Gintzler, 1980). Systemic administration of the general opioid receptor blocker, naloxone, abolishes pregnancy-induced analgesia, indicating participation of the endogenous opioid system. Strikingly, the opioid antinociception associated with pregnancy is also seen when the pregnancy blood profile of estrogen and progesterone are simulated (in the absence of physiological pregnancy). Moreover, both the antinociception induced by pregnancy and its hormonal simulation are mediated not only by the same types of opioid receptor but also the same CNS compartment, spinal cord (Dawson-Basoa & Gintzler, 1993, 1996, 1997, 1998; Medina, Dawson-Basoa, & Gintzler, 1993). This underscores that the profile of change in activity of sex steroids is not only necessary, but also sufficient for pregnancy-induced analgesia to be manifest, highlighting the potential of estrogens, acting in combination with progesterone, to unlock the powerful ability of endogenous opioids to mitigate pain.

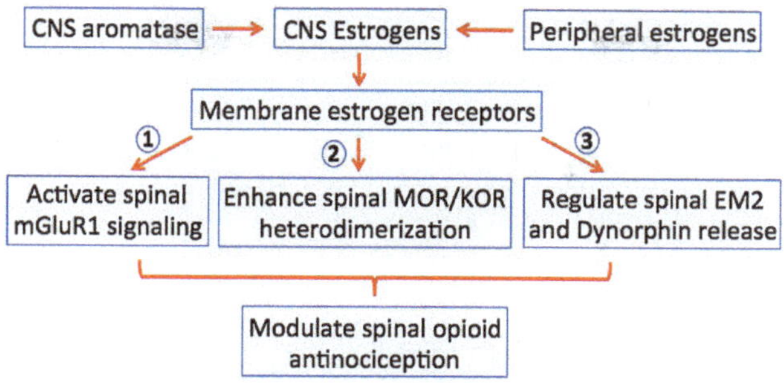

Scheme 1 CNS membrane estrogen receptors are activated by ovarian-produced and CNS-produced estrogens. Opioid parameters modulated by CNS estrogenic signaling include (1) membrane estrogen receptor alpha activation of spinal $mGluR_1$ (to inhibit spinal EM2 analgesia during diestrus), (2) enhancement of the heterodimerization of spinal MOR and KOR (as occurs during proestrus), (3) graded inhibition of spinal release of EM2 (as occurs over the reproductive cycle) and dynorphin release (as occurs during diestrus). All impact exogenous as well as endogenous opioid antinociception.

10. Conclusions

Estrogens impact multiple parameters of opioid functionality (see Scheme 1). Given the multi-faceted influence of estrogens on opioid systems, it is imperative to understand, and possibly intercede at, the points of intersection of estrogenic signaling and opioid functionality, particularly in the face of the current opioid epidemic. The propensity of aromatase to either to physically pair with or exist separately from mERα in combination with a synaptic relationship between aromatase and mERα, enables locally produced estrogens to act on multiple spatial levels, spanning the subcellular to synaptic, with a high degree of gradated regulation and plasticity. Factors that integrate estrogenic actions at subcellular, synaptic, and CNS regional levels are likely to be prime drug targets for novel pharmacotherapies designed to modulate CNS estrogen-dependent functionalities and possibly circumvent the current opioid epidemic.

Disclosures

None of the authors hold any conflicts of interest.

References

Albanito, L., Madeo, A., Lappano, R., Vivacqua, A., Rago, V., Carpino, A., et al. (2007). G protein-coupled receptor 30 (GPR30) mediates gene expression changes and growth response to 17beta-estradiol and selective GPR30 ligand G-1 in ovarian cancer cells. *Cancer Research*, *67*(4), 1859–1866. https://doi.org/10.1158/0008-5472.CAN-06-2909.

Aloisi, A. M., Affaitati, G., Ceccarelli, I., Fiorenzani, P., Lerza, R., Rossi, C., et al. (2010). Estradiol and testosterone differently affect visceral pain-related behavioural responses in male and female rats. *European Journal of Pain, 14*(6), 602–607. https://doi.org/10.1016/j.ejpain.2009.10.016.

Balthazart, J., Baillien, M., & Ball, G. F. (2001). Rapid and reversible inhibition of brain aromatase activity. *Journal of Neuroendocrinology, 13*(1), 63–73. jne598 [pii]. https://doi.org/10.1111/j.1365-2826.2001.00598.x.

Balthazart, J., Baillien, M., & Ball, G. F. (2006). Rapid control of brain aromatase activity by glutamatergic inputs. *Endocrinology, 147*(1), 359–366. https://doi.org/10.1210/en.2005-0845.

Balthazart, J., Cornil, C. A., Taziaux, M., Charlier, T. D., Baillien, M., & Ball, G. F. (2006). Rapid changes in production and behavioral action of estrogens. *Neuroscience, 138*(3), 783–791. https://doi.org/10.1016/j.neuroscience.2005.06.016.

Beyer, C., Pawlak, J., & Karolczak, M. (2003). Membrane receptors for oestrogen in the brain. *Journal of Neurochemistry, 87*(3), 545–550.

Bjornstrom, L., & Sjoberg, M. (2005). Mechanisms of estrogen receptor signaling: Convergence of genomic and nongenomic actions on target genes. *Molecular Endocrinology, 19*(4), 833–842. https://doi.org/10.1210/me.2004-0486.

Blaustein, J. D. (1992). Cytoplasmic estrogen receptors in rat brain: Immunocytochemical evidence using three antibodies with distinct epitopes. *Endocrinology, 131*(3), 1336–1342. https://doi.org/10.1210/endo.131.3.1380440.

Blaustein, J. D., Lehman, M. N., Turcotte, J. C., & Greene, G. (1992). Estrogen receptors in dendrites and axon terminals in the Guinea pig hypothalamus. *Endocrinology, 131*(1), 281–290. https://doi.org/10.1210/endo.131.1.1612006.

Boulware, M. I., & Mermelstein, P. G. (2005). The influence of estradiol on nervous system function. *Drug News & Perspectives, 18*(10), 631–637. https://doi.org/10.1358/dnp.2005.18.10.959578.

Boulware, M. I., Weick, J. P., Becklund, B. R., Kuo, S. P., Groth, R. D., & Mermelstein, P. G. (2005). Estradiol activates group I and II metabotropic glutamate receptor signaling, leading to opposing influences on cAMP response element-binding protein. *The Journal of Neuroscience, 25*(20), 5066–5078.

Bradshaw, H., Miller, J., Ling, Q., Malsnee, K., & Ruda, M. A. (2000). Sex differences and phases of the estrous cycle alter the response of spinal cord dynorphin neurons to peripheral inflammation and hyperalgesia. *Pain, 85*(1–2), 93–99.

Cao, D. Y., Ji, Y., Tang, B., & Traub, R. J. (2012). Estrogen receptor beta activation is antinociceptive in a model of visceral pain in the rat. *The Journal of Pain, 13*(7), 685–694. https://doi.org/10.1016/j.jpain.2012.04.010.

Chakrabarti, S., Liu, N. J., & Gintzler, A. R. (2010). Formation of mu-/kappa-opioid receptor heterodimer is sex-dependent and mediates female-specific opioid analgesia. *Proceedings of the National Academy of Sciences of the United States of America, 107*(46), 20115–20119. https://doi.org/10.1073/pnas.1009923107.

Charlier, T. D., Newman, A. E., Heimovics, S. A., Po, K. W., Saldanha, C. J., & Soma, K. K. (2011). Rapid effects of aggressive interactions on aromatase activity and oestradiol in discrete brain regions of wild male white-crowned sparrows. *Journal of Neuroendocrinology, 23*(8), 742–753. https://doi.org/10.1111/j.1365-2826.2011.02170.x.

Cornil, C. A., Ball, G. F., & Balthazart, J. (2006). Functional significance of the rapid regulation of brain estrogen action: Where do the estrogens come from? *Brain Research, 1126*(1), 2–26. https://doi.org/10.1016/j.brainres.2006.07.098.

Craft, R. M. (2003). Sex differences in opioid analgesia: "From mouse to man" *The Clinical Journal of Pain, 19*(3), 175–186.

Dawson-Basoa, M. E., & Gintzler, A. R. (1993). 17-b-Estradiol and progesterone modulate an intrinsic opioid analgesic system. *Brain Research, 601*, 241–245.

Dawson-Basoa, M. E., & Gintzler, A. R. (1996). Estrogen and progesterone activate spinal kappa-opiate receptor analgesic mechanisms. *Pain, 64*, 607–615.
Dawson-Basoa, M. E., & Gintzler, A. R. (1997). Involvement of spinal cord delta opiate receptors in the antinociception of gestation and its hormal simulation. *Brain Research, 757*, 37–42.
Dawson-Basoa, M. E., & Gintzler, A. R. (1998). Gestational and ovarian sex steroid antinociception: Synergy between spinal k and d opioid systems. *Brain Research, 794*, 61–67.
Dewing, P., Boulware, M. I., Sinchak, K., Christensen, A., Mermelstein, P. G., & Micevych, P. (2007). Membrane estrogen receptor-alpha interactions with metabotropic glutamate receptor 1a modulate female sexual receptivity in rats. *The Journal of Neuroscience: The Official Journal of the Society for Neuroscience, 27*(35), 9294–9300. https://doi.org/10.1523/JNEUROSCI.0592-07.2007.
Dubner, R., & Ruda, M. (1992). Activity-dependent neuronal plasticity following tissue injury and inflammation. *Trends in Neurosciences, 15*, 96–103.
Evrard, H. C. (2006). Estrogen synthesis in the spinal dorsal horn: A new central mechanism for the hormonal regulation of pain. *American Journal of Physiology. Regulatory. Integrative and Comparative Physiology, 291*(2), R291–R299. https://doi.org/10.1152/ajpregu.00930.2005.
Evrard, H., Baillien, M., Foidart, A., Absil, P., Harada, N., & Balthazart, J. (2000). Localization and controls of aromatase in the quail spinal cord. *The Journal of Comparative Neurology, 423*(4), 552–564. https://doi.org/10.1002/1096-9861(20000807)423:4<552::AID-CNE2>3.0.CO;2-S.
Evrard, H. C., & Balthazart, J. (2003). Aromatase (estrogen synthase) activity in the dorsal horn of the spinal cord: Functional implications. *Annals of the New York Academy of Sciences, 1007*, 263–271.
Evrard, H. C., & Balthazart, J. (2004). Rapid regulation of pain by estrogens synthesized in spinal dorsal horn neurons. *The Journal of Neuroscience: The Official Journal of the Society for Neuroscience, 24*(33), 7225–7229. https://doi.org/10.1523/JNEUROSCI.1638-04.2004.
Evrard, H. C., Willems, E., Harada, N., & Balthazart, J. (2003). Specific innervation of aromatase neurons by substance P fibers in the dorsal horn of the spinal cord in quail. *The Journal of Comparative Neurology, 465*(2), 309–318. https://doi.org/10.1002/cne.10854.
Filardo, E. J., Quinn, J. A., Bland, K. I., & Frackelton, A. R., Jr. (2000). Estrogen-induced activation of Erk-1 and Erk-2 requires the G protein-coupled receptor homolog, GPR30, and occurs via trans-activation of the epidermal growth factor receptor through release of HB-EGF. *Molecular Endocrinology, 14*(10), 1649–1660.
Fillingim, R. B., & Gear, R. W. (2004). Sex differences in opioid analgesia: Clinical and experimental findings. *European Journal of Pain, 8*(5), 413–425. https://doi.org/10.1016/j.ejpain.2004.01.007.
Fillingim, R. B., King, C. D., Ribeiro-Dasilva, M. C., Rahim-Williams, B., & Riley, J. L., 3rd. (2009). Sex, gender, and pain: A review of recent clinical and experimental findings. *The Journal of Pain, 10*(5), 447–485. https://doi.org/10.1016/j.jpain.2008.12.001.
Fischer, L., Torres-Chavez, K. E., Clemente-Napimoga, J. T., Jorge, D., Arsati, F., de Arruda Veiga, M. C., et al. (2008). The influence of sex and ovarian hormones on temporomandibular joint nociception in rats. *The Journal of Pain, 9*(7), 630–638. https://doi.org/10.1016/j.jpain.2008.02.006.
Fokidis, H. B., Prior, N. H., & Soma, K. K. (2013). Fasting increases aggression and differentially modulates local and systemic steroid levels in male zebra finches. *Endocrinology, 154*(11), 4328–4339. https://doi.org/10.1210/en.2013-1171.
Gaumond, I., Arsenault, P., & Marchand, S. (2002). The role of sex hormones on formalin-induced nociceptive responses. *Brain Research, 958*(1), 139–145. S0006899302036612 [pii]. https://doi.org/10.1016/S0006-8993(02)03661-2.

Gear, R. W., Gordon, N. C., Heller, P. H., Paul, S., Miaskowski, C., & Levine, J. D. (1996). Gender difference in analgesic response to the kappa-opioid pentazocine. *Neuroscience Letters*, *205*(3), 207–209.

Gear, R. W., Gordon, N. C., Miaskowski, C., Paul, S. M., Heller, P. H., & Levine, J. D. (2003). Sexual dimorphism in very low dose nalbuphine postoperative analgesia. *Neuroscience Letters*, *339*(1), 1–4.

Gear, R. W., Miaskowski, C., Gordon, N. C., Paul, S. M., Heller, P. H., & Levine, J. D. (1996). Kappa-opioids produce significantly greater analgesia in women than in men. *Nature Medicine*, *2*(11), 1248–1250.

Gear, R. W., Miaskowski, C., Gordon, N. C., Paul, S. M., Heller, P. H., & Levine, J. D. (1999). The kappa opioid nalbuphine produces gender- and dose-dependent analgesia and antianalgesia in patients with postoperative pain. *Pain*, *83*(2), 339–345.

Giamberardino, M. A., Affaitati, G., Valente, R., Iezzi, S., & Vecchiet, L. (1997). Changes in visceral pain reactivity as a function of estrous cycle in female rats with artificial ureteral calculosis. *Brain Research*, *774*(1–2), 234–238. doi: S0006-8993(97)81711-8 [pii]. https://doi.org/10.1016/s0006-8993(97)81711-8.

Gintzler, A. R. (1980). Endorphin-mediated increases in pain threshold during pregnancy. *Science*, *210*, 193–195.

Guirimand, F., Strimbu-Gozariu, M., Willer, J.-C., & Le Bars, D. (1994). Effects of mu, delta and kappa opioid antagonists on the depression of a C-fiber reflex by intrathecal morphine and DAGO in the rat. *The Journal of Pharmacology and Experimental Therapeutics*, *269*, 1007–1020.

Guyton, A., & Hall, J. E. (2010). *Guyton and Hall textbook of medical physiology* (12th ed.). Philadelphia, PA: Saunders.

Han, J. S., & Xie, C. W. (1982). Dynorphin: Potent analgesic effect in spinal cord of the rat. *Life Sciences*, *31*, 1781–1784.

Hazell, G. G., Yao, S. T., Roper, J. A., Prossnitz, E. R., O'Carroll, A. M., & Lolait, S. J. (2009). Localisation of GPR30, a novel G protein-coupled oestrogen receptor, suggests multiple functions in rodent brain and peripheral tissues. *The Journal of Endocrinology*, *202*(2), 223–236. https://doi.org/10.1677/JOE-09-0066.

Herman, B. H., & Goldstein, A. (1984). Antinociception and paralysis induced by intrathecal dynorphin a. *The Journal of Pharmacology and Experimental Therapeutics*, *232*, 27–32.

Hojo, Y., Hattori, T. A., Enami, T., Furukawa, A., Suzuki, K., Ishii, H. T., et al. (2004). Adult male rat hippocampus synthesizes estradiol from pregnenolone by cytochromes P45017alpha and P450 aromatase localized in neurons. *Proceedings of the National Academy of Sciences of the United States of America*, *101*(3), 865–870. https://doi.org/10.1073/pnas.2630225100.

Hollt, V., Haarmann, I., Millan, M. J., & Herz, A. (1987). Prodynorphin gene expression is enhanced in the spinal cord of chronic arthritic rats. *Neuroscience Letters*, *73*, 90–94.

Holtman, J. R., Jr., & Wala, E. P. (2006). Characterization of the antinociceptive effect of oxycodone in male and female rats. *Pharmacology, Biochemistry, and Behavior*, *83*(1), 100–108. https://doi.org/10.1016/j.pbb.2005.12.013.

Ibironke, G. F., & Aji, K. E. (2011). Pain threshold variations in female rats as a function of the estrus cycle. *Nigerian Journal of Physiological Sciences*, *26*(1), 67–70.

Jakab, R. L., Harada, N., & Naftolin, F. (1994). Aromatase- (estrogen synthetase) immunoreactive neurons in the rat septal area. A light and electron microscopic study. *Brain Research*, *664*(1–2), 85–93.

Ji, Y., Murphy, A. Z., & Traub, R. J. (2003). Estrogen modulates the visceromotor reflex and responses of spinal dorsal horn neurons to colorectal stimulation in the rat. *The Journal of Neuroscience: The Official Journal of the Society for Neuroscience*, *23*(9), 3908–3915. 23/9/3908 [pii]. https://doi.org/10.1523/JNEUROSCI.23-09-03908.2003.

Ji, Y., Tang, B., & Traub, R. J. (2011). Spinal estrogen receptor alpha mediates estradiol-induced pronociception in a visceral pain model in the rat. *Pain, 152*(5), 1182–1191. https://doi.org/10.1016/j.pain.2011.01.046.

Kramer, P. R., & Bellinger, L. L. (2009). The effects of cycling levels of 17beta-estradiol and progesterone on the magnitude of temporomandibular joint-induced nociception. *Endocrinology, 150*(8), 3680–3689. https://doi.org/10.1210/en.2008-1707.

Kumar, A., Storman, E. M., Liu, N. J., & Gintzler, A. R. (2015). Estrogens suppress spinal Endomorphin 2 release in female rats in phase with the estrous cycle. *Neuroendocrinology, 102*(1–2), 33–43. https://doi.org/10.1159/000430817.

Lai, J., Ossipov, M. H., Vanderah, T. W., Malan, T. P., Jr., & Porreca, F. (2001). Neuropathic pain: The paradox of dynorphin. *Molecular Interventions, 1*(3), 160–167.

Lawson, K. P., Nag, S., Thompson, A. D., & Mokha, S. S. (2010). Sex-specificity and estrogen-dependence of kappa opioid receptor-mediated antinociception and anti-hyperalgesia. *Pain, 151*(3), 806–815. https://doi.org/10.1016/j.pain.2010.09.018.

Levin, E. R. (2005). Integration of the extranuclear and nuclear actions of estrogen. *Molecular Endocrinology, 19*(8), 1951–1959. https://doi.org/10.1210/me.2004-0390.

Levin, E. R. (2008). Rapid signaling by steroid receptors. *American Journal of Physiology. Regulatory, Integrative and Comparative Physiology, 295*(5), R1425–R1430. https://doi.org/10.1152/ajpregu.90605.2008.

Levin, E. R. (2009a). G protein-coupled receptor 30: Estrogen receptor or collaborator? *Endocrinology, 150*(4), 1563–1565. https://doi.org/10.1210/en.2008-1759.

Levin, E. R. (2009b). Membrane oestrogen receptor alpha signalling to cell functions. *The Journal of Physiology, 587*(Pt. 21), 5019–5023. https://doi.org/10.1113/jphysiol.2009.177097.

Levin, E. R. (2009c). Plasma membrane estrogen receptors. *Trends in Endocrinology and Metabolism, 20*(10), 477–482. https://doi.org/10.1016/j.tem.2009.06.009.

Li, L., Fan, X., Warner, M., Xu, X. J., Gustafsson, J. A., & Wiesenfeld-Hallin, Z. (2009). Ablation of estrogen receptor alpha or beta eliminates sex differences in mechanical pain threshold in normal and inflamed mice. *Pain, 143*(1–2), 37–40. https://doi.org/10.1016/j.pain.2009.01.005.

Liu, N. J., Chakrabarti, S., Schnell, S., Wessendorf, M., & Gintzler, A. R. (2011). Spinal synthesis of estrogen and concomitant signaling by membrane estrogen receptors regulate spinal {kappa}- and {mu}-opioid receptor heterodimerization and female-specific spinal morphine antinociception. *The Journal of Neuroscience: The Official Journal of the Society for Neuroscience, 31*(33), 11836–11845. https://doi.org/10.1523/JNEUROSCI.1901-11.2011.

Liu, N. J., & Gintzler, A. R. (2013). Spinal endomorphin 2 antinociception and the mechanisms that produce it are both sex- and stage of estrus cycle-dependent in rats. *The Journal of Pain, 14*(11), 1522–1530. https://doi.org/10.1016/j.jpain.2013.09.002.

Liu, N. J., Murugaiyan, V., Storman, E. M., Schnell, S. A., Kumar, A., Wessendorf, M. W., et al. (2017a). Plasticity of signaling by spinal estrogen receptor alpha, kappa-opioid receptor and mGluRs over the rat reproductive cycle regulates spinal endomorphin 2 antinociception: Relevance of endogenous biased agonism. *The Journal of Neuroscience, 37*, 11181–11191. https://doi.org/10.1523/JNEUROSCI.1927-17.2017.

Liu, N. J., Murugaiyan, V., Storman, E. M., Schnell, S. A., Wessendorf, M. W., & Gintzler, A. R. (2017b). Estrogens synthesized and acting within a spinal oligomer suppress spinal endomorphin 2 antinociception: Ebb and flow over the rat reproductive cycle. *Pain, 158*(10), 1903–1914. https://doi.org/10.1097/j.pain.0000000000000991.

Liu, N. J., von Gizycki, H., & Gintzler, A. R. (2007). Sexually dimorphic recruitment of spinal opioid analgesic pathways by the spinal application of morphine. *The Journal of Pharmacology and Experimental Therapeutics, 322*(2), 654–660.

Loyd, D. R., Morgan, M. M., & Murphy, A. Z. (2007). Morphine preferentially activates the periaqueductal gray-rostral ventromedial medullary pathway in the male rat: A potential mechanism for sex differences in antinociception. *Neuroscience*, *147*(2), 456–468.

Lu, Y. C., Chen, C. W., Wang, S. Y., & Wu, F. S. (2009). 17Beta-estradiol mediates the sex difference in capsaicin-induced nociception in rats. *The Journal of Pharmacology and Experimental Therapeutics*, *331*(3), 1104–1110. https://doi.org/10.1124/jpet.109.158402.

Mannino, C. A., South, S. M., Quinones-Jenab, V., & Inturrisi, C. E. (2007). Estradiol replacement in ovariectomized rats is antihyperalgesic in the formalin test. *The Journal of Pain*, *8*(4), 334–342. https://doi.org/10.1016/j.jpain.2006.10.002.

Martin-Schild, S., Gerall, A. A., Kastin, A. J., & Zadina, J. E. (1999). Differential distribution of endomorphin 1- and endomorphin 2-like immunoreactivities in the CNS of the rodent. *The Journal of Comparative Neurology*, *405*, 450–471.

McEwen, B., Akama, K., Alves, S., Brake, W. G., Bulloch, K., Lee, S., et al. (2001). Tracking the estrogen receptor in neurons: Implications for estrogen-induced synapse formation. *Proceedings of the National Academy of Sciences of the United States of America*, *98*(13), 7093–7100. https://doi.org/10.1073/pnas.121146898.

McEwen, B. S., & Alves, S. E. (1999). Estrogen actions in the central nervous system. *Endocrine Reviews*, *20*(3), 279–307.

Medina, V. M., Dawson-Basoa, M. E., & Gintzler, A. R. (1993). 17-b-estradiol and progesterone positively modulate spinal cord dynorphin: Relevance to the analgesia of pregnancy. *Neuroendocrinology*, *58*, 310–315.

Menendez, L., Andres-Trelles, F., Hidalgo, A., & Baamonde, A. (1993). Involvement of spinal k opioid receptors in a type of footshock induced analgesia in mice. *Brain Research*, *611*, 264–271.

Merchenthaler, I., Lane, M. V., Numan, S., & Dellovade, T. L. (2004). Distribution of estrogen receptor alpha and beta in the mouse central nervous system: In vivo autoradiographic and immunocytochemical analyses. *The Journal of Comparative Neurology*, *473*(2), 270–291. https://doi.org/10.1002/cne.20128.

Mermelstein, P. G. (2009). Membrane-localised oestrogen receptor alpha and beta influence neuronal activity through activation of metabotropic glutamate receptors. *Journal of Neuroendocrinology*, *21*(4), 257–262. https://doi.org/10.1111/j.1365-2826.2009.01838.x.

Miaskowski, C., & Levine, J. D. (1999). Does opioid analgesia show a gender preference for females? *Pain Forum*, *8*, 34–44.

Micevych, P., & Dominguez, R. (2009). Membrane estradiol signaling in the brain. *Frontiers in Neuroendocrinology*, *30*(3), 315–327. https://doi.org/10.1016/j.yfrne.2009.04.011.

Mitra, S. W., Hoskin, E., Yudkovitz, J., Pear, L., Wilkinson, H. A., Hayashi, S., et al. (2003). Immunolocalization of estrogen receptor beta in the mouse brain: Comparison with estrogen receptor alpha. *Endocrinology*, *144*(5), 2055–2067. https://doi.org/10.1210/en.2002-221069.

Mogil, J. S., Chesler, E. J., Wilson, S. G., Juraska, J. M., & Sternberg, W. F. (2000). Sex differences in thermal nociception and morphine antinociception in rodents depend on genotype. *Neuroscience and Biobehavioral Reviews*, *24*(3), 375–389.

Naftolin, F., Horvath, T. L., Jakab, R. L., Leranth, C., Harada, N., & Balthazart, J. (1996). Aromatase immunoreactivity in axon terminals of the vertebrate brain. An immunocytochemical study on quail, rat, monkey and human tissues. *Neuroendocrinology*, *63*(2), 149–155.

Pelissier, T., Paeile, C., Soto-Moyano, R., Saavedra, H., & Hernandez, A. (1990). Analgesia is produced by intrathecal administration of the k opioid agonist, U50,488H, on formalin-evoked cutaneous pain in the rat. *European Journal of Pharmacology*, *190*, 287–293.

Perez, S. E., Chen, E. Y., & Mufson, E. J. (2003). Distribution of estrogen receptor alpha and beta immunoreactive profiles in the postnatal rat brain. *Brain Research. Developmental Brain Research*, *145*(1), 117–139.

Peterson, R. S., Yarram, L., Schlinger, B. A., & Saldanha, C. J. (2005). Aromatase is pre-synaptic and sexually dimorphic in the adult zebra finch brain. *Proceedings of the Biological Sciences*, *272*(1576), 2089–2096. https://doi.org/10.1098/rspb.2005.3181.

Phoenix, C. H., Goy, R. W., Gerall, A. A., & Young, W. C. (1959). Organizing action of prenatally administered testosterone propionate on the tissues mediating mating behavior in the female Guinea pig. *Endocrinology*, *65*, 369–382. https://doi.org/10.1210/endo-65-3-369.

Piercey, M. F., & Einspahr, F. J. (1989). Spinal analgesic actions of kappa receptor agonists, U-50488H and spiradoline (U-62066). *The Journal of Pharmacology and Experimental Therapeutics*, *251*(1), 267–271.

Przewlocka, B., Dziedzicka, M., Lason, W., & Przewlocki, R. (1992). Differential effects of opioid receptor agonists on nociception and cAMP level in the spinal cord of mon-oarthritic rats. *Life Sciences*, *50*(1), 45–54.

Przewlocki, R., Stala, L., Greczek, M., Shearmen, G. T., Przewlocka, B., & Herz, A. (1983). Analgesic effects of mu, delta, kappa-opiate agonists and in particular, dynorphin at the spinal level. *Life Sciences*, *33*(1), 649–652.

Razandi, M., Alton, G., Pedram, A., Ghonshani, S., Webb, P., & Levin, E. R. (2003). Identification of a structural determinant necessary for the localization and function of estrogen receptor alpha at the plasma membrane. *Molecular and Cellular Biology*, *23*(5), 1633–1646.

Razandi, M., Pedram, A., Greene, G. L., & Levin, E. R. (1999). Cell membrane and nuclear estrogen receptors (ERs) originate from a single transcript: Studies of ERalpha and ERbeta expressed in Chinese hamster ovary cells. *Molecular Endocrinology*, *13*(2), 307–319.

Remage-Healey, L., Maidment, N. T., & Schlinger, B. A. (2008). Forebrain steroid levels fluctuate rapidly during social interactions. *Nature Neuroscience*, *11*(11), 1327–1334. https://doi.org/10.1038/nn.2200.

Revankar, C. M., Cimino, D. F., Sklar, L. A., Arterburn, J. B., & Prossnitz, E. R. (2005). A transmembrane intracellular estrogen receptor mediates rapid cell signaling. *Science*, *307*(5715), 1625–1630. https://doi.org/10.1126/science.1106943.

Riley, J. L., 3rd, Robinson, M. E., Wise, E. A., & Price, D. D. (1999). A meta-analytic review of pain perception across the menstrual cycle. *Pain*, *81*(3), 225–235. S0304395998002589 [pii]. https://doi.org/10.1016/S0304-3959(98)00258-9.

Ruda, M. A., Iadarola, M. J., Cohen, L. V., & Young, W. S., III. (1988). In situ hybridization histochemistry and immunocytochemistry reveal an increase in spinal dynorphin biosynthesis in a rat model of peripheral inflammation and hyperalgesia. *Proceedings of the National Academy of Sciences of the United States of America*, *85*, 622–626.

Saldanha, C. J., Remage-Healey, L., & Schlinger, B. A. (2011). Synaptocrine signaling: Steroid synthesis and action at the synapse. *Endocrine Reviews*, *32*(4), 532–549. https://doi.org/10.1210/er.2011-0004.

Sanoja, R., & Cervero, F. (2005). Estrogen-dependent abdominal hyperalgesia induced by ovariectomy in adult mice: A model of functional abdominal pain. *Pain*, *118*(1–2), 243–253. https://doi.org/10.1016/j.pain.2005.08.021.

Sarajari, S., & Oblinger, M. M. (2010). Estrogen effects on pain sensitivity and neuropeptide expression in rat sensory neurons. *Experimental Neurology*, *224*(1), 163–169. https://doi.org/10.1016/j.expneurol.2010.03.006.

Schlinger, B. A., & Callard, G. V. (1989). Localization of aromatase in synaptosomal and microsomal subfractions of quail (Coturnix coturnix japonica) brain. *Neuroendocrinology*, *49*(4), 434–441.

Schlinger, B. A., Remage-Healey, L., & Rensel, M. (2014). Establishing regional specificity of neuroestrogen action. *General and Comparative Endocrinology*, *205*, 235–241. https://doi.org/10.1016/j.ygcen.2014.03.043.

Schmauss, C. (1987). Spinal k-opioid receptor-mediated antinociception is stimulus specific. *European Journal of Pharmacology*, *137*, 197–205.

Schmauss, C., & Yaksh, T. L. (1984). In vivo studies on spinal opiate receptor systems mediating antinociception. II. Pharmacological profiles suggesting a differential association of mu, delta and kappa receptors with visceral chemical and cutaneous thermal stimuli in the rat. *The Journal of Pharmacology and Experimental Therapeutics*, *228*(1), 1–12.

Shughrue, P. J., & Merchenthaler, I. (2001). Distribution of estrogen receptor beta immunoreactivity in the rat central nervous system. *The Journal of Comparative Neurology*, *436*(1), 64–81.

Stevens, C. W., & Yaksh, T. L. (1986). Dynorphin A and related peptides administered intrathecally in the rat: A search for putative kappa opiate receptor activity. *The Journal of Pharmacology and Experimental Therapeutics*, *238*(3), 833–838.

Stiller, R. U., Grubb, B. D., & Schaible, H.-G. (1993). Neurophysiological evidence for increased kappa opioidergic control of spinal cord neurons in rats with unilateral inflammation at the ankle. *The European Journal of Neuroscience*, *5*, 1520–1527.

Storman, E. M., Liu, N. J., Wessendorf, M. W., & Gintzler, A. R. (2018). Physical linkage of estrogen receptor alpha and aromatase in rat: Oligocrine and endocrine actions of CNS-produced estrogens. *Endocrinology*, *159*(7), 2683–2697. https://doi.org/10.1210/en.2018-00319.

Teepker, M., Peters, M., Vedder, H., Schepelmann, K., & Lautenbacher, S. (2010). Menstrual variation in experimental pain: Correlation with gonadal hormones. *Neuropsychobiology*, *61*(3), 131–140. https://doi.org/10.1159/000279303.

Vasudevan, N., & Pfaff, D. W. (2007). Membrane-initiated actions of estrogens in neuroendocrinology: Emerging principles. *Endocrine Reviews*, *28*(1), 1–19. https://doi.org/10.1210/er.2005-0021.

Vasudevan, N., & Pfaff, D. W. (2008). Non-genomic actions of estrogens and their interaction with genomic actions in the brain. *Frontiers in Neuroendocrinology*, *29*(2), 238–257. https://doi.org/10.1016/j.yfrne.2007.08.003.

Wang, Z., Gardell, L. R., Ossipov, M. H., Vanderah, T. W., Brennan, M. B., Hochgeschwender, U., et al. (2001). Pronociceptive actions of dynorphin maintain chronic neuropathic pain. *The Journal of Neuroscience*, *21*(5), 1779–1786.

Watkins, L. R., Wiertelak, E. P., & Maier, S. F. (1992). Kappa opiate receptors mediate tail-shock induced antinociception at spinal levels. *Brain Research*, *582*, 1–9.

Wu, M. V., Manoli, D. S., Fraser, E. J., Coats, J. K., Tollkuhn, J., Honda, S., et al. (2009). Estrogen masculinizes neural pathways and sex-specific behaviors. *Cell*, *139*(1), 61–72. https://doi.org/10.1016/j.cell.2009.07.036.

Zadina, J. E., Hackler, L., Ge, L. J., & Kastin, A. J. (1997). A potent and selective endogenous agonist for the mu-opiate receptor. *Nature*, *386*(6624), 499–502. https://doi.org/10.1038/386499a0.

CHAPTER TEN

Opioids and reproduction

Beata Seeber[a,†], Bettina Böttcher[a,*,†], Elisabeth D'Costa[b], Ludwig Wildt[a]
[a]Department of Gynecological Endocrinology and Reproductive Medicine, Medical University of Innsbruck, Innsbruck, Austria
[b]Department of Obstetrics and Gynecology, Medical University of Innsbruck, Innsbruck, Austria
*Corresponding author: e-mail address: bettina.boettcher@i-med.ac.at

Contents

Abstract

More than 40 years ago, the endogenous opioids were first described. Their role as important neuromodulators of pain and their influence on a variety of neuroendocrine control systems within the central nervous system has been recognized. More recently, endogenous opioids and their receptor have been identified in a variety of reproductive and non-reproductive tissues outside the central nervous system. What role the opioid system plays in these peripheral tissues and organs is not completely understood and thus the subjects of current research. In the central nervous system, endogenous opioids inhibit pulsatile Gonadotropin Releasing Hormone (GnRH) release, affecting the

† Authors contributed equally.

Vitamins and Hormones, Volume 111
ISSN 0083-6729
https://doi.org/10.1016/bs.vh.2019.07.001

release of gonadotropins from the pituitary, and thus mediating stress response within the central nervous-pituitary-gonadal axes in both women and men—Peripherally, endogenous opioids have been demonstrated to be present—among other organs—in the pancreas and in the ovary, where they are produced by granulosa cells and may influence oocyte maturation. In men, endogenous opioids play a role in sperm production within the testis. Opioid antagonists such as naltrexone have been used to restore cyclicity in women through improvement in insulin resistance, GnRH-pulsatility and hyperandrogenemia stemming from specific pathophysiological conditions such as hypothalamic amenorrhea, polycystic ovarian syndrome, hyperinsulinemia, ovarian hyperstimulation syndrome. Opioid antagonists have also been used to treat male sexual disorders and male infertility. In summary, endogenous opioids exert a variety of actions within the reproductive system which are reviewed in this chapter.

1. Introduction

Endogenous opioid peptides are derived from proopiomelanocortin (POMC), proenkephalin (PENK) and prodynorphin (PDYN) and exert their effects via binding to G-protein coupled receptors. While "opioids" are physiologically produced, their synthetically derived alkaloid agonists are termed "opiates." Three main classes of opioid receptors mediate the functions of the opioid peptides: mu (OPRM1), kappa (OPRK1) and delta (OPRD1), each of which has specificity for specific ligands. Endogenous opioids play a major role in stress management, cognition, mood, and pain perception in the central nervous system (CNS). In addition, opioids centrally regulate CNS reproductive physiology, by inhibiting pulsatile GnRH-secretion from the hypothalamus (Dudás & Merchenthaler, 2004). This effect leads secondarily to decreased release of gonadotropins from the pituitary, and to hypogonadism. In the periphery, the opioid system regulates and immune system functions and carbohydrate metabolism (Eyvazzadeh et al., 2009).

That the opioid system is also present in peripheral reproductive tissues in both women and men is a recent revelation. Specifically, researchers have reported the expression of opioid receptors in the granulosa cell of the ovarian follicle, the oocyte, and in the human endometrium (Agirregoitia et al., 2012; Lunger et al., 2016a; Totorikaguena et al., 2017). Concurrently, opioid receptors are found in the somatic and germ cells of the testis. The clinical correlates of these findings include effects on follicular maturation and embryo implantation, and spermatogenesis and sexual function, aspects which are being further studied.

Our group has recently summarized certain aspects of opioid action pertaining specifically to the reproductive system of women (Böttcher, Seeber, Leyendecker, & Wildt, 2017).

In this review, we aim to (1) summarize the role of endogenous opioids in female and male reproductive physiology by presenting evidence for both central and peripheral actions, (2) describe the effects of opiate agonists and antagonists on reproductive function, (3) present the pathophysiology of the opioid system as it relates to particular disease processes and (4) summarize the role of opioids in pregnancy and delivery.

2. Effects of opioids in the central nervous system on the reproductive system

The last 70 years have achieved major advances in understanding the underlying mechanisms of the endogenous opioid system (Corbett, Henderson, McKnight, & Paterson, 2006). Various opioid subsystems with specific receptors and unique functions have been revealed. All opioid receptor types belong to the family of Gi/Go protein-coupled receptors with a seven-transmembrane domain whereas the corresponding endogenous ligands are derived from different precursors. Endogenous opioids do not bind specifically to one receptor type only (Corbett et al., 2006). For example, β-endorphin, derived from the precursor peptide pro-opiomelanocortin (POMC) binds to the mu-opioid receptor and is of major importance for reproductive functioning as it inhibits the pulsatile secretion of Gonadotropin releasing hormone (GnRH) from the hypothalamus.

2.1 Central effects on the Reproductive System in Women

With regard to the menstrual cycle the following interaction between opioid tone, GnRH secretion and subsequent release of LH (luteinizing hormone) and FSH (follicle- stimulating hormone) has been proposed (Leyendecker, Waibeltreber, & Wildt, 1990): At the beginning of the menstrual cycle—in early follicular phase—the opioid tone is low resulting in an uninhibited pulsatile GnRH secretion with a high frequency. The rise of estradiol levels toward the midfollicular phase is accompanied by a rise of the opioid tone. In turn, the rise of opioid tone promotes slower GnRH pulses leading to an increase in LH pulse amplitude. The frequency of GnRH and LH pulses is reduced in the luteal phase, caused by a high opioid tone induced through high progesterone levels.

This cyclic nature of opioid tone was corroborated by a study with 131 healthy volunteers in which it was shown that β-endorphin levels increase during the follicular phase, reach a maximum approximately 4 days before ovulation, further increase in the luteal phase before decreasing again before the next menstrual bleeding. A positive linear correlation between endogenous opioids and progesterone levels during the luteal phase has been found (Ferrer et al., 1997).

In summary, the interaction between hypothalamic, pituitary hormones and the endogenous opioid system in females is illustrated in Fig. 1.

The effect of opioids on the hypothalamic-pituitary axis depends on the phase of the menstrual cycle. It has been shown that opiate antagonists such

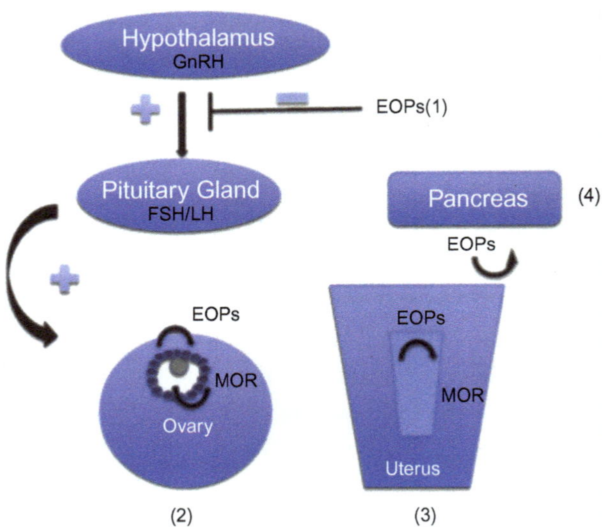

Fig. 1 Multiple levels of the female reproductive system are influenced by the opioid system. At the level of the central nervous system (CNS) endogenous opioids (EOPs) such as endorphins and dynorphin inhibit the GnRH pulse generator, resulting in a decrease of pulsatile LH and FSH secretion and a consequent decline of circulating LH and FSH levels. (1). At the ovarian level (2), endogenous opioids have been identified in follicular fluid and μ-opioid receptors have been shown in granulosa cells and the oocyte. EOPs most likely play a role in oocyte maturation and in granulosa cell-oocyte crosstalk. EOPs could be measured in uterine fluid and μ-opioid receptors have been found in the endometrium (3). The EOP system likely plays a role in embryo implantation. Within the pancreas (4), EOPs stimulate insulin secretion via paracrine and intracrine mechanisms. *With kind permission Böttcher, B., Seeber, B., Leyendecker, G., & Wildt, L. (2017) Impact of the opioid system on the reproductive axis.* Fertility and Sterility, 108 *(2), 207–213. https://doi.org/10.1016/j.fertnstert.2017.06.009 (web archive link). Epub 2017 Jun 29. Review.*

as naloxone and naltrexone lead to an increase in LH levels in the luteal phase but not in the follicular phase of the normal menstrual cycle (Petraglia et al., 1986; Rossmanith, Wirth, Sterzik, & Yen, 1989). Furthermore, the effects of opioids on LH secretion require a mature and functioning HPO Axis: opioid antagonists do not alter LH secretion during early puberty in girls and boys (Fraioli et al., 1984) nor in postmenopausal women (Reid, Quigley, & Yen, 1983). These findings underscore the association between the presence of circulating sex-steroids and the activation of endogenous opioids.

The endogenous opioid activity is stimulated by stressful psychological and physical situations. Women with eating disorders such as anorexia nervosa, or those experiencing starvation or extreme stress, demonstrate an inhibition of the GnRH pulse generator through an increase in the production and release of endogenous opioids. This leads to a consequent decreased production of LH and FSH (Hasan & Hasan, 2011). Chronic disruption of pulsatile GnRH thus leads to anovulatory cycles and is known as hypogonadotropic hypogonadism or hypothalamic ovarian insufficiency.

Several studies have simulated this activation of the endogenous opioid system by creating stressful situations through applying painful stimuli or stressful behavioral tasks (McCubbin, 1993; Smith et al., 2006; Vuong, Van Uum, O'Dell, Lutfy, & Friedman, 2010).

The relationship between female sex steroids and the production of endogenous opioids is underscored by variations in β-endorphin synthesis and opioid receptor expression according to menstrual cycle phase (Ferin, 1984; Ferin, Van Vugt, & Wardlaw, 1984; Petersen & LaFlamme, 1997). The number of mu-opioid receptors in rats varies according to the corresponding phase of the menstrual cycle. Namely, the receptors decrease with elevated estradiol levels, indicating feedback mechanisms on LH release (Piva et al., 1995). Smith (1998) examined the regulation of mu-opioid receptors in a neuroimaging study with positron emission tomography at different time points during the menstrual cycle: LH pulse number and amplitude correlated with mu-binding in the amygdala whereas negative correlations were found between estradiol levels and mu-binding affinity in the amygdala and hypothalamus during the follicular phase. No associations of peripheral β-endorphin levels with mu-binding affinity in these specific brain regions could be found. The authors concluded that mu-opioid receptors in the amygdala are involved in the regulation of GnRH pulse activity and that circulating estradiol levels influence central mu-opioid activity. An interesting link between reproductive and emotional function was suggested, as the amygdala is known to be involved in neuronal processing of emotions.

The beneficial effects of naltrexone have been used therapeutically in patients with hypothalamic amenorrhea, a centrally inducted anovulatory condition. After oral administration of the antagonist naltrexone, a resumption of pulsatile gonadotropin secretion was demonstrated, even leading to recovery of menstrual cycles with ovulations and pregnancies, as shown in Fig. 2 (Wildt & Leyendecker, 1987; Wildt, Sir-Petermann, Rabenbauer, & Wildt, 1993; Wildt, Leyendecker, Waibel-Treber, & Wildt, 1993).

2.2 Central effects on the reproductive system in men

In healthy men, opioids and their analogs decrease LH release via effects on the pulsatile GnRH release (Delitala, Giusti, et al., 1983; Delitala, Grossman, & Besser, 1983; Pende et al., 1986).

In addition, it has been shown that endogenous opioids counteract the negative feedback regulation of androgens on LH. Namely, during

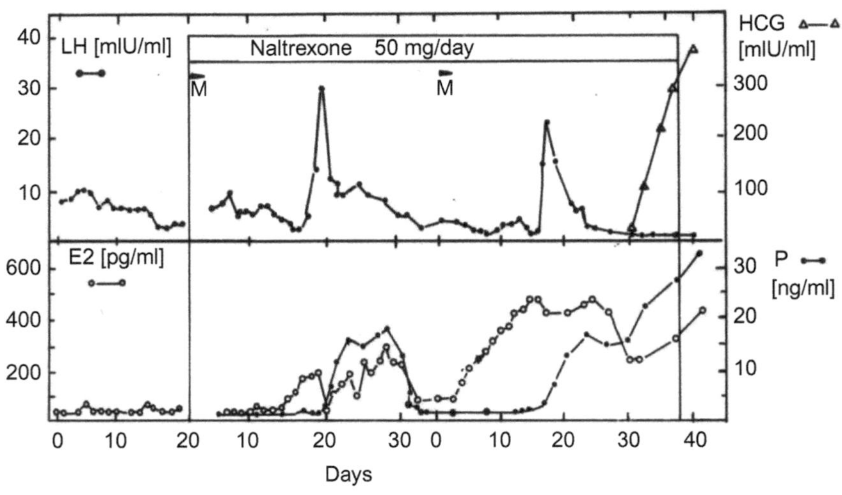

Fig. 2 Follicular maturation, ovulation, luteal function and pregnancy induced by the daily oral administration of 50 mg naltrexone in a patient suffering from secondary hypothalamic ovarian insufficiency. The left panel shows LH and E2 levels during the control period, response to treatment is shown in the right hand panel. The patient became pregnant during the second cycle. Menstruation is indicated by M, LH = luteinizing hormone, E2 = estradiol, P = progesterone, HCG = human chorionic gonadotropin. Menstruation at the beginning of treatment was induced by exogenous gestagen administration. *With kind permission Wildt, L., Leyendecker, G., Sir-Petermann T., & Waibel-Treber S., 1993. Treatment with naltrexone in hypothalamic ovarian failure: Induction of ovulation and pregnancy.* Human Reproduction, *8(3), 350–358.*

continuous infusion of 5- alpha dihydrotestosterone, LH pulse frequency is decreased but was counteracted by an opiate receptor antagonist, returning the LH pulse frequencies to baseline levels (Veldhuis, Rogol, Samojlik, & Ertel, 1984). In contrast, FSH levels are not significantly altered by opiate infusions (Delitala, Giusti, et al., 1983; Delitala, Grossman, et al., 1983; Pende et al., 1986). It has been demonstrated that the effect of opioids depends on pubertal stage, again underscoring the relationship between sex steroids and the opioid system: In late puberty naltrexone/naloxone provoked an increased LH pulse frequency in boys. Conversely, in early puberty a decreased LH pulse frequency was shown after the antagonist (Mauras, Veldhuis, & Rogol, 1986; Petraglia, Bernasconi, et al., 1986).

LH suppression secondarily leads to the suppression of sex steroids from the testes. This effect has been demonstrated after long-term intrathecal administration of opiates for pain, leading to hypogonadotropic hypogonadism in men as well as in women (Abs, 2000; Daniell, 2008; Finch, Roberts, Price, Hadlow, & Pullan, 2000; O'Rourke & Wosnitzer, 2016). Subsequently, women suffered from anovulatory irregular menstrual cycles and amenorrhea while the leading symptoms in men were loss of libido and impotency (Abs, 2000), consistent with the known effects of testosterone deficiency in men, characterized by fatigue, mood disturbances and sexual problems of erectile dysfunction and/or decreased libido (Bawor et al., 2015). Male hypogonadism can also lead to osteoporosis, and is associated with depression, anxiety and infertility (Katz & Mazer, 2009). In opiate addicts, the resulting hypogonadism is often a limitation for long-maintenance therapy (Bawor et al., 2015). Interestingly, methadone causes more sexual dysfunction than the opiate buprenorphine (Yee, Danaee, Loh, Sulaiman, & Ng, 2016). In accordance with these results, testosterone levels were higher—albeit not significantly so—in addicts treated with buprenorphine compared to methadone (Bliesener et al., 2005). As such, the opiate-mediated effect on sex steroid levels and consequently on sexual function, desire and behavior is seen in both chronic opiate users requiring pain therapy as well as in opioid addicts (Daniell, 2002; Paice, Penn, & Ryan, 1994). There is some evidence suggesting that the opioid antagonist naltrexone can improve symptoms of sexual dysfunction in otherwise healthy men although the effect ceases when patients go off treatment (Fabbri et al., 1989).

It is recommended to carefully monitor men as well as women with -opioid-induced hypogonadism and—after thorough risk-benefit-evaluation—initiate hormone replacement therapy to prevent long-term detrimental effects (Katz & Mazer, 2009; O'Rourke & Wosnitzer, 2016).

Male opiate addicts often suffer from infertility caused by low testosterone levels and poor semen quality (Bracken et al., 1990; Lafisca et al., 1985; Ragni, De Lauretis, Bestetti, Sghedoni, & Gambaro, 1988; Safarinejad et al., 2013). In fact, given the growing numbers of opiate users, opiate consumption should be on the differential diagnosis of otherwise unexplained infertility in men (Safarinejad et al., 2013). A recent meta-analysis including 17 studies confirmed significant reduced levels of testosterone in men using opiates (Bawor et al., 2015), irrespective of the type of opiate used (Bawor et al., 2015; Katz & Mazer, 2009).

While opioids and opiates induce a hypogonadotropic hypogonadism in both genders, a higher prevalence of hypogonadism and interestingly, osteoporosis has been observed in men compared to women (Fraser et al., 2009). According to the authors, underlying mechanisms explaining these differences need to be elucidated, as there was no difference in opioid dose or pain scores between male and female patients.

Naltrexone also serves as a drug for smoking cessation, treatment of alcoholism and for the prevention of weight gain and food craving in patients treated with antidepressants (Zimmermann et al., 1997). In this context, gender differences have been demonstrated in the response of the hypothalamic-pituitary-adrenal axis to naltrexone by measuring adrenocorticotropic hormone (ACTH) and cortisol levels in smokers (Roche, Childs, Epstein, & King, 2010). Compared to male smokers naltrexone increased higher levels of ACTH and cortisol in female smokers, especially at the time of ovulation. These data suggest that women might be more sensitive to opioid antagonists than men, possibly resulting in better clinical outcomes in cessation of smoking.

2.3 Central effects on other hormones involved in human reproduction

Endogenous opioids and their analogs also exert their actions on other hormones involved in female and male reproduction. Prolactin levels are increased during acute administration of opiates, presumably mediated via an effect at the hypothalamus (Dobson & Brown, 1988; Vuong et al., 2010). In pre- as well as in postmenopausal women, a single dose of morphine led to a significant increase in peripheral prolactin levels (Hemmings, Fox, & Tolis, 1982). As summarized by Vuong et al. (2010) the effects of chronic administration of opiates on prolactin levels are still unclear and require further investigation, especially with regard to the relationship between estrogen levels and opioids modulating prolactin secretion.

Significant reduced levels of dehydroepiandrosterone sulfate (DHEAS) have been found in opiate addicts and may account for common symptoms of fatigue, depression and/or weakness in this population (Daniell, 2006). The mechanism of how DHEAS levels are affected is not exactly known.

The impact of opioids on oxytocin release is of major importance with regard to pregnancy, labor and delivery, and lactation. The body produces oxytocin which is necessary for lactation, uterine contractions during labor, and bonding between mother and baby. Conversely, opiate analgesics are predominantly used for pain relief during labor. These contradictory effects are described in more detail below.

In men, only few studies on the relation between opioids and oxytocin secretion exist, with conflicting results: Two studies by the same group showed an involvement of the opioid system in oxytocin secretion (Coiro et al., 1991, 1990) whereas in another study no effect could be seen (Honer, Thompson, Lightman, Williams, & Checkley, 1986). This underscores the need for further research.

3. Peripheral action of the opioid system

As described thus far, in both women and men, endogenous opioids in the central nervous system play a regulatory role, secondarily affecting hormone production from the peripheral reproductive organs, testes and ovaries. With the growing interest in the opioid signaling system, there is accumulating evidence that opioids also exert direct effects on the reproductive organs—in animals as well as in humans.

3.1 Peripheral opioids in the reproductive system in women

Peritoneal and ovarian follicular fluid collected from women of different ages was shown to contain beta-endorphin and met-enkephalin (Petraglia et al., 1985). The concentrations of these endogenous opioids varied according to reproductive age of the woman, with high concentrations during the fertile years and nearly undetectable levels in post-menopausal women. Large mature follicles in younger women contained the highest concentrations of opioids, far higher than that measured in serum. These early studies suggested that endogenous opioids may be locally produced in the ovary and thus may play a role in oocyte maturation.

Recently, Jaschke, Lunger, Wildt, and Seeber (2018) reported that the percentage of mature oocytes collected during in vitro fertilization correlated to the concentration of follicular fluid b-endorphin. These findings provide additional indirect evidence that the endogenous opioid system plays a role in the follicular microenvironment.

Direct evidence of the presence of opioid receptors in oocytes was reported in several animal models (Dell'Aquila et al., 2008). Subsequently, Agirregoitia et al. confirmed the presence of all three opioid receptor types in human oocytes and compared their expression and distribution at various stages of oocyte maturation (Agirregoitia et al., 2012). Using western blot analysis, the researchers identified the presence of proteins corresponding to the delta (OPRD1), kappa (OPRK1), and mu (OPRM1) receptors. With immunofluorescence, the OPRD1 pattern remained the same across different stages of oocyte maturity from germinal vesicle (GV) to Metaphase II (MII) stage. Immunostaining patterns changed for the other receptors, OPRK1 and OPRM1, with localization at the periphery in the GV stage and more internally or homogenously at MI and MII. These results present additional evidence that the endogenous opioids present in the follicular fluid may exert a paracrine effect by binding to opiate receptors found on the oocyte, thus playing a role in the maturation of the oocyte.

The follicular micro-environment reflects the various secretion products of oocytes and somatic cells, as they contribute to the maturation of the oocyte (Gilchrist, Lane, & Thompson, 2008). While several factors important factors such as growth-differentiation factor 9 and bone morphogenetic protein 15 are known to be secreted from oocytes, opioid secretion has thus far not been reported. Possible effects of endogenously produced opioids on signaling pathways for gene regulation and cellular processes needs to be studied further.

Not only oocytes but also granulosa cells have been shown to possess opioid receptors. Dell'Aquilla et al. demonstrated that mural granulosa cells express the mRNA coding for the mu-opioid receptor (Dell'Aquila et al., 2002). Recently, Lunger et al. (2016a) demonstrated the presence of OPRM1 opioid receptors in non-luteinized human granulosa cell lines as well as in luteinized primary human granulosa cells. The authors also confirmed the expression of the opioid precursor POMC in the cytoplasm of granulosa cells using immunocytochemical staining (unpublished data).

The signaling pathways that are activated upon opioid binding in granulosa cells are being studied. Vascular endothelial growth factor (VEGF) secretion from cultured granulosa cells appears to decrease when the cells are incubated with the opiate antagonist naloxone (Lunger et al., 2016a). Since previous experimental evidence has shown that endogenous opioids stimulate angiogenesis in other cells (retinal epithelial cells and dermal microvascular endothelial cells) (Blebea et al., 2000; Dai et al., 2010; Singleton, Lingen, Fekete, Garcia, & Moss, 2006; Ye et al., 2012), these results likewise suggest that opioids may play a role in regulating the function of granulosa cells via a paracrine or autocrine effect.

There is evidence to suggest that the opioid system is present in the uterine cavity and plays a role in endometrial function. Furthermore, there appears to be an interaction between opioids and reproductive hormones in the uterus. The presence of the endogenous opioid peptides β-endorphin and methionine-enkephalin in the uterine fluid of fertile women was demonstrated by Petraglia and colleagues. The two peptides were not detectable in the uterine fluid of untreated postmenopausal women, but were present in those women after hormone replacement with estrogen and progesterone (Petraglia et al., 1986). Subsequently, POMC- and PDYN-derived opioid peptides were shown to be synthesized by the human endometrium epithelial cell-derived Ishikawa cell line and by primary mammalian endometrial tissues (Gravanis, Makrigiannakis, Stournaras, & Margioris, 1994; Wahlström, Laatikainen, Salminen, & Leppäluoto, 1985). Estradiol and dexamethasone could suppress the secretion of the opioid β-endorphin, but not dynorphin, by the Ishikawa cells. On the other hand, GnRH increased the secretion of only dynorphin, suggesting that each type of opioid peptide is specifically regulated and may play a distinct role in endometrial physiology.

Further evidence of an interaction between steroid hormones and opioids in the endometrium comes from Vertes et al. who showed that β-endorphin leads to an increase in the concentration of estrogen receptors in uterine epithelial cells, making the endometrium more sensitive to estrogens (Vértes, Pámer, & Garai, 1986). Recently, Totorikaguena et al. evaluated the localization of the mu opioid receptor in the human endometrium at five different time points in the menstrual cycle (Totorikaguena et al., 2017). They found that the maximum expression of OPRM1 mRNA and protein was achieved at the time of ovulation, with changes seen throughout the cycle. The authors concluded from these findings that the mu opiate receptor could play a role in the complex remodeling process of monthly endometrial cyclicity and potentially also in the early events of implantation.

Embryo development and implantation may be negatively affected by opiates as demonstrated in mice (Tang et al., 2015). It has been confirmed that preimplantation embryos express all opioid receptor subtypes. The administration of morphine derails normal embryo development and causes a delay in development to the blastocyst stage. At the same time, systemic morphine negatively impacts the receptivity of the uterus to embryo implantation in mice by impairing luminal epithelial differentiation, decreasing stromal cell proliferation and inhibiting normal angiogenesis. These effects on uterine receptivity were independent of any change in ovarian estrogen or progesterone secretion (Tang et al., 2015).

Not only endometrium but also the smooth muscle that makes up the uterine myometrium may be the target of endogenous opioids. Endometrial opioids may exert an inhibitory or relaxant effect on smooth muscle. Since myometrial contractility is essential for the transport of sperm across the uterine cavity and into the fallopian tubes where fertilization occurs, the exposure to opioids may negative influence this physiologic contractility (Gravanis et al., 1994).

The local paracrine and autocrine effects of opioids in the peripheral reproductive organs, ovary and uterus, take part in regulating their micro-environment. Future studies need to elucidate the specific roles of opioids in reproductive events, especially their influences on oocyte maturation and embryo implantation.

3.2 Peripheral opioids in the reproductive system in men

The various roles of endogenous opioids in the male reproductive system is summarized in Fig. 3. Analogous to the localization of opioids and opioid receptors in the ovary, the opioid peptides, endorphins and enkephalins, have been found in seminal fluid and cells of the testis. Their role in regulating the peripheral system of male reproduction is being unveiled.

β-Endorphin is thought to be produced locally in the male reproductive tract, as the levels of this opioid peptide in seminal plasma were reported to be 10 to 20 times higher than levels in venous plasma. In this study, no difference in seminal plasma beta-endorphin concentration was seen between fertile and infertile men (Davidson, Vermesh, Paulson, Graczykowski, & Lobo, 1989). Another study reached a different conclusion, reporting that the level of metenkephalin in the seminal fluid of semen from asthenospermic men was significantly lower than that from controls. In fact, a minimal concentration of metenkephalin was thought to be necessary for maintaining sperm motility, as demonstrated in in vitro experiments (Fujisawa, Kanzaki, Okada, Arakawa, & Kamidono, 1996).

Gerendai (1991) studied the possible physiological role of testicular opioids in the control of testicular function in rats. This researcher reported that the intratesticular administration of opiate antagonists stimulated Sertoli cell proliferation and secretion in neonatal as well as adult rats. In long-term studies, injection of opiate receptor antagonists resulted in a decrease in

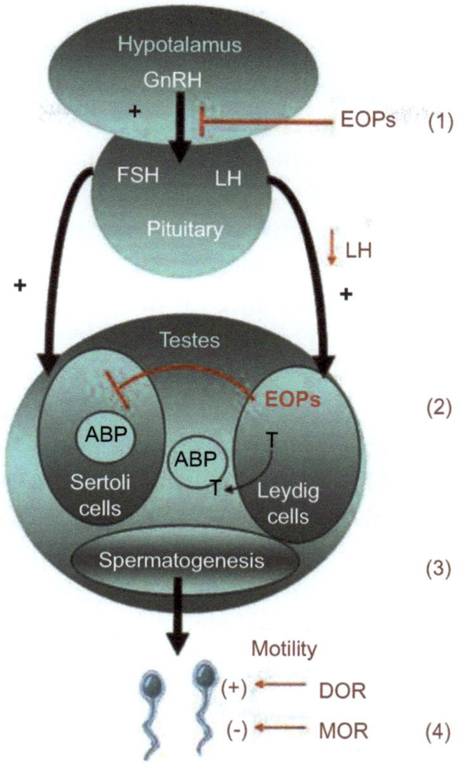

Fig. 3 Control of the male reproductive function by the opioid system at multiple levels. (1) At the level of the CNS, EOPs inhibit the secretion of GnRH, thereby suppressing the release of LH from the pituitary. (2) At the testes level, EOPs are synthesized mainly in Leydig cells after LH stimulation, and they exert an inhibitory effect on Sertoli cells. In particular, EOPs can regulate the levels of testosterone (T) indirectly, inhibiting the production of ABP (androgen binding protein) that is stimulated by FSH in Sertoli cells. (3) Genes encoding opioid peptide precursors are differentially expressed in germ cells, and somatic cells of the testes and their transcripts are not efficiently translated in spermatogenic germ cells. (4) In spermatozoa, the opioid system regulates sperm motility in a distinct manner by the activation of distinct receptors. *With kind permission Subirán, N., Casis, L., & Irazusta, J. (2011). Regulation of male fertility by the opioid system.* Molecular Medicine *(*Cambridge, Mass.*), 17(7–8), 846–853. https://doi.org/10.2119/molmed.2010.00268 (web archive link).*

steroidogenesis. Furthermore, the author demonstrated that the effects of the pharmacological agents were not seen when partial denervation of the testis was carried out, implying that local opioid actions are modulated by testicular nerves.

Opioid protein precursors are expressed differently in somatic and germ cells of the testis, implying that endogenous opioids may regulate testicular function locally by de novo synthesis (Subirán, Casis, & Irazusta, 2011). In their comprehensive review, Subiran et al. summarize the regulation of testicular function by the opioid system (Subirán et al., 2011). The two key cells of the testis, Sertoli and Leydig cells, are both influenced by the local opioid system. Specifically, LH and corticotropin-releasing factor (CRF) stimulate the production of endogenous opioid peptides from Leydig cells, the cells which are also mainly responsible for testicular testosterone production. Opioids may indirectly regulate the levels of testosterone by inhibiting the production of androgen-binding protein. Endogenous opioids exhibit an inhibitory effect on Sertoli cells, the cells which provide the structural and metabolic support for the differentiation of germ cells.

The presence of mu-, delta-, and kappa-opioid receptors in spermatogenic cells was reported in mouse germs cells by Estomba et al. (2016). Agirregoitia et al. (2006) found that human spermatozoa likewise express the three main opioid receptors, mu-, kappa-, and delta-. Furthermore, the receptors were located in different parts of the head, in the middle region and in the tail of the sperm, as demonstrated by immunofluorescence. In incubation experiments with opioid agonists and antagonists, the researchers showed that the agonist morphine increased the number of immotile sperm and that this action was reversed by naloxone. This result is consistent with clinical findings that heroin and methadone users have decreased sperm motility (Ragni et al., 1988). Thus, the low semen quality and male infertility found in opiate addicts is likely due to two separate effects: (1) direct inhibitory effects at the level of the hypothalamus and pituitary gland, resulting in decreased gonadotropin secretion and (2) direct action on sperm cells. The latter action would explain why some heroin users have abnormal semen parameters while maintaining normal hormone levels.

In order for sperm to be fertilization competent, they must undergo modifications such as activation, capacitation and hyperactivation. Principally, only capacitated sperm with sperm membrane reorganization and hyperactive sperm with increased amplitude and flagellar beat can penetrate the oocyte's zona pellucida and fertilize the egg. The role the opioid system may play in these events is poorly understood with few study in animals and no in vivo studies to date (Subirán et al., 2011).

Two additional detrimental effects on sperm integrity deserve mention. Safarinejad et al. (2013) reported a significant increase in the amount of fragmented DNA in opiate users compared to non-users. In addition, significantly decreased levels of seminal plasma anti-oxidant capacity were seen in opiate users, demonstrated through decreased levels of catalase-like and superoxide dismutase-like activity.

4. Pathophysiology of the opioid system in reproductive disorders

The endogenous opioid system is involved in the hormonal regulation of female reproduction. Therefore, opioids play a crucial role also in the pathogenesis of reproductive disorders such as hypothalamic amenorrhea or polycystic ovarian syndrome (PCOS).

4.1 Hypothalamic amenorrhea

Hypothalamic amenorrhea is characterized by a reduced or completely absent pulsatile GnRH secretion (Leyendecker & Wildt, 1983). Reduced GnRH release leads to a spectrum of problems, from corpus luteum insufficiency to anovulatory cycles, to primary or secondary amenorrhea, reflecting a pathophysiologic continuum (Leyendecker, 1979). In about 70% of women with hypothalamic amenorrhea treated with the opioid antagonist naltrexone normal menstrual cycles were restored, even leading to successful pregnancies (Leyendecker et al., 1993; Wildt, Leyendecker, et al., 1993). Pregnancy rates of 29% per treatment cycle were reported in women who received treatment with daily naltrexone doses of 50–150mg for 9 months (Wildt, Leyendecker, et al., 1993). Higher ovulation rates with naltrexone or with naltrexone in combination with clomiphene citrate have been demonstrated in anovulatory patients who did not respond to clomiphene treatment alone (Roozenburg, van Dessel, Evers, & Bots, 1997). Opioid antagonists may be considered as a therapy in patients suffering from "normogonadotropic" hypothalamic anovulation and functional hypothalamic amenorrhea as an alternative to pulsatile GnRH therapy or gonadotropin treatment, the latter being more expensive and associated with the risk of ovarian hyperstimulation and multiple pregnancies.

Naltrexone treatment leads to an increase in frequency and amplitude of gonadotropin pulses resulting most likely in an increase in GnRH secretion. This effect has been observed for up to 6 months, even after treatment discontinuation (Genazzani et al., 1995). Treatment with naltrexone is characterized by relatively few side-effects: Women reported suffering from mild nausea, headaches, and sleep disturbances (Wildt, Sir-Petermann, et al., 1993). Thus, naltrexone can be considered as a promising therapy approach in treating women with hypothalamic ovarian insufficiency.

In addition to the therapeutical approach, naltrexone offers a diagnostic tool allowing for the differentiation between hypothalamic and hyperandrogenemic causes of ovarian failure. Women with congenital adrenal hyperplasia (CAH) often present with amenorrhea caused by high androgen levels. Albeit, in some women with CAH, androgen levels are within the normal range but nonetheless amenorrhea is present. The administration of naltrexone in such women led to normal menstrual cycles in, suggesting that hypothalamic disorders due to alterations of the opioid system may also contribute to the pathology seen in this condition (Böttcher, Dörr, & Wildt, 2014).

4.2 Polycystic ovary syndrome (PCOS)

Polycystic ovary syndrome (PCOS) is a common cause of ovulatory dysfunction in reproductive age women, defined by published criteria (Azziz et al., 2009; Rotterdam ESHRE/ASRM-Sponsored PCOS Consensus Workshop Group, 2004). In addition, there is a high prevalence of insulin resistance (IR) and other metabolic disorders, affecting up to 50% of PCOS women. In overweight POCS women, glucose intolerance and signs of the metabolic syndrome are prevalent.

It has been shown that patients with PCOS present with elevated levels of β-endorphins (Wortsman, Wehrenberg, Gavin, & Allen, 1984) independent of the individual body mass index (Chapron et al., 2011; Pasquali & Vicennati, 2000). Most of the pituitary cells that express insulin receptors also show immunoreactions for the mu-opioid receptor agonist β-endorphin (Unger & Lange, 1997). Typical brain areas for opioid neurotransmission are the amygdala and nucleus accumbens which influence the regulation of appetite and mood. An interesting finding is the relationship between insulin resistance and mu-opioid neurotransmission in PCOS patients (Berent-Spillson et al., 2011): Insulin- resistant patients had a greater limbic availability of mu-opioid receptors than controls on PET. A correlation of receptor availability to the grade of insulin resistance was found. The opioid receptor availability normalized after metformin therapy. These findings suggest that an improvement of insulin sensitivity can lead to an alteration of dysregulated opioid neurotransmission centrally.

Pilot studies administering 50 mg daily naltrexone to women with PCOS and IR showed a dramatic decline in insulin levels after 3–7 weeks of treatment (Wildt, Sir-Petermann, et al., 1993). Subsequent studies corroborated these findings with a significant decline in serum androgens independent of any change in body mass index, as shown in Fig. 4 (Hadziomerović, Rabenbauer, & Wildt, 2006). As treatment durations did not exceed

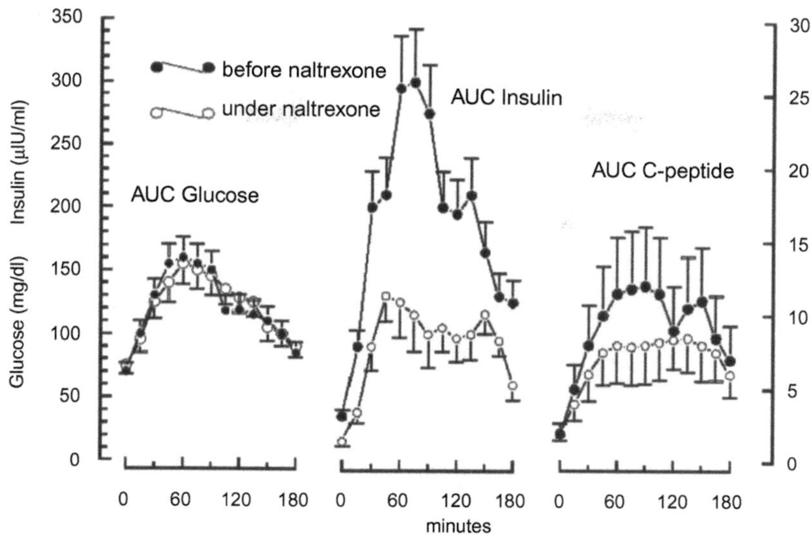

Fig. 4 Time course of glucose, insulin and C-peptide during a 75 g glucose load in 39 women suffering from hyperandrogenemia and insulin resistance before and after >3 weeks of administration of 50 mg naltrexone daily. Data are given as mean ± SD. Under naltrexone treatment, glucose concentrations were not different from those before treatment, however, concentrations of insulin dramatically declined ($P \leq 0.01$) when compared to the response to glucose load at baseline, indicating amelioration of insulin resistance. C-peptide levels were less dramatically lowered during naltrexone treatment. *With kind permission Hadziomerović, D., Rabenbauer, B. & Wildt, L. (2006). Normalization of hyperinsulinemia by chronic opioid receptor blockade in hyperandrogenemic women.* Fertility and Sterility, 86 *(3), 651–657.*

9 weeks, no significant changes in clinical parameters like hirsutism, acne or cycle irregularities were noted and could not account for the decrease in androgens. Fruzzetti, Bersi, Parrini, Ricci, and Genazzani (2002) continued the treatment with naltrexone for 6 months and showed a significant decline in body mass index, cycle irregularities, as well as a decrease in androgen levels in a cohort of obese patients with a BMI >25 kg/m^2. In this case, the improvement in metabolic parameters was attributed to both the weight loss as well as the change in opioid response. These studies confirm that opioids play a central role in the pathogenesis of insulin resistance in PCOS. Although an increase of opioid activity leads to only minimal reduction in pulse frequency of the GnRH pulse generator, there is a consequent elevation in LH levels and a subsequent increase in the LH/FSH, prompting an increase in androgen levels. It should be noted that opioids are generally regarded as inhibitory neurotransmitters and thus the stimulatory effect on LH in PCOS patients may be considered as paradox. It may be explained by the

small inhibitory effect of androgens on pulsatile GnRH secretion, resulting in higher LH-pulse amplitudes, as demonstrated by the effect of the antiandrogen flutamide on pulsatile LH secretion (Eyvazzadeh et al., 2009; Sir-Petermann, Rabenbauer, & Wildt, 1993).

Beta endorphin containing cells have been identified in areas surrounding the pancreatic β-cells and opioids have been shown to stimulate insulin secretion (Hadziomerović et al., 2006; Sir-Petermann et al., 1998). It is therefore tempting to speculate that locally elevated opioid activity stimulates insulin secretion from the pancreas. Insulin resistance develops as a consequence of the downregulation of hepatic insulin receptors, thereby increasing insulin levels in peripheral blood (Hadziomerović et al., 2006; Wildt, Leyendecker, et al., 1993).

The interaction between opiates and insulin secretion has been validated in a study of PCOS patients who were treated with pulsatile administration of GnRH in combination with naltrexone. This treatment resulted in an improvement of the LH response and ovulation rates. A decrease in insulin secretion was also shown and, interestingly, mainly observed in hyperinsulinemic patients (Fulghesu et al., 2001). A similar effect was demonstrated by Lanzone et al. (1993): PCOS patients showed no significant changes of the LH/FSH response using a GnRH-Test after naltrexone treatment for 4 weeks compared to baseline GnRH-Test. No differences in LH responses compared to healthy controls were evident after naltrexone treatment. FSH levels remained unchanged, suggesting a selective sensitizing effect of opioids on LH (Lanzone et al., 1993).

In another study, women with PCOS received one of three different therapeutic treatments (metformin, naltrexone, or a combination of an oral antiandrogenic contraceptive in combination with prednisolone) over 3 months. The researchers demonstrated significant improvements in ovulation rates as well as in hyperandrogenemia in all groups. No concurrent change in metabolic parameters was seen, leading to the conclusion that insulin resistance is only one component influencing hyperandrogenemic anovulation (Hadžiomerović-Pekić et al., 2010).

The presence of opioid receptors on granulosa cells and the link to granulosa cell VEGF secretion has been discussed above. In subsequent work, Lunger et al. challenged granulosa cells obtained from women with PCOS and from controls in vitro with the opiate antagonist naloxone, insulin, or both and measured VEGF in culture supernatant (Lunger et al., 2016b). They found that baseline VEGF levels were higher in PCOS than controls, could be increased by insulin, and could be reduced by exposing the cells to naloxone. This study further implicates the regulatory effects of opioids on granulosa cell function, but also demonstrates inherent differences in granulosa cell stemming from PCOS and non-PCOS women. Potential future therapeutic options for treating ovarian hyperstimulation syndrome, a common complication of ovarian stimulation in PCOS women mediated by high VEGF levels, by targeting the opiate signaling system deserve further study.

5. Opioids in pregnancy

In pregnancy endogenous opioids mediate important changes which support pregnancy and prepare women for childbirth and the postpartum period. Endogenous opioids play a key roles in the regulation of the stress axis, the oxytocin system, the prolactin system and the neural circuits controlling pain and maternal behavior (Brunton, 2019).

5.1 Oxytocin

Oxytocin is produced in magnocellular neurosecretory cells of the supraoptic and paraventricular nucleus in the hypothalamus whose axons reach the posterior lobe of the pituitary gland. Upon activation, oxytocin is secreted from the nerve terminals in the posterior pituitary into the systemic circulation (Russell & Brunton, 2009).

Oxytocin is the major hormone during parturition which stimulates uterine contractions and supports lactation via the milk ejection reflex. Douglas et al. postulated that the synthesis of oxytocin per se is not increased in pregnancy, but instead its release is increased (Douglas et al., 1998). The oxytocin neurons stay in a quiescent state that limits oxytocin secretion and permits the accumulation of oxytocin stores in the posterior pituitary (which expand by around 75% in rats), for when it is required at parturition and lactation (Douglas, Dye, Leng, Russell, & Bicknell, 1993).

Oxytocin is involved in a positive feedback loop during labor, where oxytocin causes uterine contractions, which in turn lead to more oxytocin release until the baby is delivered. When the infant suckles at the nipple and stimulates the mammary glands, oxytocin is released, causing milk ejection and milk letdown, also from the contralateral breast. Furthermore, oxytocin is responsible for fetomaternal bonding.

The opiate morphine is used clinically in selected obstetrical patients to provide rest and sedation during the stressful prodromal stages of labor when insufficient cervical dilation occurs with early contractions. In view of the importance of the endogenous opioids system in modulating oxytocin release, the use of exogenous opioids—opiate analgesics—for pain relief during human parturition, seems is contradictory to the physiological effects of opioids during parturition (Morris, Domino, & Domino, 2013).

Generally, endogenous opioids inhibit oxytocin release in animals and humans (Hartman, Rosella-Dampman, Emmert, & Summy-Long, 1986). Both mu and kappa opioid receptor agonists binding to receptors in the-posterior pituitary have been shown to inhibit posterior pituitary hormone release.

Specifically, oxytocin secretion by the posterior pituitary gland is inhibited by both mu and kappa agonists acting at the hypothalamus or directly at the posterior pituitary via kappa receptors (Morris et al., 2013).

Animal studies showed that opioids suppress the functional activity of oxytocin neurons in the hypothalamus and inhibit the oxytocin release from the axon terminal in the neural lobe. When given intracerebroventricularly and intravenously, morphine depresses the firing rate of supraoptic oxytocin neurons. After intracerebroventricular morphine infusion given for 5 days, tolerance and dependence developed in magnocellular oxytocin neurons. Hence Pumford et al. suggested that tolerance and dependence to endogenous opioids is important for oxytocin neurons during pregnancy, in preparation for parturition (Pumford, Leng, & Russell, 1991).

In rats, morphine treatment for 1 week showed decreased oxytocin content and mRNA expression in the supraoptic nucleus and nucleus accumbens. Based on these findings the authors postulated that morphine treatment in rats affect the synthesis and secretion of oxytocin (You, Li, Song, Wang, & Lu, 2000).

Bicknell et al postulated that opioids act directly at the neural lobe of the pituitary (Bicknell, Ingram, & Leng, 1983) to inhibit oxytocin release in a variety of situations. After an isolated rat neural lobe had been electrically stimulated administered morphine inhibited oxytocin levels (Clarke & Patrick, 1983). The hydration state of the animal did not affect the ability of mu- and kappa-opioid receptor agonists to suppress oxytocin levels (Evans, Olley, Rice, & Abrahams, 1989).

Maternal morphine administration during parturition in rats inhibited the levels of oxytocin with a following interruption of parturition and impaired maternal behavior (Russell et al., 1989). Morphine given intracerebroventricularly on days 17–18 of pregnancy delayed the start of parturition by 4 h. Once parturition was established, however, it was not affected by morphine, indicating tolerance to morphine inhibition of oxytocin secretion.

Several authors suggest an interaction between the lactation state and the effects of opioids on oxytocin levels. Hartman et al. proposed that endogenous opioid peptides inhibit oxytocin so that the pituitary can store and conserve oxytocin needed for lactation. In dehydrated rats during lactation opioids also inhibited oxytocin release (Hartman, Rosella-Dampman, & Summy-Long, 1987).

Concerning human studies, women have been studied for the effects of opioids on oxytocin, especially the influence of analgesic opioids in pregnancy and parturition.

Lindow et al. showed that during late pregnancy, morphine and naloxone administration had no effect on the maternal oxytocin concentration (Lindow, van der Spuy, Hendricks, Nugent, & Dunne, 1993). In contrast to these findings the same authors found that, during the first stage of labor, morphine significantly inhibited plasma oxytocin levels. The opioid-antagonist naloxone had no effect (Lindow et al., 1992). The fetal oxytocin production was not affected by the administration of morphine to the mother. Furthermore, after delivery, morphine inhibited the pulsatile oxytocin secretion in breast-feeding women. Naloxone again had no effect (Lindow, Hendricks, Nugent, Dunne, & van der Spuy, 1999; Lindow, Hendricks, Thompson, & van der Spuy, 1998).

Other opioid analgesics, such as fentanyl or sufentanil, did not suppress oxytocin levels before labor, but suppressed oxytocin levels during the onset of labor (Shibli et al., 2001; Stocche, Klamt, Antunes-Rodrigues, Garcia, & Moreira, 2001).

A biphasic effect of opioids on oxytocin release could have important clinical implications related to pregnancy and parturition. The clinical effect of initial decreased oxytocin levels may be exploited in pregnant women with a dysfunctional pattern of labor contractions. Obstetricians give morphine for a period of rest, obtaining temporary pain relief. When the effects of the drug wear off, normal uterine contractions often follow perhaps due to increased endogenous oxytocin release. Theoretically, a mu agonist such as morphine would be preferred because of the dysphoric and psychotomimetic properties of kappa-opioid receptor agonists (Grell, Fjalland, & Christensen, 1989; Trescot, Datta, Lee, & Hansen, 2008).

In summary opioids inhibit oxytocin secretion in animals and humans. Both kappa- and mu-opioid receptor agonists can be used to decrease oxytocin levels. Opioids act directly at the level of the pituitary and at the level of the hypothalamus. Chronic morphine treatment affects the synthesis and the secretion of oxytocin. The clinical observation that morphine inhibits contractions and may delay delivery is due to this effect on inhibition of oxytocin release. As the drug wears off or as opioid receptors are down-regulated, oxytocin levels rebound significantly, and the patient finds herself in active labor or return to baseline uterine activity without active phase of labor. As such, morphine can safely be administered to a patient in the early stage of labor. If necessary, the opioid antagonist naloxone can be administered to newborns to counteract residual agonist effects after delivery (Morris et al., 2013).

5.2 Prolactin

As mentioned above, the prolactin system is likewise influenced by endogenous opioids during pregnancy and has important functions to support pregnancy.

In non-pregnant animals the anterior pituitary lactotrophs secrete prolactin under tonic inhibitory control by dopamine. Dopamine is produced and secreted by hypothalamic tuberoinfundibular dopamine (TIDA) neurons in the arcuate nucleus neurons that project to the median eminence and secrete dopamine into the hypothalamo-hypophysial portal blood system (Reymond & Porter, 1985). In the anterior pituitary dopamine acts on D2 receptors expressed on lactotrophs to inhibit prolactin production (Caron et al., 1978; Guillou et al., 2015).

Animal studies postulated that prolactin plays an important role in preventing regression of the corpus luteum in the first trimester, ensuring progesterone secretion to sustain the pregnancy (Ormandy, Binart, & Kelly, 1997). This role of prolactin, however, seems questionable in primates. In interaction with estrogen, progesterone and oxytocin, prolactin primes the brain to promote maternal behavior after birth and prepares for lactation by stimulating milk production and secretion into alveoli of the mammary gland (Bridges & Ronsheim, 1990; Lucas, Ormandy, Binart, Bridges, & Kelly, 1998).

Beyond these effects, rising prolactin in the maternal brain leads to depression of basal and stress-induced activity of the oxytocin system and the hypothalamo-pituitary-adrenal (HPA) axis, which is the major neuroendocrine system that responds to stress (Gustafson, Bunn, & Grattan, 2017; Torner & Neumann, 2002). Gunnet and Freeman showed that pairing in rodents stimulates the onset of rising prolactin secretion twice daily, which persist for the first days of pregnancy (Gunnet & Freeman, 1983). These surges are suppressed when the secretion of placental lactogen begins to increase around the second trimester. Placental lactogen stimulates dopamine release from TIDA neurons and thereby inhibits prolactin secretion from the lactotrophs (Lee & Voogt, 1999).

Endogenous opioids play a key role in regulating prolactin secretion via inhibition of TIDA neurons and consequently dopamine secretion. This effect promotes prolactin surge in pregnancy and suckling-induced prolactin secretion during lactation (Andrews & Grattan, 2002, 2003).

Administration of naloxone the night before birth increases dopamine turnover and inhibits prolactin surge in rodents (Andrews & Grattan, 2002).

In contrast, administration of naloxone some days earlier in pregnancy has no effect on prolactin secretion or dopaminergic activity unless progesterone actions are blocked. This effect decreases TIDA neuronal activity and permits an increase in prolactin secretion, supporting the idea that progesterone reduction at the end of pregnancy plays a key role in reducing activity of the TIDA neurons and facilitating increased prolactin secretion (Andrews & Grattan, 2002; Soaje, Valdez, Bregonzio, Penissi, & Deis, 2006).

The administration of naloxone during pregnancy has no greater effect on the activity of the dopamine neurons than anti-progesterone given alone (Soaje et al., 2006). Based on these findings, the authors suggest that endogenous opioids may regulate prolactin secretion through another mechanism in the presence of high levels of progesterone, possibly via oxytocin, which stimulates prolactin secretion (Gonzalez-Iglesias et al., 2015). As a matter of fact, recent findings support a role for oxytocin in facilitating prolactin secretion in late pregnancy but this effect is only evident when progesterone action is blocked (Villegas-Gabutti, Pennacchio, Vivas, Jahn, & Soaje, 2018).

5.3 Endorphins

Besides their effects on oxytocin and prolactin, endorphins are of special interest because of their key role in maternal pain perception during pregnancy and delivery. As previously discussed, endorphins are derived from proopiomelanocortin (POMC) precursor and bind to mu- and kappa-receptors (Russell & Douglas, 2000).

Measurements of maternal beta- endorphin concentrations in plasma during pregnancy showed higher concentrations in pregnant women, with a nadir during the second trimester (Goland, Wardlaw, Stark, & Frantz, 1981; Guillemin et al., 1977). Maternal plasma endorphin concentrations rise during labor and remain high during the early postpartum period. This conforms to the increase of ACTH during labor and its peak at delivery (Abboud, 1988).

Plasma endorphin concentrations fall in response to epidural analgesia during labor and delivery as well as after intrathecal morphine analgesia. The decline in maternal plasma beta-endorphin concentrations after maternal analgesia occurs despite uterine contractions and progression in labor, indicating that pain, stress associated with pain, or both, are major stimuli of pituitary β-endorphin release and not labor itself (Abboud, 1988).

To conclude, endogenous opioids play key roles in adaptations among others in the regulation of the oxytocin and prolactin system and in endorphin effects.

These changes are important for preservation and development of pregnancy, providing sufficient oxytocin for parturition and lactation. In light of these physiological actions, the use of opioids during this sensible time should be used with knowledge of their potential side effects.

6. Conclusion

As reviewed in this manuscript, endogenous opioids play an important role in both female as well as male reproductive function. In the central nervous system, they are involved in controlling the release of GnRH and thus the gonadotropins LH and FSH. In females, this affects ovarian hormonal production and follicular growth and ovulation. In males, opioids' central effects affect spermatogenesis and testicular hormone production.

Peripherally, endogenous opioids act directly as important neuromodulators as well as signaling peptides within several reproductive organs and tissues. These include the endocrine pancreas, and in women the different compartments of the ovary, including the ovarian follicles and the oocyte, as well as the endometrium. In men, they exert effects on the Leydig and Sertoli cells of the testes, affecting testosterone production and sperm function.

In females, we have shown that endogenous opioids may play an important role in hypothalamic amenorrhea and in insulin resistance associated with hyperandrogenemia and PCOS. In men, opioids and opiates may contribute to sexual dysfunction and pathospermia, a condition that may be reversed upon discontinuation or antagonism.

The endogenous opioid system has an important role during pregnancy and parturition. In addition to its direct roles, opioids also affect the prolactin and oxytocin systems, exerting additional indirect effects.

The availability of specific antagonists such as naloxone or naltrexone offers unique therapeutic options to modulate the pathophysiological conditions. The efficacy of these agents in hypothalamic amenorrhea as well as in insulin resistance has been demonstrated in research studies which we summarized in this review. Additional potential clinical uses such as for the prevention or treatment of ovarian hyperstimulation syndrome and other disorders of female reproduction, deserve further study. Local application of opiates may deserve further study as non-hormonal male contraceptives. Altogether, future research into the role of endogenous opioids in reproductive disorders may lead to a better understanding of the pathophysiology of female and male reproduction as well as to novel treatments options.

References

Abboud, T. K. (1988). Maternal and fetal beta endorphin: Effects of pregnancy and labour. *Archives of Disease in Childhood*, *63*, 707–709. BMJ Publishing Group. 7 Spec No.

Abs, R. (2000). Endocrine consequences of long-term intrathecal administration of opioids. *The Journal of Clinical Endocrinology and Metabolism*, *85*(6), 2215–2222. https://doi.org/10.1210/jc.85.6.2215.

Agirregoitia, E., Peralta, L., Mendoza, R., Expósito, A., Ereño, E. D., Matorras, R., et al. (2012). Expression and localization of opioid receptors during the maturation of human oocytes. *Reproductive Biomedicine Online*, *24*(5), 550–557. https://doi.org/10.1016/j.rbmo.2012.02.007.

Agirregoitia, E., Valdivia, A., Carracedo, A., Casis, L., Gil, J., Subirán, N., et al. (2006). Expression and localization of delta-, kappa-, and mu-opioid receptors in human spermatozoa and implications for sperm motility. *The Journal of Clinical Endocrinology and Metabolism*, *91*(12), 4969–4975. https://doi.org/10.1210/jc.2006-0599.

Andrews, Z. B., & Grattan, D. R. (2002). Opioid control of prolactin secretion in late pregnant rats is mediated by tuberoinfundibular dopamine neurons. *Neuroscience Letters*, *328*(1), 60–64.

Andrews, Z. B., & Grattan, D. R. (2003). Opioid receptor subtypes involved in the regulation of prolactin secretion during pregnancy and lactation. *Journal of Neuroendocrinology*, *15*(3), 227–236.

Azziz, R., Carmina, E., Dewailly, D., Diamanti-Kandarakis, E., Escobar-Morreale, H. F., Futterweit, W., et al. (2009). The androgen excess and PCOS society criteria for the polycystic ovary syndrome: The complete task force report. *Fertility and Sterility*, *91*, 456–488. https://doi.org/10.1016/j.fertnstert.2008.06.035.

Bawor, M., Bami, H., Dennis, B. B., Plater, C., Worster, A., Varenbut, M., et al. (2015). Testosterone suppression in opioid users: A systematic review and meta-analysis. *Drug and Alcohol Dependence*, *149*, 1–9. https://doi.org/10.1016/j.drugalcdep.2015.01.038.

Berent-Spillson, A., Love, T., Pop-Busui, R., Sowers, M., Persad, C. C., Pennington, K. P., et al. (2011). Insulin resistance influences central opioid activity in polycystic ovary syndrome. *Fertility and Sterility*, *95*(8), 2494–2498. https://doi.org/10.1016/j.fertnstert.2011.03.031.

Bicknell, R. J., Ingram, C. D., & Leng, G. (1983). Oxytocin release is inhibited by opiates from the neural lobe, not those from the intermediate lobe. *Neuroscience Letters*, *43*(2-3), 227–230.

Blebea, J., Mazo, J. E., Kihara, T. K., Vu, J. H., McLaughlin, P. J., Atnip, R. G., et al. (2000). Opioid growth factor modulates angiogenesis. *Journal of Vascular Surgery*, *32*(2), 364–373. https://doi.org/10.1067/mva.2000.107763b.

Bliesener, N., Albrecht, S., Schwager, A., Weckbecker, K., Lichtermann, D., & Klingmüller, D. (2005). Plasma testosterone and sexual function in men receiving buprenorphine maintenance for opioid dependence. *The Journal of Clinical Endocrinology and Metabolism*, *90*(1), 203–206. https://doi.org/10.1210/jc.2004-0929.

Böttcher, B., Dörr, H. G., & Wildt, L. (2014). Naltrexone as a diagnostic tool to distinguish between hyperandrogenemic and hypothalamic ovarian failure in females with congenital adrenal hyperplasia due to 21-hydroxylase-deficiency (CAH). *European Journal of Obstetrics, Gynecology, and Reproductive Biology*, *180*, 153–156. https://doi.org/10.1016/j.ejogrb.2014.06.007.

Böttcher, B., Seeber, B., Leyendecker, G., & Wildt, L. (2017). Impact of the opioid system on the reproductive axis. *Fertility and Sterility*, *108*(2), 207–213. https://doi.org/10.1016/j.fertnstert.2017.06.009.

Bracken, M. B., Eskenazi, B., Sachse, K., McSharry, J. E., Hellenbrand, K., & Leo-Summers, L. (1990). Association of cocaine use with sperm concentration, motility, and morphology. *Fertility and Sterility*, *53*(2), 315–322.

Bridges, R. S., & Ronsheim, P. M. (1990). Prolactin (PRL) regulation of maternal behavior in rats: Bromocriptine treatment delays and PRL promotes the rapid onset of behavior. *Endocrinology*, *126*(2), 837–848. https://doi.org/10.1210/endo-126-2-837.

Brunton, P. J. (2019). Endogenous opioid signalling in the brain during pregnancy and lactation. *Cell and Tissue Research*, *375*(1), 69–83. https://doi.org/10.1007/s00441-018-2948-1.

Caron, M. G., Beaulieu, M., Raymond, V., Gagné, B., Drouin, J., Lefkowitz, R. J., et al. (1978). Dopaminergic receptors in the anterior pituitary gland. Correlation of [3H] dihydroergocryptine binding with the dopaminergic control of prolactin release. *The Journal of Biological Chemistry*, *253*(7), 2244–2253.

Chapron, C., Souza, C., Borghese, B., Lafay-Pillet, M.-C., Santulli, P., Bijaoui, G., et al. (2011). Oral contraceptives and endometriosis: The past use of oral contraceptives for treating severe primary dysmenorrhea is associated with endometriosis, especially deep infiltrating endometriosis. *Human Reproduction (Oxford, England)*, *26*(8), 2028–2035. https://doi.org/10.1093/humrep/der156.

Clarke, G., & Patrick, G. (1983). Differential inhibitory action by morphine on the release of oxytocin and vasopressin from the isolated neural lobe. *Neuroscience Letters*, *39*(2), 175–180.

Coiro, V., Capretti, L., Davoli, C., Speronij, G., Bianconi, L., Cavazzini, U., et al. (1991). Endogenous opioids modulate the oxytocin response to insulin-induced hypoglycaemia and partially mediate the inhibitory effect of ethanol in man. *Journal of Neuroendocrinology*, *3*(4), 401–405. John Wiley & Sons, Ltd (10.1111). https://doi.org/10.1111/j.1365-2826.1991.tb00294.x.

Coiro, V., Capretti, L., Speroni, G., Castelli, A., Bianconi, L., Cavazzini, U., et al. (1990). Increase by naloxone of arginine vasopressin and oxytocin responses to insulin-induced hypoglycemia in obese men. *Journal of Endocrinological Investigation*, *13*(9), 757–763. Springer International Publishing. https://doi.org/10.1007/BF03349617.

Corbett, A. D., Henderson, G., McKnight, A. T., & Paterson, S. J. (2006). 75 years of opioid research: The exciting but vain quest for the Holy Grail. *British Journal of Pharmacology*, *147*(Suppl1), S153–S162. John Wiley & Sons, Ltd (10.1111). https://doi.org/10.1038/sj.bjp.0706435.

Dai, X., Song, H.-J., Cui, S.-G., Wang, T., Liu, Q., & Wang, R. (2010). The stimulative effects of endogenous opioids on endothelial cell proliferation, migration and angiogenesis in vitro. *European Journal of Pharmacology*, *628*(1-3), 42–50. https://doi.org/10.1016/j.ejphar.2009.11.035.

Daniell, H. W. (2002). Hypogonadism in men consuming sustained-action oral opioids. *The journal of pain: official journal of the American Pain Society*, *3*(5), 377–384.
Daniell, H. W. (2006). DHEAS deficiency during consumption of sustained-action prescribed opioids: Evidence for opioid-induced inhibition of adrenal androgen production. *The journal of pain: Official journal of the American Pain Society*, 7(12), 901–907. https://doi.org/10.1016/j.jpain.2006.04.011.
Daniell, H. W. (2008). Opioid endocrinopathy in women consuming prescribed sustained-action opioids for control of nonmalignant pain. *The journal of pain: Official journal of the American Pain Society*, *9*(1), 28–36. https://doi.org/10.1016/j.jpain.2007.08.005.
Davidson, A., Vermesh, M., Paulson, R. J., Graczykowski, J. W., & Lobo, R. A. (1989). Presence of immunoreactive beta-endorphin and calcitonin in human seminal plasma, and their relation to sperm physiology. *Fertility and Sterility*, *51*(5), 878–880.
Delitala, G., Giusti, M., Mazzocchi, G., Granziera, L., Tarditi, W., & Giordano, G. (1983). Participation of endogenous opiates in regulation of the hypothalamic-pituitary-testicular axis in normal men. *The Journal of Clinical Endocrinology and Metabolism*, *57*(6), 1277–1281. https://doi.org/10.1210/jcem-57-6-1277.
Delitala, G., Grossman, A., & Besser, M. (1983). Differential effects of opiate peptides and alkaloids on anterior pituitary hormone secretion. *Neuroendocrinology*, *37*(4), 275–279. https://doi.org/10.1159/000123558Karger Publishers.
Dell'Aquila, M. E., Albrizio, M., Guaricci, A. C., De Santis, T., Maritato, F., Tremoleda, J. L., et al. (2008). Expression and localization of the mu-opioid receptor (MOR) in the equine cumulus-oocyte complex and its involvement in the seasonal regulation of oocyte meiotic competence. *Molecular Reproduction and Development*, *75*(8), 1229–1246. https://doi.org/10.1002/mrd.20869.
Dell'Aquila, M. E., Casavola, V., Reshkin, S. J., Albrizio, M., Guerra, L., Maritato, F., et al. (2002). Effects of beta-endorphin and Naloxone on in vitro maturation of bovine oocytes. *Molecular Reproduction and Development*, *63*(2), 210–222. https://doi.org/10.1002/mrd.10163.
Dobson, P. R., & Brown, B. L. (1988). Involvement of the hypothalamus in opiate-stimulated prolactin secretion. *Regulatory Peptides*, *20*(4), 305–310.
Douglas, A. J., Dye, S., Leng, G., Russell, J. A., & Bicknell, R. J. (1993). Endogenous opioid regulation of oxytocin secretion through pregnancy in the rat. *Journal of Neuroendocrinology*, *5*(3), 307–314.
Douglas, A. J., Meeren, H. K., Johnstone, L. E., Pfaff, D. W., Russell, J. A., & Brooks, P. J. (1998). Stimulation of expression of the oxytocin gene in rat supraoptic neurons at parturition. *Brain Research*, *782*(1-2), 167–174.
Dudás, B., & Merchenthaler, I. (2004). Close anatomical associations between beta-endorphin and luteinizing hormone-releasing hormone neuronal systems in the human diencephalon. *Neuroscience*, *124*(1), 221–229. https://doi.org/10.1016/j.neuroscience.2003.11.021.
Estomba, H., Muñoa-Hoyos, I., Gianzo, M., Urizar-Arenaza, I., Casis, L., Irazusta, J., et al. (2016). Expression and localization of opioid receptors in male germ cells and the Implication for Mouse Spermatogenesis. *PLoS One*, *11*(3), e0152162. Edited by S. Schlatt. Public Library of Science. https://doi.org/10.1371/journal.pone.0152162.
Evans, R. G., Olley, J. E., Rice, G. E., & Abrahams, J. M. (1989). mu- and K-opiate receptor agonists reduce plasma neurohypophysial hormone concentrations in water-deprived and normally hydrated rats. *Clinical and Experimental Pharmacology & Physiology*, *16*(3), 191–197.
Eyvazzadeh, A. D., Pennington, K. P., Pop-Busui, R., Sowers, M., Zubieta, J.-K., & Smith, Y. R. (2009). The role of the endogenous opioid system in polycystic ovary syndrome. *Fertility and Sterility*, *92*(1), 1–12. https://doi.org/10.1016/j.fertnstert.2009.05.012.
Fabbri, A., Jannini, E. A., Gnessi, L., Moretti, C., Ulisse, S., Franzese, A., et al. (1989). Endorphins in male impotence: Evidence for naltrexone stimulation of erectile activity in patient therapy. *Psychoneuroendocrinology*, *14*(1-2), 103–111.

Ferin, M. (1984). Endogenous opioid peptides and the menstrual cycle. *Trends in Neurosciences*, 7(6), 194–196. https://doi.org/10.1016/S0166-2236(84)80009-0.
Ferin, M., Van Vugt, D., & Wardlaw, S. (1984). The hypothalamic control of the menstrual cycle and the role of endogenous opioid peptides. *Recent Progress in Hormone Research*, *40*, 441–485.
Ferrer, J., Mtnez-Guisasola, J., Díaz, F., Alonso, F., Guerrero, M., & Marín, B. (1997). Plasma levels of beta-endorphin during the menstrual cycle. *Gynecological endocrinology: The official journal of the International Society of Gynecological Endocrinology*, *11*(2), 75–82.
Finch, P. M., Roberts, L. J., Price, L., Hadlow, N. C., & Pullan, P. T. (2000). Hypogonadism in patients treated with intrathecal morphine. *The Clinical Journal of Pain*, *16*(3), 251–254.
Fraioli, F., Cappa, M., Fabbri, A., Gnessi, L., Moretti, C., Borrelli, P., et al. (1984). Lack of endogenous opioid inhibitory tone on LH secretion in early puberty. *Clinical Endocrinology*, *20*(3), 299–305.
Fraser, L.-A., Morrison, D., Morley-Forster, P., Paul, T. L., Tokmakejian, S., Larry Nicholson, R., et al. (2009). Oral opioids for chronic non-cancer pain: Higher prevalence of hypogonadism in men than in women. *Experimental and clinical endocrinology & diabetes: Official journal, German Society of Endocrinology [and] German Diabetes Association*, *117*(1), 38–43. https://doi.org/10.1055/s-2008-1076715.
Fruzzetti, F., Bersi, C., Parrini, D., Ricci, C., & Genazzani, A. R. (2002). Effect of long-term naltrexone treatment on endocrine profile, clinical features, and insulin sensitivity in obese women with polycystic ovary syndrome. *Fertility and Sterility*, 77(5), 936–944.
Fujisawa, M., Kanzaki, M., Okada, H., Arakawa, S., & Kamidono, S. (1996). Metenkephalin in seminal plasma of infertile men. *International Journal of Urology: Official Journal of the Japanese Urological Association*, *3*(4), 297–300.
Fulghesu, A. M., Ciampelli, M., Belosi, C., Apa, R., Guido, M., Caruso, A., et al. (2001). Naltrexone effect on pulsatile GnRH therapy for ovulation induction in polycystic ovary syndrome: A pilot prospective study. *Journal of Endocrinological Investigation*, *24*(7), 483–490.
Genazzani, A. D., Petraglia, F., Gastaldi, M., Volpogni, C., Gamba, O., & Genazzani, A. R. (1995). Naltrexone treatment restores menstrual cycles in patients with weight loss-related amenorrhea. *Fertility and Sterility*, *64*(5), 951–956.
Gerendai, I. (1991). Modulation of testicular functions by testicular opioid peptides. *Journal of physiology and pharmacology: An official journal of the Polish Physiological Society*, *42*(4), 427–437.
Gilchrist, R. B., Lane, M., & Thompson, J. G. (2008). Oocyte-secreted factors: Regulators of cumulus cell function and oocyte quality. *Human Reproduction Update*, *14*(2), 159–177. https://doi.org/10.1093/humupd/dmm040.
Goland, R. S., Wardlaw, S. L., Stark, R. I., & Frantz, A. G. (1981). Human plasma beta-endorphin during pregnancy, labor, and delivery. *The Journal of Clinical Endocrinology and Metabolism*, *52*(1), 74–78. https://doi.org/10.1210/jcem-52-1-74.
Gonzalez-Iglesias, A. E., Fletcher, P. A., Arias-Cristancho, J. A., Cristancho-Gordo, R., Helena, C. V., Bertram, R., et al. (2015). Direct stimulatory effects of oxytocin in female rat gonadotrophs and somatotrophs in vitro: Comparison with lactotrophs. *Endocrinology*, *156*(2), 600–612. https://doi.org/10.1210/en.2014-1543.
Gravanis, A., Makrigiannakis, A., Stournaras, C., & Margioris, A. N. (1994). Interaction between steroid hormones and endometrial opioids. *Annals of the New York Academy of Sciences*, *734*, 245–256.
Grell, S., Fjalland, B., & Christensen, J. D. (1989). Biphasic effect of a kappa-opioid receptor agonist on plasma oxytocin levels in rats. *European Journal of Pharmacology*, *166*(1), 41–48.
Guillemin, R., Vargo, T., Rossier, J., Minick, S., Ling, N., Rivier, C., et al. (1977). Beta-Endorphin and adrenocorticotropin are selected concomitantly by the pituitary gland. *Science (New York, N.Y.)*, *197*(4311), 1367–1369.

Guillou, A., Romanò, N., Steyn, F., Abitbol, K., Le Tissier, P., Bonnefont, X., et al. (2015). Assessment of lactotroph axis functionality in mice: Longitudinal monitoring of PRL secretion by ultrasensitive-ELISA. *Endocrinology, 156*(5), 1924–1930. https://doi.org/10.1210/en.2014-1571.

Gunnet, J. W., & Freeman, M. E. (1983). The mating-induced release of prolactin: A unique neuroendocrine response. *Endocrine Reviews, 4*(1), 44–61. https://doi.org/10.1210/edrv-4-1-44.

Gustafson, P., Bunn, S. J., & Grattan, D. R. (2017). The role of prolactin in the suppression of Crh mRNA expression during pregnancy and lactation in the mouse. *Journal of Neuroendocrinology, 29*(9), e12511. https://doi.org/10.1111/jne.12511. 4th ed.

Hadziomerović, D., Rabenbauer, B., & Wildt, L. (2006). Normalization of hyperinsulinemia by chronic opioid receptor blockade in hyperandrogenemic women. *Fertility and Sterility, 86*(3), 651–657. https://doi.org/10.1016/j.fertnstert.2006.01.039.

Hadžiomerović-Pekić, D., Wildt, L., Weiss, J. M., Moeller, K., Mattle, V., & Seeber, B. E. (2010). Metformin, naltrexone, or the combination of prednisolone and antiandrogenic oral contraceptives as first-line therapy in hyperinsulinemic women with polycystic ovary syndrome. *Fertility and Sterility, 94*(6), 2385–2388. https://doi.org/10.1016/j.fertnstert.2010.02.041.

Hartman, R. D., Rosella-Dampman, L. M., Emmert, S. E., & Summy-Long, J. Y. (1986). Ontogeny of opioid inhibition of vasopressin and oxytocin release in response to osmotic stimulation. *Endocrinology, 119*(1), 1–11. https://doi.org/10.1210/endo-119-1-1.

Hartman, R. D., Rosella-Dampman, L. M., & Summy-Long, J. Y. (1987). Endogenous opioid peptides inhibit oxytocin release in the lactating rat after dehydration and urethane. *Endocrinology, 121*(2), 536–543. https://doi.org/10.1210/endo-121-2-536.

Hasan, T. F., & Hasan, H. (2011). Anorexia nervosa: A unified neurological perspective. *International Journal of Medical Sciences, 8*(8), 679–703. Ivyspring International Publisher.

Hemmings, R., Fox, G., & Tolis, G. (1982). Effect of morphine on the hypothalamic-pituitary axis in postmenopausal women. *Fertility and Sterility, 37*(3), 389–391.

Honer, W. G., Thompson, C., Lightman, S. L., Williams, T. D., & Checkley, S. A. (1986). No effect of naloxone on plasma oxytocin in normal men. *Psychoneuroendocrinology, 11*(2), 245–248.

Jaschke, N., Lunger, F., Wildt, L., & Seeber, B. (2018). Beta endorphin in serum and follicular fluid of PCOS- and non-PCOS women. *Archives of Gynecology and Obstetrics, 298*(1), 217–222. Springer Berlin Heidelberg. https://doi.org/10.1007/s00404-018-4793-6.

Katz, N., & Mazer, N. A. (2009). The impact of opioids on the endocrine system. *The Clinical Journal of Pain, 25*(2), 170–175. https://doi.org/10.1097/AJP.0b013e3181850df6.

Lafisca, S., Bolelli, G., Franceschetti, F., Danieli, A., Tagliaro, F., Marigo, M., et al. (1985). Free and bound testosterone in male heroin addicts. *Archives of toxicology. Supplement. = Archiv fur Toxikologie. Supplement, 8*, 394–397.

Lanzone, A., Apa, R., Fulghesu, A. M., Cutillo, G., Caruso, A., & Mancuso, S. (1993). Long-term naltrexone treatment normalizes the pituitary response to gonadotropin-releasing hormone in polycystic ovarian syndrome. *Fertility and Sterility, 59*(4), 734–737.

Lee, Y., & Voogt, J. L. (1999). Feedback effects of placental lactogens on prolactin levels and Fos-related antigen immunoreactivity of tuberoinfundibular dopaminergic neurons in the arcuate nucleus during pregnancy in the rat. *Endocrinology, 140*(5), 2159–2166. https://doi.org/10.1210/endo.140.5.6730.

Leyendecker, G. (1979). The pathophysiology of hypothalamic ovarian failure. Diagnostic and therapeutical considerations. *European Journal of Obstetrics, Gynecology, and Reproductive Biology, 9*(3), 175–186.

Leyendecker, G., Waibeltreber, S., & Wildt, L. (1990). The central control of follicular maturation and ovulation in the human. *Oxford Reviews of Reproductive Biology, 12*, 93–146.

Leyendecker, G., Waibel-Treber, S., & Wildt, L. (1993). Pulsatile administration of gonadotrophin releasing hormone and oral administration of naltrexone in hypothalamic amenorrhoea. *Human Reproduction*, *8*(Suppl. 2), 184–188. https://doi.org/10.1093/humrep/8.suppl_2.184.

Leyendecker, G., & Wildt, L. (1983). Induction of ovulation with chronic intermittent (pulsatile) administration of Gn-RH in women with hypothalamic amenorrhoea. *Reproduction*, *69*(1), 397–409. https://doi.org/10.1530/jrf.0.0690397.

Lindow, S. W., Hendricks, M. S., Nugent, F. A., Dunne, T. T., & van der Spuy, Z. M. (1999). Morphine suppresses the oxytocin response in breast-feeding women. *Gynecologic and Obstetric Investigation*, *48*(1), 33–37. Karger Publishers. https://doi.org/10.1159/000010130.

Lindow, S. W., Hendricks, M. S., Thompson, J. W., & van der Spuy, Z. M. (1998). Effects of morphine administration on the fetal production of oxytocin in labour. *Clinical Science (London, England: 1979)*, *95*(1), 91–95.

Lindow, S. W., van der Spuy, Z. M., Hendricks, M. S., Nugent, F. A., & Dunne, T. T. (1993). The effect of morphine and naloxone administration on maternal oxytocin concentration in late pregnancy. *Clinical Endocrinology*, *39*(6), 671–675.

Lindow, S. W., van der Spuy, Z. M., Hendricks, M. S., Rosselli, A. P., Lombard, C., & Leng, G. (1992). The effect of morphine and naloxone administration on plasma oxytocin concentrations in the first stage of labour. *Clinical Endocrinology*, *37*(4), 349–353.

Lucas, B. K., Ormandy, C. J., Binart, N., Bridges, R. S., & Kelly, P. A. (1998). Null mutation of the prolactin receptor gene produces a defect in maternal behavior. *Endocrinology*, *139*(10), 4102–4107. https://doi.org/10.1210/endo.139.10.6243.

Lunger, F., Vehmas, A. P., Fürnrohr, B. G., Sopper, S., Wildt, L., & Seeber, B. (2016a). Opiate receptor blockade on human granulosa cells inhibits VEGF release. *Reproductive Biomedicine Online*, *32*(3), 316–322. https://doi.org/10.1016/j.rbmo.2015.12.006.

Lunger, F., Vehmas, A. P., Fürnrohr, B. G., Sopper, S., Wildt, L., & Seeber, B. (2016b). The opiate antagonist naloxone reduces the release of VEGF from primary granulosa cells of PCOS women. *Human Reproduction*, *31*(Suppl. 1), i444.

Mauras, N., Veldhuis, J. D., & Rogol, A. D. (1986). Role of endogenous opiates in pubertal maturation: Opposing actions of naltrexone in prepubertal and late pubertal boys. *The Journal of Clinical Endocrinology and Metabolism*, *62*(6), 1256–1263. https://doi.org/10.1210/jcem-62-6-1256.

McCubbin, J. A. (1993). Stress and endogenous opioids: Behavioral and circulatory interactions. *Biological Psychology*, *35*(2), 91–122.

Morris, M. S., Domino, E. F., & Domino, S. E. (2013). Opioid modulation of oxytocin release. *The Journal of Clinical Pharmacology*, *50*(10), 1112–1117. https://doi.org/10.1177/0091270010361256.

Ormandy, C. J., Binart, N., & Kelly, P. A. (1997). Mammary gland development in prolactin receptor knockout mice. *Journal of Mammary Gland Biology and Neoplasia*, *2*(4), 355–364.

O'Rourke, T. K., & Wosnitzer, M. S. (2016). Opioid-induced androgen deficiency (OPIAD): Diagnosis, management, and literature review. *Current Urology Reports*, *17*(10), 76. Springer US. https://doi.org/10.1007/s11934-016-0634-y.

Paice, J. A., Penn, R. D., & Ryan, W. G. (1994). Altered sexual function and decreased testosterone in patients receiving intraspinal opioids. *Journal of Pain and Symptom Management*, *9*(2), 126–131.

Pasquali, R., & Vicennati, V. (2000). Activity of the hypothalamic-pituitary-adrenal axis in different obesity phenotypes. *International Journal of Obesity and Related Metabolic Disorders: Journal of the International Association for the Study of Obesity*, *24*(Suppl. 2), S47–S49.

Pende, A., Musso, N. R., Montaldi, M. L., Pastorino, G., Arzese, M., & Devilla, L. (1986). Evaluation of the effects induced by four opiate drugs, with different affinities to opioid receptor subtypes, on anterior pituitary LH, TSH, PRL and GH secretion and on cortisol secretion in normal men. *Biomedicine & pharmacotherapy = Biomedecine & pharmacotherapie*, *40*(5), 178–182.

Petersen, S. L., & LaFlamme, K. D. (1997). Progesterone increases levels of μ-opioid receptor mRNA in the preoptic area and arcuate nucleus of ovariectomized, estradiol-treated female rats. *Molecular Brain Research, 52*(1), 32–37. https://doi.org/10.1016/S0169-328X(97)00194-0.

Petraglia, F., Bernasconi, S., Iughetti, L., Loche, S., Romanini, F., Facchinetti, F., et al. (1986). Naloxone-induced luteinizing hormone secretion in normal, precocious, and delayed puberty. *The Journal of Clinical Endocrinology and Metabolism, 63*(5), 1112–1116. https://doi.org/10.1210/jcem-63-5-1112.

Petraglia, F., Facchinetti, F., M'Futa, K., Ruspa, M., Bonavera, J. J., Gandolfi, F., et al. (1986). Endogenous opioid peptides in uterine fluid. *Fertility and Sterility, 46*(2), 247–251.

Petraglia, F., Segre, A., Facchinetti, F., Campanini, D., Ruspa, M., & Genazzani, A. R. (1985). Beta-endorphin and met-enkephalin in peritoneal and ovarian follicular fluids of fertile and postmenopausal women. *Fertility and Sterility, 44*(5), 615–621.

Piva, F., Limonta, P., Dondi, D., Pimpinelli, F., Martini, L., & Maggi, R. (1995). Effects of steroids on the brain opioid system. *The Journal of Steroid Biochemistry and Molecular Biology, 53*(1-6), 343–348. https://doi.org/10.1016/0960-0760(95)00072-8.

Pumford, K. M., Leng, G., & Russell, J. A. (1991). Morphine actions on supraoptic oxytocin neurones in anaesthetized rats: Tolerance after i.c.v. morphine infusion. *The Journal of Physiology, 440*, 437–454. Wiley-Blackwell.

Ragni, G., De Lauretis, L., Bestetti, O., Sghedoni, D., & Gambaro, V. (1988). Gonadal function in male heroin and methadone addicts. *International Journal of Andrology, 11*(2), 93–100.

Reid, R. L., Quigley, M. E., & Yen, S. S. (1983). The disappearance of opioidergic regulation of gonadotropin secretion in postmenopausal women. *The Journal of Clinical Endocrinology and Metabolism, 57*(6), 1107–1110. https://doi.org/10.1210/jcem-57-6-1107.

Reymond, M. J., & Porter, J. C. (1985). Involvement of hypothalamic dopamine in the regulation of prolactin secretion. *Hormone Research, 22*(3), 142–152. https://doi.org/10.1159/000180088.

Roche, D. J. O., Childs, E., Epstein, A. M., & King, A. C. (2010). Acute HPA axis response to naltrexone differs in female vs. male smokers. *Psychoneuroendocrinology, 35*(4), 596–606. https://doi.org/10.1016/j.psyneuen.2009.09.016.

Roozenburg, B. J., van Dessel, H. J., Evers, J. L., & Bots, R. S. (1997). Successful induction of ovulation in normogonadotrophic clomiphene resistant anovulatory women by combined naltrexone and clomiphene citrate treatment. *Human Reproduction, 12*(8), 1720–1722.

Rossmanith, W. G., Wirth, U., Sterzik, K., & Yen, S. S. (1989). The effects of prolonged opioidergic blockade on LH pulsatile secretion during the menstrual cycle. *Journal of Endocrinological Investigation, 12*(4), 245–252.

Rotterdam ESHRE/ASRM-Sponsored PCOS Consensus Workshop Group. (2004). Revised 2003 consensus on diagnostic criteria and long-term health risks related to polycystic ovary syndrome. *Fertility and Sterility, 81*, 19–25.

Russell, J. A., & Brunton, P. J. (2009). Oxytocin (Peripheral/Central actions and their regulation). In *Encyclopedia of neuroscience* (pp. 337–347). Elsevier. https://doi.org/10.1016/B978-008045046-9.01192-X.

Russell, J. A., & Douglas, A. J. (2000). Opioids. In G. Fink (Ed.), *Encyclopedia of stress* (pp. 86–98). San Diego: Academic Press.

Russell, J. A., Gosden, R. G., Humphreys, E. M., Cutting, R., Fitzsimons, N., Johnston, V., et al. (1989). Interruption of parturition in rats by morphine: A result of inhibition of oxytocin secretion. *The Journal of Endocrinology, 121*(3), 521–536.

Safarinejad, M. R., Asgari, S. A., Farshi, A., Ghaedi, G., Kolahi, A. A., Iravani, S., et al. (2013). The effects of opiate consumption on serum reproductive hormone levels, sperm parameters, seminal plasma antioxidant capacity and sperm DNA integrity. *Reproductive toxicology (Elmsford, N.Y.), 36*, 18–23. https://doi.org/10.1016/j.reprotox.2012.11.010.

Shibli, K. U., Dhillon, A. R., Goode, J. A., Gilbert, C. L., Thompson, J. W., Russell, I. F., et al. (2001). Effect of intrathecal fentanyl on oxytocin secretion in pregnant women not in labour. *Clinical Science (London, England: 1979)*, *101*(4), 415–419.

Singleton, P. A., Lingen, M. W., Fekete, M. J., Garcia, J. G. N., & Moss, J. (2006). Methylnaltrexone inhibits opiate and VEGF-induced angiogenesis: role of receptor transactivation. *Microvascular Research*, *72*(1-2), 3–11. https://doi.org/10.1016/j.mvr.2006.04.004.

Sir-Petermann, T., López, G., Castillo, T., Calvillán, M., Rabenbauer, B., & Wildt, L. (1998). Naltrexone effects on insulin sensitivity and insulin secretion in hyperandrogenic women. *Experimental and clinical endocrinology & diabetes: Official journal, German Society of Endocrinology [and] German Diabetes Association*, *106*(5), 389–394. https://doi.org/10.1055/s-0029-1212004. © J. A. Barth Verlag in Georg Thieme Verlag KG Stuttgart · New York.

Sir-Petermann, T., Rabenbauer, B., & Wildt, L. (1993). The effect of flutamide on pulsatile gonadotrophin secretion in hyperandrogenaemic women. *Human Reproduction*, *8*(11), 1807–1812.

Smith, Y. R. (1998). Brain opioid receptor measurements by positron emission tomography in normal cycling women: Relationship to luteinizing hormone pulsatility and gonadal steroid hormones. *The Journal of Clinical Endocrinology and Metabolism*, *83*(12), 4498–4505. https://doi.org/10.1210/jc.83.12.4498.

Smith, Y. R., Stohler, C. S., Nichols, T. E., Bueller, J. A., Koeppe, R. A., & Zubieta, J.-K. (2006). Pronociceptive and antinociceptive effects of estradiol through endogenous opioid neurotransmission in women. *The Journal of neuroscience: The official journal of the Society for Neuroscience*, *26*(21), 5777–5785. https://doi.org/10.1523/JNEUROSCI.5223-05.2006.

Soaje, M., Valdez, S., Bregonzio, C., Penissi, A., & Deis, R. P. (2006). Dopaminergic mechanisms involved in prolactin release after mifepristone and naloxone treatment during late pregnancy in the rat. *Neuroendocrinology*, *84*(1), 58–67. Karger Publishers. https://doi.org/10.1159/000096825.

Stocche, R. M., Klamt, J. G., Antunes-Rodrigues, J., Garcia, L. V., & Moreira, A. C. (2001). Effects of intrathecal sufentanil on plasma oxytocin and cortisol concentrations in women during the first stage of labor. *Regional Anesthesia and Pain Medicine*, *26*(6), 545–550.

Subirán, N., Casis, L., & Irazusta, J. (2011). Regulation of male fertility by the opioid system. *Molecular medicine (Cambridge, Mass.)*, *17*(7-8), 846–853. https://doi.org/10.2119/molmed.2010.00268.

Tang, X., Chen, Y., Ran, H., Jiang, Y., He, B., Wang, B., et al. (2015). Systemic morphine treatment derails normal uterine receptivity, leading to embryo implantation failure in mice. *Biology of Reproduction*, *92*(5), 118. https://doi.org/10.1095/biolreprod.115.128686.

Torner, L., & Neumann, I. D. (2002). The brain prolactin system: Involvement in stress response adaptations in lactation. *Stress (Amsterdam, Netherlands)*, *5*(4), 249–257. https://doi.org/10.1080/1025389021000048638.

Totorikaguena, L., Olabarrieta, E., Matorras, R., Alonso, E., Agirregoitia, E., & Agirregoitia, N. (2017). Mu opioid receptor in the human endometrium: Dynamics of its expression and localization during the menstrual cycle. *Fertility and Sterility*, *107*(4), 1070–1077.e1. https://doi.org/10.1016/j.fertnstert.2017.01.020.

Trescot, A. M., Datta, S., Lee, M., & Hansen, H. (2008). Opioid pharmacology. *Pain Physician*, *11*(Suppl 2), S133–S153.

Unger, J. W., & Lange, W. (1997). Insulin receptors in the pituitary gland: Morphological evidence for influence on opioid peptide-synthesizing cells. *Cell and Tissue Research*, *288*(3), 471–483.

Veldhuis, J. D., Rogol, A. D., Samojlik, E., & Ertel, N. H. (1984). Role of endogenous opiates in the expression of negative feedback actions of androgen and estrogen on pulsatile properties of luteinizing hormone secretion in man. *The Journal of Clinical Investigation, 74*(1), 47–55. American Society for Clinical Investigation. https://doi.org/10.1172/JCI111417.

Vértes, M., Pámer, Z., & Garai, J. (1986). On the mechanism of opioid-oestradiol interactions. *Journal of Steroid Biochemistry, 24*(1), 235–238.

Villegas-Gabutti, C. M., Pennacchio, G. E., Vivas, L., Jahn, G. A., & Soaje, M. (2018). Role of oxytocin in prolactin secretion during late pregnancy. *Neuroendocrinology, 106*(4), 324–334. https://doi.org/10.1159/000480669.

Vuong, C., Van Uum, S. H. M., O'Dell, L. E., Lutfy, K., & Friedman, T. C. (2010). The effects of opioids and opioid analogs on animal and human endocrine systems. *Endocrine Reviews, 31*(1), 98–132. https://doi.org/10.1210/er.2009-0009.

Wahlström, T., Laatikainen, T., Salminen, K., & Leppäluoto, J. (1985). Immunoreactive beta-endorphin is demonstrable in the secretory but not in the proliferative endometrium. *Life Sciences, 36*(10), 987–990.

Wildt, L., & Leyendecker, G. (1987). Induction of ovulation by the chronic administration of naltrexone in hypothalamic amenorrhea. *The Journal of Clinical Endocrinology and Metabolism, 64*(6), 1334–1335. https://doi.org/10.1210/jcem-64-6-1334.

Wildt, L., Leyendecker, G., Sir-Petermann, T., & Waibel-Treber, S. (1993). Treatment with naltrexone in hypothalamic ovarian failure: Induction of ovulation and pregnancy. *Human Reproduction, 8*(3), 350–358.

Wildt, L., Sir-Petermann, T., Leyendecker, G., Waibel-Treber, S., & Rabenbauer, B. (1993). Opiate antagonist treatment of ovarian failure. *Human Reproduction, 8*(Suppl. 2), 168–174.

Wortsman, J., Wehrenberg, W. B., Gavin, J. R., & Allen, J. P. (1984). Elevated levels of plasma beta-endorphin and gamma 3-melanocyte stimulating hormone in the polycystic ovary syndrome. *Obstetrics and Gynecology, 63*(5), 630–634.

Ye, X., Ren, H., Zhang, M., Sun, Z., Jiang, A. C., & Xu, G. (2012). ERK1/2 signaling pathway in the release of VEGF from Müller cells in diabetes. *Investigative Ophthalmology & Visual Science, 53*(7), 3481–3489. https://doi.org/10.1167/iovs.11-9076.

Yee, A., Danaee, M., Loh, H. S., Sulaiman, A. H., & Ng, C. G. (2016). Sexual dysfunction in heroin dependents: A comparison between methadone and buprenorphine maintenance treatment. *PLoS One, 11*(1). Edited by J. M. Dominguez. e0147852. https://doi.org/10.1371/journal.pone.0147852.

You, Z. D., Li, J. H., Song, C. Y., Wang, C. H., & Lu, C. L. (2000). Chronic morphine treatment inhibits oxytocin synthesis in rats. *Neuroreport, 11*(14), 3113–3116.

Zimmermann, U., Rechlin, T., Plaskacewicz, G. J., Barocka, A., Wildt, L., & Kaschka, W. P. (1997). Effect of naltrexone on weight gain and food craving induced by tricyclic antidepressants and lithium: An open study. *Biological Psychiatry, 41*(6), 747–749. https://doi.org/10.1016/S0006-3223(96)00523-9.

CHAPTER ELEVEN

Kappa opioid receptor and oligodendrocyte remyelination

Fei Wang, Feng Mei*
Department of Histology and Embryology, Brain Science Key Laboratory of Chongqing Education Commission, Third Military Medical University, Chongqing, China
*Corresponding author: e-mail address: meif@tmmu.edu.cn

Contents

Abstract

Adult oligodendrocyte precursor cells (OPCs) maintain the abilities to differentiate and myelinate denuded axons in demyelinating diseases, such as Multiple Sclerosis (MS), albert often inefficiently. Remyelination therapies seek to enhance endogenous remyelination and represent a promising approach to achieve functional and cellular architectural recovery against neuronal deficits. Recent findings indicate that the kappa opioid receptor (KOR), a G-protein coupled receptor (GPCR), plays an important role in regulating oligodendrocyte differentiation and myelination. In this chapter, we reviewed (1) current knowledge of the functional importance of remyelination in demyelination diseases; (2) the opioids that can alter oligodendroglial proliferation and differentiation; (3) the endogenous KOR signaling in regulating oligodendrocyte myelination.

Vitamins and Hormones, Volume 111
ISSN 0083-6729
https://doi.org/10.1016/bs.vh.2019.05.004

1. Oligodendrocyte and remyelination

1.1 Myelin and oligodendrocytes

Myelin is a fatty insulation that ensheaths axons in the vertebrate CNS. Each mature oligodendrocytes (OLs) generates multiple myelin sheets that repeatedly and compactly wrap axons, known as myelin sheaths (Hartline & Colman, 2007; Zalc, Goujet, & Colman, 2008). Myelin sheaths are discontinued and gapped by segments of naked axons; so-called the Node of Ranvier (Eshed-Eisenbach & Peles, 2013; Normand & Rasband, 2015). The nodal axolemma is in direct contact with the extracellular environment and enriched with ion channels. It has been well established that myelin sheaths insult axons and ensure fast and salutatory transmission of active potentials (Baumann & Pham-Dinh, 2001; Smith, Blakemore, & McDonald, 1981; Utzschneider, Archer, Kocsis, Waxman, & Duncan, 1994). Additionally, OLs also provide metabolic supports to meet the energy demands of axons by transporting lactate via monocarboxylate transporter 1 (MCT1) (Lee et al., 2012; Nave, 2010). Indeed, OLs can dictate to the axonal needs by sensing the spiking amplitude and adaptively "feed" axons in a NMDA-dependent manner (Bergles, Roberts, Somogyi, & Jahr, 2000; Saab et al., 2016). Recent evidence also indicates that myelination may facilitate synaptogenesis in developing brains, suggesting a potential role in establishing the neural circuitries (Wang et al., 2018). Together, myelin sheaths have been well recognized as a constitutive component that is essentially required for neuronal functions and survival.

1.2 Demyelination and axonal loss

Demyelination insults directly impair neuronal functions by disrupting the action potential conduction along axons in demyelinating diseases, such as Multiple Sclerosis (MS) (Dutta & Trapp, 2011; Franklin, 2002). Immunosuppression is the principal therapeutic strategy for the autoimmune induced demyelination, regarding that overactivated immune cells are the primary reason that causes OL death and demyelination in MS (Haghikia, Hohlfeld, Gold, & Fugger, 2013; International Multiple Sclerosis Genetics Consortium, et al., 2011). More importantly, the inflammation may further damage and degrade axons, and even ultimately result in permanent neuronal function deficits (Bjartmar & Trapp, 2003; Dutta & Trapp, 2011; Trapp et al., 1998). It is noticeable that the immunoregulatory efforts have successfully slowed down the relapses of MS but fail to cease the progression of the disease nor restore CNS functions (Bitsch, Schuchardt, Bunkowski, Kuhlmann, & Bruck, 2000; Franklin, ffrench-Constant, Edgar, & Smith, 2012; Kuhlmann, Lingfeld, Bitsch, Schuchardt, & Bruck, 2002; Lovas, Szilagyi, Majtenyi, Palkovits, & Komoly, 2000; Plemel, Liu, & Yong, 2017; Reich, Lucchinetti, & Calabresi, 2018; Tallantyre et al., 2009; Trapp & Stys, 2009). The fact that remyelination is often incomplete in the lesions and no remyelination therapies are available in clinic to date, begs the question: is successful remyelination required for preserving axonal integrity and restoring neuronal functions in MS? Evidence from the material of autopsies has shown that the axon density is greatly decreased in the lesions with poor remyelination (Dutta & Trapp, 2011; Kornek et al., 2000). Consistently, better preserved axons always accompany with better remyelination, suggesting that the denuded axons may be extremely vulnerable to inflammatory insults and remyelination may protect axons against degeneration (Kornek et al., 2000; Nikic et al., 2011). A plausible mechanism is that remyelination could benefit axonal survival by reconstructing the energy supply and shielding axons from the microenvironmental inflammation (Irvine & Blakemore, 2008; Zambonin et al., 2011). Given that active inflammation always accompanies with demyelination, axonal injuries and possible remyelination, it is important to understand when remyelination could occur and if enhancing remyelination is effective to preserve axonal integrity? Since longitudinal observations are not feasible by using histologic examinations in demyelinating diseases, cell-lineage specific tracing is a feasible approach to identify newly generated myelin sheaths from pre-existing myelin (McKenzie et al., 2014; Tripathi et al., 2017; Xiao et al., 2016; Young et al., 2013). Recent evidence indicates that newly formed myelin sheaths could be observed at the peak of disease in the EAE model, an animal model of MS (Mei et al., 2016). This finding is in line with the observations that relatively early remyelination can be spotted in the MS lesions (Chang, Tourtellotte, Rudick, & Trapp, 2002; Prineas, Barnard, Kwon, Sharer, & Cho, 1993). More importantly, oligodendroglia-specific deletion of the muscarinic receptor 1, a negative regulator for oligodendroglial differentiation, enhances remyelination and subsequently preserves axonal survivals and decreases neuronal function deficits in the EAE mice (Mei et al., 2016). Therefore, these findings demonstrate that enhancing remyelination may represent a promising strategy to preserve axonal integrity in demyelination diseases.

1.3 Remyelination therapies

Remyelination therapies seek to reconstruct the cellular architecture and restore neuronal functions, more importantly, to preserve axons from degeneration in demyelinating diseases. Despite that remyelination can spontaneously occurs in the lesions, it often remains incomplete and insufficient (Franklin, 2002; Gilson & Blakemore, 1993). The pre-existing mature OLs are unlikely to contribute to remyelination, given that these cells are stable and stop adding new myelin segments (Keirstead & Blakemore, 1997). Rather, oligodendrocyte precursor cells (OPCs) are responsible for remyelinating the denuded axons via differentiation into mature OLs (Goldschmidt, Antel, Konig, Bruck, & Kuhlmann, 2009; Nunes et al., 2003; Patrikios et al., 2006). Evidence has shown that oligodendroglial cells are present but arrested from terminal maturation in the lesions of MS brains (Chang et al., 2002; Kuhlmann et al., 2008; Wolswijk, 1998, 2000). Indeed, abundant OPCs are evenly present in adult CNS throughout the life span and they are capable of migrating and proliferating (Hughes, Kang, Fukaya, & Bergles, 2013). Several recent lines of evidence have shown life-long myelinogenesis in the healthy mouse CNS, suggesting OPCs maintain their full abilities of differentiation, albert gradually declining while aging (Cantuti-Castelvetri & Fitzner, 2018; Hill, Li, & Grutzendler, 2018; Hughes & Orthmann-Murphy, 2018; Safaiyan et al., 2016). The access to OPCs may not the major reason that causes the failure of remyelination, since a number of OPCs are present within and adjacent to the lesions (Wolswijk, 1998). In support with this notion, addition of platelet-derived growth factor (PDGF), a mitogen to promote OPC proliferation, did not successfully accelerate remyelination (Woodruff, Fruttiger, Richardson, & Franklin, 2004). It has been hypothesized that the inhibitory signals in the microenvironment niche may block oligodendroglial differentiation, thus deactivating the inhibitory signals may represent an effective remyelination approach. One remarkable example is the utilization of anti-LINGO-1 antibody as a remyelination reagent. LINGO-1 is a potent inhibitory signal that blocks oligodendroglial differentiation and it is widely expressed by axons and myelin sheaths. It has been demonstrated that the anti-LINGO-1 antibody is efficient in promoting remyelination in the demyelination mouse models (Mi et al., 2007, 2004, 2005; Sun, Ren, Xu, & Zhang, 2015). A second strategy is to enhance endogenous oligodendroglial differentiation and myelination by using small molecule compounds. With this regard, screening and identification of remyelination-enhancing compounds may be a feasible and realistic approach (Deshmukh et al., 2013; Hubler et al., 2018; Mei et al., 2014; Najm et al., 2015). Recent efforts in screening myelination-enhancing drugs have yielded a number of potent compounds that can accelerate oligodendrocyte remyelination, and some of them are FDA-approved drugs, such as clemastine and benzatropine (Cree et al., 2018; Deshmukh et al., 2013; Jensen et al., 2018; Li, He, Fan, & Sun, 2015; Liu & Dupree, 2016; Mei et al., 2014). As a proof of concept, a recent stage-two clinical trial reveals that clemastine, an anti-allergy drug, can improve neuronal functional recovery in MS patients (Cohen & Tesar, 2017; Green et al., 2017). Together, these findings indicate that enhancing myelination may be promising and realistic approach to preserve axonal integrity and promote functional recovery in demyelination diseases.

2. Opioids in regulating oligodendroglial differentiation

Opioids have been widely used to treat nerve pain and addictive disorders in clinic (Besson, 1999; Kreek, LaForge, & Butelman, 2002). Natural and synthetic opioids exert their functions by binding to opioid receptors (Cox, 2013). The opioid receptors are a group of G-protein coupled receptors (GPCRs), which mainly consists of three classic families of receptors: μ (MOR), δ (λ, DOR), and κ (KOR) (Cox, 2013; Katritch, Cherezov, & Stevens, 2013). The opioid receptors are widely expressed by most nerve cells in the CNS, including oligodendroglia, neurons and astrocytes (Du & Duan, 2016; Petrillo, Tavani, Verotta, Robson, & Kosterlitz, 1987; Stiene-Martin et al., 2001). Numerous literatures have shown that quite a few opioids can impact oligodendroglial proliferation and differentiation (Table 1).

2.1 KOR agonists

Recent advance in high-throughput screening has developed an *in vitro* platform, so-called Binary Indicant for myelination using Micropillar Array (BIMA), which allows for rapid and efficient screening for remyelination compounds (Mei et al., 2014). After a screen of 250 GPCR compounds, a cluster of KOR agonists are identified as potent myelination-enhancing compounds, which includes (±)-U50488, ICI-199441, U54494, Matrine, N-MPPP, BRL52537, GR89696, dynorphin B, 6-GNTI, and salvinorin A (Mei & Mayoral, 2016). In a consistent manner, KOR antagonists, such as nor-binaltorphimine (nor-BNI), DIPPA and GNTI exhibit inhibitory effects on

Table 1 Opioids in regulating oligodendroglial differentiation and proliferation.

Compounds	Pharmacodynamics	Effects	References
(±)-U50488, ICI-199441, U54494, Matrine, et al.	Selective KOR agonism	Promote OPCs differentiation and myelination in cultures. (±)-U50488 improves remyelination and functional recovery in demyelination mice and hypoxic mice	Du and Duan (2016), Mei and Mayoral (2016), and Wang et al. (2018)
Asimadoline	Selective KOR agonism	Improves functional recovery in EAE mice	Du and Duan (2016)
nor-BNI, DIPPA, GNTI	Selective KOR antagonism	Inhibits OPCs differentiation in cultures	Mei and Mayoral (2016)
PL017	Selective MOR agonism	Promotes proliferation in immature oligodendrocytes in cultures	Knapp, Maderspach, and Hauser (1998)
Morphine	MOR agonism	Promotes oligodendroglial proliferation *in vitro*. Inhibit oligodendroglial differentiation and cause OL death *in vivo*	Hauser et al. (2009), Stokes and Lee (1976), and Ujcikova, Eckhardt, Kagan, Roubalova, and Svoboda (2014)
Oxycodone	MOR agonism	Decreases myelination *in vivo*	Fan et al. (2018)
Buprenorphine	MOR agonism and partial antagonism to DOR and KOR	Accelerates myelination at a low dosage while inhibits myelination at a high dosage in the brains of newborn rodents	Sanchez, Bigbee, Fobbs, Robinson, and Sato-Bigbee (2008)
Methadone	MOR agonism and lower affinity to DOR or KOR	Promotes oligodendrocyte differentiation and myelination both *in vivo* and *in vitro*, and increases oligodendroglia proliferation in cultures	Vestal-Laborde, Eschenroeder, Bigbee, Robinson, and Sato-Bigbee (2014)
Nalmefene	Selective MOR antagonism and partial KOR agonism	Increases myelination in the spinal cord injury	Faden, Sacksen, and Noble (1988) and Puniak et al. (1991)

oligodendroglial differentiation (Mei & Mayoral, 2016). It is noticeable that, among these KOR agonists, (±)-U50488 is a highly selective KOR agonist and displays as the most potent compound in promoting oligodendroglial differentiation. *In vitro* evidence indicates that the activation of KOR by (±)-U50488 or other KOR agonists can promote oligodendroglial differentiation and myelination in purified rat OPC cultures (Mei & Mayoral, 2016). In addition, (±)-U50488 promotes the differentiation of human induced pluripotent stem cells (iPSC) derived OPCs *in vitro* (Mei & Mayoral, 2016). Further, in the demyelination model induced by lysolecithin or cuprizone, the administration of (±)-U50488 accelerates oligodendrocyte remyelination (Du & Duan, 2016; Mei & Mayoral, 2016). (±)-U50488 treatment also enhances myelinogenesis in the hypoxia induced hypomyelination mouse model and consequently promoted functional recovery (Wang et al., 2018), suggesting (±)-U50488 represents a powerful compound to promote oligodendroglial differentiation and myelination. Finally, on the EAE mouse model, KOR agonism by (±)-U50488 decreases the clinical score and rescues neuronal deficits by accelerating remyelination (Du & Duan, 2016). These findings indicate that KOR agonists may be promising potential drugs for treating demyelination diseases.

2.2 MOR agonists

MOR agonists have been used for the treatment of addiction and pain (Kieffer & Gaveriaux-Ruff, 2002). Clinical evidence has shown that the opioid-dependent individuals exhibit impaired white matter besides a cohort of psychiatry disorders (Upadhyay et al., 2010). As a representative of MOR agonists, morphine has been widely studied for its pharmaceutical functions and mechanisms in neurons. Administration of morphine induces a significant decline of neuron cell numbers in adult rat brains, along with a decrease of OL density (Hauser et al., 2009; Ujcikova et al., 2014). The decrease of OL number could be partially explained by elevated activation of caspase 3 in mature OLs upon the exposure to morphine, which subsequently causes cell apoptosis (Hauser et al., 2009; Ujcikova et al., 2014). In line with these findings, oxycodone, another MOR agonist, also leads to demyelination and axonal degeneration in adult rats when treated for a prolonged period of time at a high dosage (Fan et al., 2018). The *in vivo* evidence suggests that the over-activation of MOR signaling may impair myelin maintenance and OL survival. In support this notion, a recent study by using a single-cell sequencing method shows that morphine exposure changes the landscape of OL gene profiles, including the genes related to oligodendroglial differentiation and maturation, and the unfold gene response (Avey et al., 2018). Regarding that MOR is widely expressed by neurons and glial cells, if it is possible that MOR agonism indirectly cause OL apoptosis or death? In purified OPC culture, morphine only alters oligodendroglial RNA synthesis and the activation of MOR by PL017 increases oligodendroglial proliferation (Knapp et al., 1998; Stokes & Lee, 1976). Moreover, global MOR deletion does not significantly alter the clinical score of EAE model (Du & Duan, 2016). Thus, it seems that MOR agonism may not directly induce OL death and demyelination.

2.3 Opioid receptor modulator

Opioid receptor modulators are the drugs that display mixed agonism and/or antagonism with comparable affinity at different opioid receptors. A number of the opioid modulators have been demonstrated effective in regulating oligodendroglial differentiation and myelination. For example, buprenorphine is a partial agonist for MOR with DOR and KOR antagonism. This drug has been used to treat opioid addiction during pregnancy and pain. It is interesting that a lower dosage of buprenorphine increases myelin basic protein (MBP) expression in neonatal rats, while a higher dosage reverses its effects, namely, inhibits oligodendrocyte myelination (Eschenroeder, Vestal-Laborde, Sanchez, Robinson, & Sato-Bigbee, 2012; Sanchez et al., 2008). Similar to buprenorphine, methadone is a substitutive drug used for opioid addiction during pregnancy. As an agonist for all three opioid receptors, methadone exposure during pregnancy and early postnatal period accelerates myelination in neonatal mice (Vestal-Laborde et al., 2014). In line with the *in vivo* finding, methadone exposure promotes oligodendroglial differentiation in a dose-dependent manner *in vitro* and also increases OPC proliferation revealed by [^{3}H] thymidine incorporation (Vestal-Laborde et al., 2014). Nalmefene acts as a selective antagonist for MOR and DOR and a partial agonist for KOR. Administration of nalmefene to a classical spinal cord injury model resulted in better preserved of axonal integrity and myelin, which may consequently promote the motor function recovery (Faden et al., 1988; Puniak et al., 1991). Regarding the unselective binding affinity to multiple opioid receptors, it is difficult to determine which receptor(s) is responsible for the effects on oligodendroglial differentiation.

3. Endogenous opioid signaling in oligodendroglia

The effects of the compounds that target to opioid receptors indicate that endogenous opioid signaling may play a role in regulating oligodendroglial differentiation and myelination. These opioid receptors can be activated by endogenous opioid peptides that are widely expressed in the CNS and PNS (Zollner & Stein, 2007). For example, proenkephalin and prodynorphin precursors are detectable in a mixed neuron-glia cultures. Indeed, a previous study has shown that the KOR expression is significantly decreased in the oligodendroglia of the *Jimpy* mice. The *Jimpy* mouse carries an X chromosome-linked recessive gene defect and exhibits severe hypomyelination in the CNS, hinting KOR may be functional in regulating oligodendroglial differentiation (Knapp, Adjan, & Hauser, 2009). Genetic deletion of opioid receptor provides direct evidence for understanding the function of endogenous opioid signaling. A recent literature has shown that the deletion of KOR rather than MOR or DOR greatly changes the clinical score of EAE mice (Du & Duan, 2016), suggesting the KOR signaling may play an important role in regulating oligodendroglial differentiation and myelination.

3.1 KOR in regulating oligodendrocyte differentiation

The effects of KOR agonists on myelination suggest a potential role of endogenous KOR signaling in regulating oligodendroglial differentiation. It is noticeable that (±)-U50488 is a highly selective to KOR without activating MOR or DOR. The KOR mutant and conditional knockout mice provide direct evidence to understand the function of KOR in oligodendroglial differentiation. The OPCs purified from the KOR knockout mice, display inhibited differentiation and myelination in cultures (Du & Duan, 2016; Mei & Mayoral, 2016). Consistently, conditional deletion of KOR in OPCs also results in a severe delay of myelination *in vivo*, suggesting endogenous KOR signaling is important for oligodendroglial differentiation and myelination. Since the natural KOR ligands, such as prodynorphin and dynorphin A, are widely expressed by oligodendroglia, astroglia and neuron, it is plausible the endogenous KOR signaling may regulate oligodendroglial differentiation and myelination through an auto-/paracrine manner (Alvarez-Bolado, Fairen, Douglass, & Naranjo, 1990; Knapp et al., 2001; Wahlert, Funkelstein, Fitzsimmons, Yaksh, & Hook, 2013; Zamir, Quirion, & Segal, 1985) (Fig. 1). In demyelination animal models, KOR deletion severely delays remyelination in the cuprizone

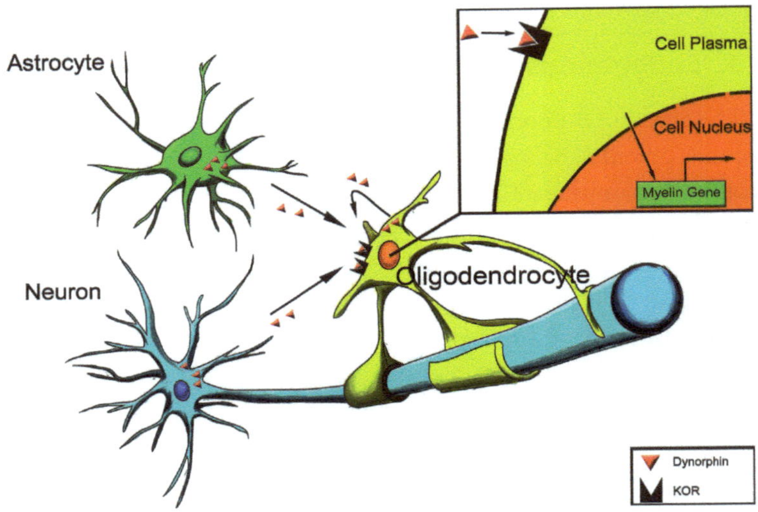

Fig. 1 KOR signaling in oligodendroglia. Endogenous KOR ligands, like dynorphin, are expressed by oligodendrocyte, neuron and astrocyte and KOR agonism by dynorphin or (±)-U50488 may activate MEK/ERK1/2 pathways to enhance oligodendroglial differentiation and myelination.

induced demyelination model (Du & Duan, 2016). OPC-specific ablation of KOR also results in inhibited oligodendroglial differentiation and remyelination in the lysolecithin-induced demyelination model (Mei & Mayoral, 2016). In the EAE model, the KOR deletion deteriorates the clinic scores by inhibiting remyelination rather than immunoreactions, and also attenuates the effect of (±)-U50488 (Du & Duan, 2016). These findings indicate that endogenous KOR signaling is essential for oligodendroglial differentiation and myelination.

3.2 KOR downstream signaling

Admittedly, little is known about the KOR downstream signal pathways in oligodendroglia so far. Understanding the KOR downstream is critical for translating KOR agonism into a clinical approach, given that KOR activation also causes some side effects, such as hallucinations and dysphoria (Bruchas & Chavkin, 2010). It has been demonstrated that the KOR activation is coupled to the G protein Gi/G0 and subsequently changes a number of downstream pathways, including the MAPK-ERK1/2 pathways (Kim et al., 2006). The MAPK-ERK1/2 signaling has been well recognized as the most important signaling for oligodendroglial differentiation and myelination (Fyffe-Maricich, Schott, Karl, Krasno, & Miller, 2013; Ishii, Furusho, & Dupree, 2014; Jeffries, Urbanek, Torres, & Wendell, 2016). Several lines of evidence have shown that the activation of ERK1/2 signaling is essential for maintaining the integrity of myelin and axon, as well as remyelination (Ishii et al., 2014). Oligodendroglia-special deletion of ERK1/2 results in decreased myelin gene expression including MBP, PLP, MAG and CNPase (Ishii et al., 2014). Consistently, overactivated ERK1/2 signaling in OLs results in a higher density of myelinated axons with thicker myelin sheathes (Ishii, Furusho, & Bansal, 2013). Given that KOR agonism also causes an increase of cAMP, a secondary messenger, and phosphorylation of ERK1/2 in embryonic stem cells, it is plausible that KOR agonism may promote oligodendrocyte remyelination through activating ERK1/2 signaling (Fig. 1) (Kim et al., 2006). Since the KOR signaling activates several downstream cascades, it holds promise to develop novel KOR agonists with maximal potency in promoting myelination and minimal side effects by biasing the KOR downstream cascades.

Altogether, recent evidence has clearly shown that the endogenous KOR signaling plays an important role in regulating oligodendroglial differentiation and myelination. Agonizing KOR represents a potential strategy to promote remyelination and developing novel KOR agonists with minimal side effects will pave the way to remyelination therapies for demyelination diseases.

Acknowledgments

We would like to thank Mr. Nan-Xing Huang for his assistance in drafting Fig. 1. This work was supported by the National Natural Science Foundation of China (31771120).

References

Alvarez-Bolado, G., Fairen, A., Douglass, J., & Naranjo, J. R. (1990). Expression of the prodynorphin gene in the developing and adult cerebral cortex of the rat: An in situ hybridization study. *The Journal of Comparative Neurology*, *300*(3), 287–300. https://doi.org/10.1002/cne.903000302.

Avey, D., Sankararaman, S., Yim, A. K. Y., Barve, R., Milbrandt, J., & Mitra, R. D. (2018). Single-cell RNA-Seq uncovers a robust transcriptional response to morphine by glia. *Cell Reports*, *24*(13), 3619–3629. e3614. https://doi.org/10.1016/j.celrep.2018.08.080.

Baumann, N., & Pham-Dinh, D. (2001). Biology of oligodendrocyte and myelin in the mammalian central nervous system. *Physiological Reviews*, *81*(2), 871–927. https://doi.org/10.1152/physrev.2001.81.2.871.

Bergles, D. E., Roberts, J. D., Somogyi, P., & Jahr, C. E. (2000). Glutamatergic synapses on oligodendrocyte precursor cells in the hippocampus. *Nature*, *405*(6783), 187–191. https://doi.org/10.1038/35012083.

Besson, J. M. (1999). The neurobiology of pain. *Lancet*, *353*(9164), 1610–1615.

Bitsch, A., Schuchardt, J., Bunkowski, S., Kuhlmann, T., & Bruck, W. (2000). Acute axonal injury in multiple sclerosis. Correlation with demyelination and inflammation. *Brain*, *123*(Pt. 6), 1174–1183.

Bjartmar, C., & Trapp, B. D. (2003). Axonal degeneration and progressive neurologic disability in multiple sclerosis. *Neurotoxicity Research*, *5*(1–2), 157–164.

Bruchas, M. R., & Chavkin, C. (2010). Kinase cascades and ligand-directed signaling at the kappa opioid receptor. *Psychopharmacology*, *210*(2), 137–147. https://doi.org/10.1007/s00213-010-1806-y.

Cantuti-Castelvetri, L., & Fitzner, D. (2018). Defective cholesterol clearance limits remyelination in the aged central nervous system. *359*(6376), 684–688. https://doi.org/10.1126/science.aan4183.

Chang, A., Tourtellotte, W. W., Rudick, R., & Trapp, B. D. (2002). Premyelinating oligodendrocytes in chronic lesions of multiple sclerosis. *The New England Journal of Medicine*, *346*(3), 165–173. https://doi.org/10.1056/NEJMoa010994.

Cohen, J. A., & Tesar, P. J. (2017). Clemastine fumarate for promotion of optic nerve remyelination. *Lancet*, *390*(10111), 2421–2422. https://doi.org/10.1016/s0140-6736(17)32639-9.

Cox, B. M. (2013). Recent developments in the study of opioid receptors. *Molecular Pharmacology*, *83*(4), 723–728. https://doi.org/10.1124/mol.112.083279.

Cree, B. A. C., Niu, J., Hoi, K. K., Zhao, C., Caganap, S. D., Henry, R. G., et al. (2018). Clemastine rescues myelination defects and promotes functional recovery in hypoxic brain injury. *Brain*, *141*(1), 85–98. https://doi.org/10.1093/brain/awx312.

Deshmukh, V. A., Tardif, V., Lyssiotis, C. A., Green, C. C., Kerman, B., Kim, H. J., et al. (2013). A regenerative approach to the treatment of multiple sclerosis. *Nature*, *502*(7471), 327–332. https://doi.org/10.1038/nature12647.

Du, C., & Duan, Y. (2016). Kappa opioid receptor activation alleviates experimental autoimmune encephalomyelitis and promotes oligodendrocyte-mediated remyelination. *Nature Communications*, 7, 11120. https://doi.org/10.1038/ncomms11120.

Dutta, R., & Trapp, B. D. (2011). Mechanisms of neuronal dysfunction and degeneration in multiple sclerosis. *Progress in Neurobiology*, *93*(1), 1–12. https://doi.org/10.1016/j.pneurobio.2010.09.005.

Eschenroeder, A. C., Vestal-Laborde, A. A., Sanchez, E. S., Robinson, S. E., & Sato-Bigbee, C. (2012). Oligodendrocyte responses to buprenorphine uncover novel and opposing roles of mu-opioid- and nociceptin/orphanin FQ receptors in cell development: Implications for drug addiction treatment during pregnancy. *Glia*, *60*(1), 125–136. https://doi.org/10.1002/glia.21253.

Eshed-Eisenbach, Y., & Peles, E. (2013). The making of a node: A co-production of neurons and glia. *Current Opinion in Neurobiology*, *23*(6), 1049–1056. https://doi.org/10.1016/j.conb.2013.06.003.

Faden, A. I., Sacksen, I., & Noble, L. J. (1988). Opiate-receptor antagonist nalmefene improves neurological recovery after traumatic spinal cord injury in rats through a central mechanism. *The Journal of Pharmacology and Experimental Therapeutics*, *245*(2), 742–748.

Fan, R., Schrott, L. M., Arnold, T., Snelling, S., Rao, M., Graham, D., et al. (2018). Chronic oxycodone induces axonal degeneration in rat. *Brain*, *19*(1), 15. https://doi.org/10.1186/s12868-018-0417-0.

Franklin, R. J. M. (2002). Why does remyelination fail in multiple sclerosis? *Nature Reviews Neuroscience*, *3*, 705. https://doi.org/10.1038/nrn917.

Franklin, R. J., ffrench-Constant, C., Edgar, J. M., & Smith, K. J. (2012). Neuroprotection and repair in multiple sclerosis. *Nature Reviews. Neurology*, *8*(11), 624–634. https://doi.org/10.1038/nrneurol.2012.200.

Fyffe-Maricich, S. L., Schott, A., Karl, M., Krasno, J., & Miller, R. H. (2013). Signaling through ERK1/2 controls myelin thickness during myelin repair in the adult central nervous system. *Journal of Neuroscience, 33*(47), 18402–18408. https://doi.org/10.1523/jneurosci.2381-13.2013.
Gilson, J., & Blakemore, W. F. (1993). Failure of remyelination in areas of demyelination produced in the spinal cord of old rats. *Neuropathology and Applied Neurobiology, 19*(2), 173–181.
Goldschmidt, T., Antel, J., Konig, F. B., Bruck, W., & Kuhlmann, T. (2009). Remyelination capacity of the MS brain decreases with disease chronicity. *Neurology, 72*(22), 1914–1921. https://doi.org/10.1212/WNL.0b013e3181a8260a.
Green, A. J., Gelfand, J. M., Cree, B. A., Bevan, C., Boscardin, W. J., Mei, F., et al. (2017). Clemastine fumarate as a remyelinating therapy for multiple sclerosis (ReBUILD): A randomised, controlled, double-blind, crossover trial. *Lancet, 390*(10111), 2481–2489. https://doi.org/10.1016/s0140-6736(17)32346-2.
Haghikia, A., Hohlfeld, R., Gold, R., & Fugger, L. (2013). Therapies for multiple sclerosis: Translational achievements and outstanding needs. *Trends in Molecular Medicine, 19*(5), 309–319. https://doi.org/10.1016/j.molmed.2013.03.004.
Hartline, D. K., & Colman, D. R. (2007). Rapid conduction and the evolution of giant axons and myelinated fibers. *Current Biology, 17*(1), R29–R35. https://doi.org/10.1016/j.cub.2006.11.042.
Hauser, K. F., Hahn, Y. K., Adjan, V. V., Zou, S., Buch, S. K., Nath, A., et al. (2009). HIV-1 tat and morphine have interactive effects on oligodendrocyte survival and morphology. *Glia, 57*(2), 194–206. https://doi.org/10.1002/glia.20746.
Hill, R. A., Li, A. M., & Grutzendler, J. (2018). Lifelong cortical myelin plasticity and age-related degeneration in the live mammalian brain. *Nature Neuroscience, 21*(5), 683–695. https://doi.org/10.1038/s41593-018-0120-6.
Hubler, Z., Allimuthu, D., Bederman, I., Elitt, M. S., Madhavan, M., Allan, K. C., et al. (2018). Accumulation of 8,9-unsaturated sterols drives oligodendrocyte formation and remyelination. *Nature, 560*(7718), 372–376. https://doi.org/10.1038/s41586-018-0360-3.
Hughes, E. G., Kang, S. H., Fukaya, M., & Bergles, D. E. (2013). Oligodendrocyte progenitors balance growth with self-repulsion to achieve homeostasis in the adult brain. *Nature Neuroscience, 16*, 668–676.
Hughes, E. G., & Orthmann-Murphy, J. L. (2018). Myelin remodeling through experience-dependent oligodendrogenesis in the adult somatosensory cortex. *Nature Neuroscience, 21*(5), 696–706. https://doi.org/10.1038/s41593-018-0121-5.
International Multiple Sclerosis Genetics Consortium, Wellcome Trust Case Control, C, Sawcer, S., Hellenthal, G., Pirinen, M., Spencer, C. C., et al. (2011). Genetic risk and a primary role for cell-mediated immune mechanisms in multiple sclerosis. *Nature, 476*(7359), 214–219. https://doi.org/10.1038/nature10251.
Irvine, K. A., & Blakemore, W. F. (2008). Remyelination protects axons from demyelination-associated axon degeneration. *Brain, 131*(Pt. 6), 1464–1477. https://doi.org/10.1093/brain/awn080.
Ishii, A., Furusho, M., & Bansal, R. (2013). Sustained activation of ERK1/2 MAPK in oligodendrocytes and schwann cells enhances myelin growth and stimulates oligodendrocyte progenitor expansion. *Journal of Neuroscience, 33*(1), 175–186. https://doi.org/10.1523/jneurosci.4403-12.2013.
Ishii, A., Furusho, M., & Dupree, J. L. (2014). Role of ERK1/2 MAPK signaling in the maintenance of myelin and axonal integrity in the adult CNS. *Journal of Neuroscience, 34*(48), 16031–16045. https://doi.org/10.1523/jneurosci.3360-14. 2014.

Jeffries, M. A., Urbanek, K., Torres, L., & Wendell, S. G. (2016). ERK1/2 activation in preexisting oligodendrocytes of adult mice drives new Myelin synthesis and enhanced CNS function. *Journal of Neuroscience*, *36*(35), 9186–9200. https://doi.org/10.1523/jneurosci.1444-16.2016.

Jensen, S. K., Michaels, N. J., Ilyntskyy, S., Keough, M. B., Kovalchuk, O., & Yong, V. W. (2018). Multimodal enhancement of Remyelination by exercise with a pivotal role for Oligodendroglial PGC1alpha. *Cell Reports*, *24*(12), 3167–3179. https://doi.org/10.1016/j.celrep.2018.08.060.

Katritch, V., Cherezov, V., & Stevens, R. C. (2013). Structure-function of the G protein-coupled receptor superfamily. *Annual Review of Pharmacology and Toxicology*, *53*, 531–556. https://doi.org/10.1146/annurev-pharmtox-032112-135923.

Keirstead, H. S., & Blakemore, W. F. (1997). Identification of post-mitotic oligodendrocytes incapable of remyelination within the demyelinated adult spinal cord. *Journal of Neuropathology and Experimental Neurology*, *56*(11), 1191–1201.

Kieffer, B. L., & Gaveriaux-Ruff, C. (2002). Exploring the opioid system by gene knockout. *Progress in Neurobiology*, *66*(5), 285–306.

Kim, E., Clark, A. L., Kiss, A., Hahn, J. W., Wesselschmidt, R., Coscia, C. J., et al. (2006). Mu- and kappa-opioids induce the differentiation of embryonic stem cells to neural progenitors. *The Journal of Biological Chemistry*, *281*(44), 33749–33760. https://doi.org/10.1074/jbc.M603862200.

Knapp, P. E., Adjan, V. V., & Hauser, K. F. (2009). Cell-specific loss of kappa-opioid receptors in oligodendrocytes of the dysmyelinating jimpy mouse. *Neuroscience Letters*, *451*(2), 114–118. https://doi.org/10.1016/j.neulet.2008.12.022.

Knapp, P. E., Itkis, O. S., Zhang, L., Spruce, B. A., Bakalkin, G., & Hauser, K. F. (2001). Endogenous opioids and oligodendroglial function: Possible autocrine/paracrine effects on cell survival and development. *Glia*, *35*(2), 156–165.

Knapp, P. E., Maderspach, K., & Hauser, K. F. (1998). Endogenous opioid system in developing normal and jimpy oligodendrocytes: Mu and kappa opioid receptors mediate differential mitogenic and growth responses. *Glia*, *22*(2), 189–201.

Kornek, B., Storch, M. K., Weissert, R., Wallstroem, E., Stefferl, A., Olsson, T., et al. (2000). Multiple sclerosis and chronic autoimmune encephalomyelitis: A comparative quantitative study of axonal injury in active, inactive, and remyelinated lesions. *American Journal of Pathology*, *157*(1), 267–276. https://doi.org/10.1016/s0002-9440(10)64537-3.

Kreek, M. J., LaForge, K. S., & Butelman, E. (2002). Pharmacotherapy of addictions. *Nature Reviews. Drug Discovery*, *1*(9), 710–726. https://doi.org/10.1038/nrd897.

Kuhlmann, T., Lingfeld, G., Bitsch, A., Schuchardt, J., & Bruck, W. (2002). Acute axonal damage in multiple sclerosis is most extensive in early disease stages and decreases over time. *Brain*, *125*(Pt. 10), 2202–2212.

Kuhlmann, T., Miron, V., Cui, Q., Wegner, C., Antel, J., & Bruck, W. (2008). Differentiation block of oligodendroglial progenitor cells as a cause for remyelination failure in chronic multiple sclerosis. *Brain*, *131*(Pt. 7), 1749–1758. https://doi.org/10.1093/brain/awn096.

Lee, Y., Morrison, B. M., Li, Y., Lengacher, S., Farah, M. H., Hoffman, P. N., et al. (2012). Oligodendroglia metabolically support axons and contribute to neurodegeneration. *Nature*, *487*(7408), 443–448. https://doi.org/10.1038/nature11314.

Li, Z., He, Y., Fan, S., & Sun, B. (2015). Clemastine rescues behavioral changes and enhances remyelination in the cuprizone mouse model of demyelination. *Neuroscience Bulletin*, *31*(5), 617–625. https://doi.org/10.1007/s12264-015-1555-3.

Liu, J., & Dupree, J. L. (2016). Clemastine enhances myelination in the prefrontal cortex and rescues behavioral changes in socially isolated mice. *Journal of Neuroscience*, *36*(3), 957–962. https://doi.org/10.1523/jneurosci.3608-15.2016.

Lovas, G., Szilagyi, N., Majtenyi, K., Palkovits, M., & Komoly, S. (2000). Axonal changes in chronic demyelinated cervical spinal cord plaques. *Brain*, *123*(Pt. 2), 308–317.

McKenzie, I. A., Ohayon, D., Li, H., de Faria, J. P., Emery, B., Tohyama, K., et al. (2014). Motor skill learning requires active central myelination. *Science*, *346*(6207), 318–322. https://doi.org/10.1126/science.1254960.

Mei, F., Fancy, S. P., Shen, Y. A., Niu, J., Zhao, C., Presley, B., et al. (2014). Micropillar arrays as a high-throughput screening platform for therapeutics in multiple sclerosis. *Nature Medicine*, *20*(8), 954–960. https://doi.org/10.1038/nm.3618.

Mei, F., Lehmann-Horn, K., Shen, Y. A., Rankin, K. A., Stebbins, K. J., Lorrain, D. S., et al. (2016). Accelerated remyelination during inflammatory demyelination prevents axonal loss and improves functional recovery. *eLife*, *5*, e18246. https://doi.org/10.7554/eLife.18246.

Mei, F., & Mayoral, S. R. (2016). Identification of the Kappa-opioid receptor as a therapeutic target for oligodendrocyte remyelination. *Journal of Neuroscience*, *36*(30), 7925–7935. https://doi.org/10.1523/jneurosci.1493-16.2016.

Mi, S., Hu, B., Hahm, K., Luo, Y., Kam Hui, E. S., Yuan, Q., et al. (2007). LINGO-1 antagonist promotes spinal cord remyelination and axonal integrity in MOG-induced experimental autoimmune encephalomyelitis. *Nature Medicine*, *13*(10), 1228–1233. https://doi.org/10.1038/nm1664.

Mi, S., Lee, X., Shao, Z., Thill, G., Ji, B., Relton, J., et al. (2004). LINGO-1 is a component of the Nogo-66 receptor/p75 signaling complex. *Nature Neuroscience*, 7(3), 221–228. https://doi.org/10.1038/nn1188.

Mi, S., Miller, R. H., Lee, X., Scott, M. L., Shulag-Morskaya, S., Shao, Z., et al. (2005). LINGO-1 negatively regulates myelination by oligodendrocytes. *Nature Neuroscience*, *8*(6), 745–751. https://doi.org/10.1038/nn1460.

Najm, F. J., Madhavan, M., Zaremba, A., Shick, E., Karl, R. T., Factor, D. C., et al. (2015). Drug-based modulation of endogenous stem cells promotes functional remyelination in vivo. *Nature*, *522*(7555), 216–220. https://doi.org/10.1038/nature14335.

Nave, K. A. (2010). Myelination and the trophic support of long axons. *Nature Reviews. Neuroscience*, *11*(4), 275–283. https://doi.org/10.1038/nrn2797.

Nikic, I., Merkler, D., Sorbara, C., Brinkoetter, M., Kreutzfeldt, M., Bareyre, F. M., et al. (2011). A reversible form of axon damage in experimental autoimmune encephalomyelitis and multiple sclerosis. *Nature Medicine*, *17*(4), 495–499. https://doi.org/10.1038/nm.2324.

Normand, E. A., & Rasband, M. N. (2015). Subcellular patterning: Axonal domains with specialized structure and function. *Developmental Cell*, *32*(4), 459–468. https://doi.org/10.1016/j.devcel.2015.01.017.

Nunes, M. C., Roy, N. S., Keyoung, H. M., Goodman, R. R., McKhann, G., 2nd, Jiang, L., et al. (2003). Identification and isolation of multipotential neural progenitor cells from the subcortical white matter of the adult human brain. *Nature Medicine*, *9*(4), 439–447. https://doi.org/10.1038/nm837.

Patrikios, P., Stadelmann, C., Kutzelnigg, A., Rauschka, H., Schmidbauer, M., Laursen, H., et al. (2006). Remyelination is extensive in a subset of multiple sclerosis patients. *Brain*, *129*(Pt. 12), 3165–3172. https://doi.org/10.1093/brain/awl217.

Petrillo, P., Tavani, A., Verotta, D., Robson, L. E., & Kosterlitz, H. W. (1987). Differential postnatal development of mu-, delta- and kappa-opioid binding sites in rat brain. *Brain Research*, *428*(1), 53–58.

Plemel, J. R., Liu, W. Q., & Yong, V. W. (2017). Remyelination therapies: A new direction and challenge in multiple sclerosis. *Nature Reviews. Drug Discovery*, *16*(9), 617–634. https://doi.org/10.1038/nrd.2017.115.

Prineas, J. W., Barnard, R. O., Kwon, E. E., Sharer, L. R., & Cho, E. S. (1993). Multiple sclerosis: Remyelination of nascent lesions. *Annals of Neurology, 33*(2), 137–151. https://doi.org/10.1002/ana.410330203.

Puniak, M. A., Freeman, G. M., Agresta, C. A., Van Newkirk, L., Barone, C. A., & Salzman, S. K. (1991). Comparison of a serotonin antagonist, opioid antagonist, and TRH analog for the acute treatment of experimental spinal trauma. *Journal of Neurotrauma, 8*(3), 193–203. https://doi.org/10.1089/neu.1991.8.193.

Reich, D. S., Lucchinetti, C. F., & Calabresi, P. A. (2018). Multiple Sclerosis. *The New England Journal of Medicine, 378*(2), 169–180. https://doi.org/10.1056/NEJMra1401483.

Saab, A. S., Tzvetavona, I. D., Trevisiol, A., Baltan, S., Dibaj, P., Kusch, K., et al. (2016). Oligodendroglial NMDA receptors regulate glucose import and axonal energy metabolism. *Neuron, 91*(1), 119–132. https://doi.org/10.1016/j.neuron.2016.05.016.

Safaiyan, S., Kannaiyan, N., Snaidero, N., Brioschi, S., Biber, K., Yona, S., et al. (2016). Age-related myelin degradation burdens the clearance function of microglia during aging. *Nature Neuroscience, 19*(8), 995–998. https://doi.org/10.1038/nn.4325.

Sanchez, E. S., Bigbee, J. W., Fobbs, W., Robinson, S. E., & Sato-Bigbee, C. (2008). Opioid addiction and pregnancy: Perinatal exposure to buprenorphine affects myelination in the developing brain. *Glia, 56*(9), 1017–1027. https://doi.org/10.1002/glia.20675.

Smith, K. J., Blakemore, W. F., & McDonald, W. I. (1981). The restoration of conduction by central remyelination. *Brain, 104*(2), 383–404.

Stiene-Martin, A., Knapp, P. E., Martin, K., Gurwell, J. A., Ryan, S., Thornton, S. R., et al. (2001). Opioid system diversity in developing neurons, astroglia, and oligodendroglia in the subventricular zone and striatum: Impact on gliogenesis in vivo. *Glia, 36*(1), 78–88.

Stokes, K. B., & Lee, N. M. (1976). Morphine effects of RNA synthesis in purified oligodendroglial nuclei. *Proceedings of the Western Pharmacology Society, 19*, 48–54.

Sun, J. J., Ren, Q. G., Xu, L., & Zhang, Z. J. (2015). LINGO-1 antibody ameliorates myelin impairment and spatial memory deficits in experimental autoimmune encephalomyelitis mice. *Scientific Reports, 5*, 14235. https://doi.org/10.1038/srep14235.

Tallantyre, E. C., Bo, L., Al-Rawashdeh, O., Owens, T., Polman, C. H., Lowe, J., et al. (2009). Greater loss of axons in primary progressive multiple sclerosis plaques compared to secondary progressive disease. *Brain, 132*(Pt. 5), 1190–1199. https://doi.org/10.1093/brain/awp106.

Trapp, B. D., Peterson, J., Ransohoff, R. M., Rudick, R., Mork, S., & Bo, L. (1998). Axonal transection in the lesions of multiple sclerosis. *The New England Journal of Medicine, 338*(5), 278–285. https://doi.org/10.1056/nejm199801293380502.

Trapp, B. D., & Stys, P. K. (2009). Virtual hypoxia and chronic necrosis of demyelinated axons in multiple sclerosis. *Lancet Neurology, 8*(3), 280–291. https://doi.org/10.1016/s1474-4422(09)70043-2.

Tripathi, R. B., Jackiewicz, M., McKenzie, I. A., Kougioumtzidou, E., Grist, M., & Richardson, W. D. (2017). Remarkable stability of myelinating oligodendrocytes in mice. *Cell Reports, 21*(2), 316–323. https://doi.org/10.1016/j.celrep.2017.09.050.

Ujcikova, H., Eckhardt, A., Kagan, D., Roubalova, L., & Svoboda, P. (2014). Proteomic analysis of post-nuclear supernatant fraction and percoll-purified membranes prepared from brain cortex of rats exposed to increasing doses of morphine. *Proteome Science, 12*(1), 11. https://doi.org/10.1186/1477-5956-12-11.

Upadhyay, J., Maleki, N., Potter, J., Elman, I., Rudrauf, D., Knudsen, J., et al. (2010). Alterations in brain structure and functional connectivity in prescription opioid-dependent patients. *Brain, 133*(Pt. 7), 2098–2114. https://doi.org/10.1093/brain/awq138.

Utzschneider, D. A., Archer, D. R., Kocsis, J. D., Waxman, S. G., & Duncan, I. D. (1994). Transplantation of glial cells enhances action potential conduction of amyelinated spinal cord axons in the myelin-deficient rat. *Proceedings of the National Academy of Sciences of the United States of America, 91*(1), 53–57.

Vestal-Laborde, A. A., Eschenroeder, A. C., Bigbee, J. W., Robinson, S. E., & Sato-Bigbee, C. (2014). The opioid system and brain development: Effects of methadone on the oligodendrocyte lineage and the early stages of myelination. *Developmental Neuroscience*, *36*(5), 409–421. https://doi.org/10.1159/000365074.

Wahlert, A., Funkelstein, L., Fitzsimmons, B., Yaksh, T., & Hook, V. (2013). Spinal astrocytes produce and secrete dynorphin neuropeptides. *Neuropeptides*, *47*(2), 109–115. https://doi.org/10.1016/j.npep.2012.10.006.

Wang, F., Yang, Y. J., Yang, N., Chen, X. J., Huang, N. X., Zhang, J., et al. (2018). Enhancing oligodendrocyte myelination rescues synaptic loss and improves functional recovery after chronic hypoxia. *Neuron*, *99*(4), 689–701. e685. https://doi.org/10.1016/j.neuron.2018.07.017.

Wolswijk, G. (1998). Chronic stage multiple sclerosis lesions contain a relatively quiescent population of oligodendrocyte precursor cells. *Journal of Neuroscience*, *18*(2), 601–609.

Wolswijk, G. (2000). Oligodendrocyte survival, loss and birth in lesions of chronic-stage multiple sclerosis. *Brain*, *123*(Pt. 1), 105–115.

Woodruff, R. H., Fruttiger, M., Richardson, W. D., & Franklin, R. J. (2004). Platelet-derived growth factor regulates oligodendrocyte progenitor numbers in adult CNS and their response following CNS demyelination. *Molecular and Cellular Neurosciences*, *25*(2), 252–262. https://doi.org/10.1016/j.mcn.2003.10.014.

Xiao, L., Ohayon, D., McKenzie, I. A., Sinclair-Wilson, A., Wright, J. L., Fudge, A. D., et al. (2016). Rapid production of new oligodendrocytes is required in the earliest stages of motor-skill learning. *Nature Neuroscience*, *19*(9), 1210–1217. https://doi.org/10.1038/nn.4351.

Young, K. M., Psachoulia, K., Tripathi, R. B., Dunn, S. J., Cossell, L., Attwell, D., et al. (2013). Oligodendrocyte dynamics in the healthy adult CNS: Evidence for myelin remodeling. *Neuron*, 77(5), 873–885. https://doi.org/10.1016/j.neuron.2013.01.006.

Zalc, B., Goujet, D., & Colman, D. (2008). The origin of the myelination program in vertebrates. *Current Biology*, *18*(12), R511–R512. https://doi.org/10.1016/j.cub.2008.04.010.

Zambonin, J. L., Zhao, C., Ohno, N., Campbell, G. R., Engeham, S., Ziabreva, I., et al. (2011). Increased mitochondrial content in remyelinated axons: Implications for multiple sclerosis. *Brain*, *134*(Pt. 7), 1901–1913. https://doi.org/10.1093/brain/awr110.

Zamir, N., Quirion, R., & Segal, M. (1985). Ontogeny and regional distribution of proenkephalin- and prodynorphin-derived peptides and opioid receptors in rat hippocampus. *Neuroscience*, *15*(4), 1025–1034.

Zollner, C., & Stein, C. (2007). Opioids. *Handbook of Experimental Pharmacology*, (177), 31–63.

CHAPTER TWELVE

Inhibitory role of G_i-coupled receptors on cAMP-driven cancers with focus on opioid receptors in lung adenocarcinoma and its stem cells

Hildegard M. Schuller*
Department of Biomedical & Diagnostic Science, College of Veterinary Medicine, The University of Tennessee, Knoxville, TN, United States
*Corresponding author: e-mail address: schullerhildegard@gmail.com

Contents

Abstract

The development, progression, metastasis and drug resistance of the most common human cancers are driven by cyclic adenosine monophosphate (cAMP)—signaling downstream of beta-adrenergic receptors (β-Ars) coupled to the stimulatory G-protein G_s. Receptors coupled to the inhibitory G-protein G_i inhibit this signaling cascade by blocking the activation of the enzyme adenylyl cyclase that catalyzes the formation of cAMP and function as the physiological inhibitors of this signaling cascade. Members of the G_i-coupled receptor family widely expressed in the mammalian organism are GABA B receptors ($GABA_B$-Rs) for the inhibitory neurotransmitter γ-aminobutyric acid (GABA), opioid receptors for endogenous opioid peptides and cannabinoid receptors for endogenous cannabinoids. This review summarizes current evidence for the concept that the activation of G_i-receptor signaling by pharmacological and psychological means is a promising tool for the long-term management of cAMP-driven cancers with special emphasis on the inhibitory effects of opioids on lung adenocarcinoma and its stem cells.

Vitamins and Hormones, Volume 111
ISSN 0083-6729
https://doi.org/10.1016/bs.vh.2019.05.009

1. Introduction

Lung cancer is the leading cause of cancer deaths and despite of tremendous progress in preclinical and clinical lung cancer research, 5-year survivals continue to be rare (Meza, Meernik, Jeon, & Cote, 2015). There are four major histological types of lung cancer: adenocarcinoma, squamous cell carcinoma, large cell carcinoma and small cell carcinoma, with adenocarcinoma predominating and increasing in incidence world-wide (Nakamura & Saji, 2014). Due to profound differences in therapy responsiveness between small cell lung cancer and the remaining lung cancer types, clinical oncologists have traditionally classified lung cancers into small cell lung cancer (SCLC) versus non-small cell lung cancer (NSCLC). SCLC is a neuroendocrine cancer with characteristics of pulmonary neuroendocrine cells (PNECs), including the nicotinic acetylcholine receptor (nAChR)-regulated synthesis and release of the autocrine growth factors serotonin (Cattaneo, Codignola, Vicentini, Clementi, & Sher, 1993; Jull, Plummer, & Schuller, 2001) and neuropeptides (Moody, Ramos-Alvarez, & Jensen, 2018; Nylen et al., 1993). Squamous cell carcinoma demonstrates features of squamous cells with and without evidence of keratin formation. Adenocarcinoma is a cancer with gland-like histological features with and without evidence of mucin production. Lung cancers are generally classified as large cell carcinomas when none of the characteristics of SCLC adenocarcinoma or squamous cell carcinoma can be identified (Travis et al., 2015). The leading type of lung cancer, adenocarcinoma, is regulated by the antagonistic forces of receptors coupled to the stimulatory G protein G_s and receptors coupled to the inhibitory G protein G_i and is the focus of this review.

Cancer stem cells (CSCs) are a small population of cells with stem cell characteristics present in all cancers (Tirino et al., 2013). CSCs in lung adenocarcinomas express a variety of stem cell markers, including sonic hedgehog (SHH) and its downstream effector, the zinc finger protein Gli1, and aldehyde dehydrogenase-1 (ALDH-1) (Singh & Chellappan, 2014; Zhang et al., 2012). CSCs divide unevenly, giving rise to self-renewal and formation of more differentiated cancer cells, and are thought to be the driving force of cancer growth, progression, metastasis and drug resistance (Tirino et al., 2013). CSCs can be isolated by cell sorting or by selective culture conditions that take advantage of the ability of these cells to grow as floating cell aggregates (spheroids) in serum-free culture medium (Eramo et al., 2008). In order to cause lung cancer, etiological factors for this malignancy must have the ability to activate lung cancer stem cells. On the other hand, lung cancer therapy will only be successful if it not only eradicates the vast majority of differentiated cells present in a tumor but also the small population of CSCs.

Traditional in vitro and in vivo preclinical investigations typically assess the cancer therapeutic potency of potential new anti-cancer agents in human cancer cell lines in vitro and in mouse xenografts from such cell lines in vivo. However, these cell lines are mixtures of predominantly differentiated cancer cells and only small populations of CSCs. Unless such studies are accompanied by investigations in CSCs enriched by flow cytometry or selective culture conditions, they can therefore be grossly misleading. This may be one reason, why so many agents that have shown great promise as cancer therapeutics in preclinical studies, later disappoint in clinical trials.

A second reason why promising preclinical data often disappoint in the clinical oncology setting is the strong cancer promoting effect of psychological stress in cancer patients. Any effects of stress are completely absent in the in vitro tests unless they are conducted in the presence of the stress neurotransmitters epinephrine (Epi) and norepinephrine (Nor) and are greatly minimized in laboratory animals due to strict regulations for the humane care of laboratory animals aimed at keeping the animals in a species appropriate stress-free and happy environment.

2. Antagonistic effects of G_s- and G_i-coupled receptors on pulmonary adenocarcinoma and its stem cells

The mammalian body responds to psychological stress by the nicotinic acetylcholine receptor (nAChR)-mediated release of the stress neurotransmitters Epi and Nor from the adrenal gland and from nerves of the sympathetic branch of the autonomic nervous system (McEwen, 2000). Nicotine and the carcinogenic nicotine-derived nitrosamines *N*-nitroso-nor-nicotine (NNN) and 4(methylnitrosamino)-1-(3pyridyl)-1-butanone (*N*-nitroso-nicotine ketone, NNK) also stimulate the release of Epi and Nor by binding as agonists to nAChRs (Al-Wadei & Schuller, 2009; Haass & Kubler, 1997; Schuller & Orloff, 1998). In turn, Epi and Nor bind as agonists to alpha -and beta-adrenergic receptors (α-ARs, β-ARs), which are expressed universally in mammalian cells (Lefkowitz, 2007). NNK is additionally a high affinity agonist for β-ARs, thus directly activating signaling downstream of these receptors (Schuller, Tithof, Williams, & Plummer, 1999).

Beta-adrenergic receptors are seven trans-membrane receptors coupled to the stimulatory G protein G_s that activates the enzyme adenylyl cyclase (AC) upon binding of an agonist to the receptor (Lefkowitz, 2007). Activated AC catalyzes the formation of intracellular cyclic adenosine monophosphate (cAMP), which activates protein kinase A (PKA) and the transcription factor cAMP response element binding protein (CREB) (Tang, Li, Lu, & Cho, 2013; Wallukat, 2002). This signaling cascade causes the release of multiple growth factors from pulmonary adenocarcinoma cells and from the tumor environment, including the epidermal growth factor (EGF) (Grau, Soley, & Ramirez, 1997; Schuller et al., 1999), vascular endothelial growth factor (VEGF) (Madden, Szpunar, & Brown, 2011), arachidonic acid (AA) (Schuller et al., 1999) and pro-inflammatory cytokines (Madden et al., 2011), all of which stimulate the growth, progression, metastasis and drug resistance of numerous cancers, including lung adenocarcinoma. The stimulatory signaling pathway downstream of G_s-coupled receptors in mammalian cells is counteracted by receptors coupled to the inhibitory G protein G_i that inhibits the activation of AC (El-Armouche, Zolk, Rau, & Eschenhagen, 2003; Lefkowitz, 2000). G_i-coupled receptors expressed in lung adenocarcinomas and normal airway epithelia are mu opioid receptor (MORs), kappa-opioid receptors (KORs) and delta-opioid receptors (DORs) and for endogenous opioid peptides (Banerjee, Papu John, & Schuller, 2015; Kuzumaki et al., 2012; Maneckjee & Minna, 1990), $GABA_B$-receptors ($GABA_B$-Rs) for the inhibitory neurotransmitter γ-aminobutyric acid (GABA) (Schuller, Al-Wadei, & Majidi, 2008; Zhang et al., 2013) and the cannabinoid receptors CB1 and CB2 for endocannabinoids (Ramer et al., 2012; Ravi, Elbaz, Wani, Nasser, & Ganju, 2016).

Numerous preclinical studies have shown that the activation of beta-adrenergic receptor signaling by agonists stimulates the proliferation, migration and drug resistance of pulmonary adenocarcinoma cells in vitro and in animal models (Schuller, 2008, 2013; Schuller & Al-Wadei, 2012; Tang et al., 2013). In addition to systemic increases in stress neurotransmitters caused by smoking or psychological stress, many cancer cells can also self-stimulate their own proliferation, migration and drug resistance by synthesizing and releasing Epi and Nor in response to nAChR activation, a phenomenon that has been documented by in vitro studies for adenocarcinomas of the lungs (Schuller & Al-Wadei, 2012), stomach (Shin et al., 2007), colon (Wong et al., 2007), pancreas (Al-Wadei, Al-Wadei, & Schuller, 2012) and mammary gland (Cakir, Plummer, Tithof, & Schuller, 2002).

Preclinical in vivo investigations by our laboratory have shown that experimentally induced social stress significantly increased the growth of xenografts from human lung adenocarcinoma cell lines in nude mice (Al-Wadei, Plummer, et al., 2012). This effect was accompanied by increased systemic and tumor levels of Epi, Nor, cAMP and the activated signaling proteins CREB and ERK whereas the levels of the inhibitory neurotransmitter GABA and the expression of its synthesizing enzymes glutamate decarboxylase (GAD) 65 and GAD 67 were reduced. Treatment of the animals with GABA in the drinking water reversed all of these tumor-promoting effects (Al-Wadei, Plummer, et al., 2012). The reduction of cAMP observed in this study supports the interpretation that the anti-cancer effects of GABA were mediated by G_i-coupled $GABA_B$-Rs.

While these findings identify GABA as a powerful inhibitor of stress-associated lung adenocarcinoma progression, they also suggest that a reduction in psychological stress may have similar cancer inhibiting effects. This hypothesis was tested in a subsequent experiment using mouse xenografts from the same lung adenocarcinoma cell lines (Banerjee et al., 2015). Stress reduction was achieved by species-appropriate environmental enrichment (Gurfein et al., 2012) and was evidenced by significantly decreased systemic levels of Epi, Nor and corticosterone whereas the levels of GABA almost doubled. Because stress impairs the endogenous opioid system (Knoll & Carlezon, 2010), whereas pleasurable emotions are mediated by endorphins (Esch & Stefano, 2004), we also determined the systemic levels of dynorphins A and B and met-enkephalin. The levels of all three endogenous opioid peptides increased significantly in mice maintained in the stress-reducing environment. Compared with control mice not maintained in a stress-free environment, the number of xenograft tumors that developed after subcutaneous injection with cancer cells was reduced by 50% with volumes of the xenografts that did develop significantly reduced. The level of cAMP was significantly reduced in xenograft tissues from mice in the stress-reducing environment, indicating that the observed tumor inhibiting effect was mediated by G_i-coupled receptors, an interpretation supported by the increased systemic levels of GABA and opioid peptides in response to stress reduction (Banerjee et al., 2015). Moreover, the xenograft levels of VEGF and the cancer stem cell markers SHH, Gli1 and ALDH-1 were significantly reduced by stress reduction, suggesting that the tumor inhibiting mechanisms of stress reduction included a decrease in angiogenesis and a significant inhibition of the self-renewal of CSCs. Moreover, the expression of multiple phosphorylated signaling proteins associated with cell proliferation was significantly reduced by stress reduction whereas the expression of proteins that mediate apoptosis was increased (Banerjee et al., 2015). The observed inhibitory effects of stress reduction on CSCs were further supported by in vitro experiments with CSCs isolated by selective culture conditions. These experiments showed that the characteristic CSC spheroids significantly increased in number in response to Epi exposure, a response accompanied by more than threefold increases in intracellular cAMP and significant increases in the CSC markers Notch-1, SHH, Gli1 and ALDH-1 (Banerjee et al., 2015). All of these responses were completely suppressed below control levels by treatment with dynorphin B or GABA (Banerjee et al., 2015). Collectively, these findings identify a strong stimulation of CSC self-renewal by the stress neurotransmitters via G_s coupled adrenergic receptors that triggered the cAMP-driven activation of multiple CSC markers and their associated signaling pathways. Moreover, these data show, for the first time, that opioid peptides and GABA have powerful direct inhibitory effects on pulmonary adenocarcinoma CSCs via their G_i-coupled receptors that inhibited the formation of intracellular cAMP, leading to the suppression of CSC self-renewal and all investigated CSC markers.

Early in vitro investigations by another laboratory have reported strong growth inhibiting effects of morphin via opioid receptor mediated decrease in intracellular cAMP in a diverse panel of NSCLC cell lines, including adenocarcinoma, squamous cell carcinoma and large cell carcinoma, and in several SCLC cell lines (Maneckjee & Minna, 1990). These studies were further extended to show that methadone, a synthetic opioid widely used for the management of opioid addiction (Krantz & Mehler, 2004) and as a cancer analgesic (Michalska, Katzenwadel, & Wolf, 2017), induced apoptosis in the same panel of lung cancer cell lines. An independent second laboratory later identified the mechanism of opioid receptor-induced growth inhibition in an SCLC cell line as suppression of the Ca^{2+}-dependent release of the autocrine growth factor serotonin, an effect that was reversed by treatment of the cells with cAMP (Sher et al., 1996). These findings are in accord with the physiological function of intracellular cAMP that cannot only activate PKA/CREB but additionally increases cellular Ca^{2+} levels by the PKA-induced up-regulation of Ca^{2+}-channels in the cell membrane or via the release of Ca^{2+} from intracellular stores induced by activated PKA or by exchange protein activated by cAMP (Epac) (Pereira et al., 2007). In turn, the resulting increase in intracellular Ca^{2+} causes the release of cellular secretory products, including that of growth factors, by exocytosis (Jang, Boo, Lee, Min, & Lee, 2016; Messenger, Falkowski, & Groblewski, 2014).

3. Anti-carcinogenic effects of methadone

More recently, preclinical studies on the potential antineoplastic effects of methadone have been conducted in a variety of cancers. In vitro studies have thus shown that methadone induced a decrease in intracellular cAMP, thereby sensitizing bladder cancer cells to cisplatin therapy (Michalska, Schultze-Seemann, Kuckuck, Katzenwadel, & Wolf, 2018). Another laboratory reported methadone—induced apoptosis in leukemia cell lines (Friesen, Roscher, Alt, & Miltner, 2008; Friesen et al., 2013; Singh et al., 2011) and inhibition of the growth of glioblastoma cells in vitro and in mouse xenografts while increasing tumor responsiveness to doxorubicin (Friesen et al., 2014). However, the potential inhibitory effect of methadone on cancer stem cells has not been investigated to date. Nevertheless, the observed sensitization of glioblastoma by methadone to doxorubicin indicates inhibitory effects of this agent on CSCs, because CSCs are the driving forces of drug resistance in cancer cells.

4. Anti-carcinogenic effects of cannabinoids

Potential anti-cancer effects of cannabinoids on a variety of cancers have been tested in preclinical investigations and observational clinical studies. It has thus been shown that overexpression of the cannabinoid receptor 2 (CB2) was associated with improved recurrence-free survival of breast cancer patients (Elbaz, Ahirwar, Ravi, Nasser, & Ganju, 2017). Mechanistic experiments then revealed that the synthetic selective CB2 agonist JWH-015 inhibited the proliferation of breast cancer cells in vitro and in mouse xenografts by inhibiting signaling downstream of the epidermal growth factor receptor (EGFR) and insulin-like growth factor-I receptor (IGF-IR) (Elbaz et al., 2017). These findings are in line with the established regulation of EGF and IGF release by cAMP signaling downstream of G_s-coupled β-Ars (Grau et al., 1997; Jang et al., 2016) as well as the transactivation of the EGFR by β-ARs (Askari, Tsao, & Schuller, 2005) and indicate that the observed inhibitory effects of cannabinoids were mediated by suppression of cAMP formation by the G_i-coupled CB2 receptor. Cannabinoids also significantly decreased the viability of lung cancer cells, neuroblastoma cells and testicular cancer cells while increasing apoptosis (Muller, Radtke, Decker, Koch, & Belge, 2017). Moreover, in vitro studies have shown that cannabidiol inhibited the self-renewal of CSCs isolated from glioblastomas and overcame the resistance of these cells to cancer therapy with BCNU (Nabissi et al., 2015). These findings reinforce the notion that the G_i-coupled CB1 and CB2 receptors could potentially be targeted as powerful adjuvant therapy for numerous cAMP driven cancers. However, clinical trials with cannabinoids alone or in combination with cancer therapeutics have not been conducted to date. Based on their analgesic and anti-emetic effects, along with their ability to positively impact the mood, cannabinoids may prove a valid alternative to opioids in the clinical oncology setting.

5. Conclusions

The most common human cancers are stimulated in their development, progression and resistance to therapy by cAMP-dependent signaling downstream of G_s-coupled β-ARs and include adenocarcinomas of the lungs, breast, stomach, pancreas, prostate, colon, rectum and ovary as well as esophageal and nasopharyngeal cancers, melanomas, leukemias, hemangiomas and hemangiosarcomas (Tang et al., 2013). In addition, the responsiveness of squamous cell carcinoma and SCLC of the lungs as well as bladder cancer, glioblastoma and neuroblastoma to the anti-carcinogenic effects of opioids or cannabinoids associated with suppression of cAMP summarized in this review identify these malignancies as cAMP-driven cancers as well.

There has been considerable excitement about the re-purposing of beta-blockers commonly used as cardio-vascular therapeutics (Powe et al., 2010; Schuller, 2010) and of synthetic opioids (Friesen et al., 2008, 2013, 2014) and cannabinoids (Chakravarti, Ravi, & Ganju, 2014) used as analgesics and/or anti-emetics as adjuvant therapy for these cancers. However, there is a caveat for this approach: The chronic exposure of any G_i-coupled receptor to agonist can desensitize the receptor, resulting in tolerance (Lefkowitz, 2007). In the case of opioid receptors and cannabinoid receptors this would cause addiction, while at the same time reversing their agonists from tumor inhibitors to tumor promoters because the desensitized receptors do no longer respond to their physiological agonists. In addition synthetic receptor agonists generally have a higher affinity to the receptor than physiological agonists, a factor that additionally favors the development of receptor desensitization/tolerance. A good example for this phenomenon is a comparative study on the effects of GABA and the synthetic selective $GABA_B$-R agonist baclofen on nicotine-induced gemcitabine resistance in pancreatic cancer cells in vitro and in mouse xenografts (Banerjee, Al-Wadei, Al-Wadei,

Dagnon, & Schuller, 2014). Both agents were administered for 7 days and yielded completely opposite responses: GABA inhibited intracellular cAMP and signaling proteins associated with cell proliferation while increasing the expression of activated proteins associated with apoptosis and significantly inhibited tumor growth in vitro and in vivo. By contrast, baclofen increased cAMP and activated proteins associated with cell proliferation while decreasing apoptotic signaling proteins and promoted tumor growth in vitro and in the mouse xenografts (Banerjee et al., 2014). Even though this experiment encourages the use of GABA for cancer intervention and the regular consumption of nutritional GABA supplements reportedly has no adverse effects (Thorne Research, 2007), GABA supplementation over several years carries the risk of desensitizing the GABA receptors. The chronic treatment with beta-blockers can create a similar but mechanistically different dilemma: long-term exposure to antagonist causes reactive sensitization and upregulation of G_s-coupled receptors (Lefkowitz, 2007). In turn, this phenomenon would render β-ARs more sensitive to their physiological agonists, thereby promoting cancer development, progression and resistance to therapy. These problems can be avoided by using positive allosteric modulators (PAMS) instead of synthetic agonists for G_i-coupled receptors. PAMS bind at a non-agonist site to the receptor and enhance its response to even low levels of agonist without desensitizing the receptor under long-term treatment protocols (Filip et al., 2015). PAMS for the $GABA_B$-R are used as therapeutics for addiction (Filip et al., 2015). PAMs for the cannabinoid receptors are currently in preclinical testing because they function as analgesics without having psychoactive effects (Ignatowska-Jankowska et al., 2015). Similarly, PAMS for opioid receptors are under development for the treatment of pain and depression (Livingston & Traynor, 2018). In recognition of the powerful impact of the mood on cancer, these pharmacological approaches need to be accompanied by methods to reduce psychological stress and increase pleasurable emotions.

References

Al-Wadei, M. H., Al-Wadei, H. A., & Schuller, H. M. (2012). Pancreatic cancer cells and normal pancreatic duct epithelial cells express an autocrine catecholamine loop that is activated by nicotinic acetylcholine receptors alpha3, alpha5, and alpha7. *Molecular Cancer Research*, *10*(2), 239–249. 1541-7786.MCR-11-0332 [pii], https://doi.org/10.1158/1541-7786.MCR-11-0332.

Al-Wadei, H. A., Plummer, H. K., 3rd, Ullah, M. F., Unger, B., Brody, J. R., & Schuller, H. M. (2012). Social stress promotes and gamma-aminobutyric acid inhibits tumor growth in mouse models of non-small cell lung cancer. *Cancer Prevention Research (Philadelphia, Pa)*, *5*(2), 189–196. https://doi.org/10.1158/1940-6207.CAPR-11-0177.

Al-Wadei, H. A., & Schuller, H. M. (2009). Nicotinic receptor-associated modulation of stimulatory and inhibitory neurotransmitters in NNK-induced adenocarcinoma of the lungs and pancreas. *The Journal of Pathology, 218*(4), 437–445. https://doi.org/10.1002/path.2542.

Askari, M. D., Tsao, M. S., & Schuller, H. M. (2005). The tobacco-specific carcinogen, 4-(methylnitrosamino)-1-(3-pyridyl)-1-butanone stimulates proliferation of immortalized human pancreatic duct epithelia through beta-adrenergic transactivation of EGF receptors. *Journal of Cancer Research and Clinical Oncology, 131*(10), 639–648. https://doi.org/10.1007/s00432-005-0002-7.

Banerjee, J., Al-Wadei, H. A., Al-Wadei, M. H., Dagnon, K., & Schuller, H. M. (2014). Differential modulation of nicotine-induced gemcitabine resistance by GABA receptor agonists in pancreatic cancer cell xenografts and in vitro. *BMC Cancer, 14*, 725. https://doi.org/10.1186/1471-2407-14-725.

Banerjee, J., Papu John, A. M., & Schuller, H. M. (2015). Regulation of nonsmall-cell lung cancer stem cell like cells by neurotransmitters and opioid peptides. *International Journal of Cancer, 137*(12), 2815–2824. https://doi.org/10.1002/ijc.29646.

Cakir, Y., Plummer, H. K., 3rd, Tithof, P. K., & Schuller, H. M. (2002). Beta-adrenergic and arachidonic acid-mediated growth regulation of human breast cancer cell lines. *International Journal of Oncology, 21*(1), 153–157.

Cattaneo, M. G., Codignola, A., Vicentini, L. M., Clementi, F., & Sher, E. (1993). Nicotine stimulates a serotonergic autocrine loop in human small-cell lung carcinoma. *Cancer Research, 53*(22), 5566–5568.

Chakravarti, B., Ravi, J., & Ganju, R. K. (2014). Cannabinoids as therapeutic agents in cancer: Current status and future implications. *Oncotarget, 5*(15), 5852–5872. https://doi.org/10.18632/oncotarget.2233.

El-Armouche, A., Zolk, O., Rau, T., & Eschenhagen, T. (2003). Inhibitory G-proteins and their role in desensitization of the adenylyl cyclase pathway in heart failure. *Cardiovascular Research, 60*(3), 478–487[doi:S0008636303006308 [pii].

Elbaz, M., Ahirwar, D., Ravi, J., Nasser, M. W., & Ganju, R. K. (2017). Novel role of cannabinoid receptor 2 in inhibiting EGF/EGFR and IGF-I/IGF-IR pathways in breast cancer. *Oncotarget, 8*(18), 29668–29678. https://doi.org/10.18632/oncotarget.9408.

Eramo, A., Lotti, F., Sette, G., Pilozzi, E., Biffoni, M., Di Virgilio, A., et al. (2008). Identification and expansion of the tumorigenic lung cancer stem cell population. *Cell Death and Differentiation, 15*(3), 504–514. https://doi.org/10.1038/sj.cdd.4402283.

Esch, T., & Stefano, G. B. (2004). The neurobiology of pleasure, reward processes, addiction and their health implications. *Neuro Endocrinology Letters, 25*(4), 235–251.

Filip, M., Frankowska, M., Sadakierska-Chudy, A., Suder, A., Szumiec, L., Mierzejewski, P., et al. (2015). GABAB receptors as a therapeutic strategy in substance use disorders: Focus on positive allosteric modulators. *Neuropharmacology, 88*, 36–47. https://doi.org/10.1016/j.neuropharm.2014.06.016.

Friesen, C., Hormann, I., Roscher, M., Fichtner, I., Alt, A., Hilger, R., et al. (2014). Opioid receptor activation triggering downregulation of cAMP improves effectiveness of anti-cancer drugs in treatment of glioblastoma. *Cell Cycle, 13*(10), 1560–1570. https://doi.org/10.4161/cc.28493.

Friesen, C., Roscher, M., Alt, A., & Miltner, E. (2008). Methadone, commonly used as maintenance medication for outpatient treatment of opioid dependence, kills leukemia cells and overcomes chemoresistance. *Cancer Research, 68*(15), 6059–6064. https://doi.org/10.1158/0008-5472.CAN-08-1227.

Friesen, C., Roscher, M., Hormann, I., Fichtner, I., Alt, A., Hilger, R. A., et al. (2013). Cell death sensitization of leukemia cells by opioid receptor activation. *Oncotarget, 4*(5), 677–690. https://doi.org/10.18632/oncotarget.952.

Grau, M., Soley, M., & Ramirez, I. (1997). Interaction between adrenaline and epidermal growth factor in the control of liver glycogenolysis in mouse. *Endocrinology, 138*(6), 2601–2609.

Gurfein, B. T., Stamm, A. W., Bacchetti, P., Dallman, M. F., Nadkarni, N. A., Milush, J. M., et al. (2012). The calm mouse: An animal model of stress reduction. *Molecular Medicine, 18*(1), 606–617. molmed.2012.00053 [pii]. https://doi.org/10.2119/molmed.2012.00053.

Haass, M., & Kubler, W. (1997). Nicotine and sympathetic neurotransmission. *Cardiovascular Drugs and Therapy, 10*(6), 657–665.

Ignatowska-Jankowska, B. M., Baillie, G. L., Kinsey, S., Crowe, M., Ghosh, S., Owens, R. A., et al. (2015). A cannabinoid CB1 receptor-positive allosteric modulator reduces neuropathic pain in the mouse with no psychoactive effects. *Neuropsychopharmacology, 40*(13), 2948–2959. https://doi.org/10.1038/npp.2015.148.

Jang, H. J., Boo, H. J., Lee, H. J., Min, H. Y., & Lee, H. Y. (2016). Chronic stress facilitates lung tumorigenesis by promoting exocytosis of IGF2 in lung epithelial cells. *Cancer Research, 76*(22), 6607–6619. https://doi.org/10.1158/0008-5472.CAN-16-0990.

Jull, B. A., Plummer, H. K., 3rd, & Schuller, H. M. (2001). Nicotinic receptor-mediated activation by the tobacco-specific nitrosamine NNK of a Raf-1/MAP kinase pathway, resulting in phosphorylation of c-myc in human small cell lung carcinoma cells and pulmonary neuroendocrine cells. *Journal of Cancer Research and Clinical Oncology, 127*(12), 707–717.

Knoll, A. T., & Carlezon, W. A., Jr. (2010). Dynorphin, stress, and depression. *Brain Research, 1314*, 56–73. doi:S0006-8993(09)02005-8 [pii]. https://doi.org/10.1016/j.brainres.2009.09.074.

Krantz, M. J., & Mehler, P. S. (2004). Treating opioid dependence. Growing implications for primary care. *Archives of Internal Medicine, 164*(3), 277–288. https://doi.org/10.1001/archinte.164.3.277.

Kuzumaki, N., Suzuki, A., Narita, M., Hosoya, T., Nagasawa, A., Imai, S., et al. (2012). Effect of kappa-opioid receptor agonist on the growth of non-small cell lung cancer (NSCLC) cells. *British Journal of Cancer, 106*(6), 1148–1152. https://doi.org/10.1038/bjc.2011.574.

Lefkowitz, R. J. (2000). The superfamily of heptahelical receptors. *Nature Cell Biology, 2*(7), E133–E136. https://doi.org/10.1038/35017152.

Lefkowitz, R. J. (2007). Seven transmembrane receptors: Something old, something new. *Acta Physiologica (Oxford, England), 190*(1), 9–19. https://doi.org/10.1111/j.1365-201X.2007.01693.x. APS1693 [pii].

Livingston, K. E., & Traynor, J. R. (2018). Allostery at opioid receptors: Modulation with small molecule ligands. *British Journal of Pharmacology, 175*(14), 2846–2856. https://doi.org/10.1111/bph.13823.

Madden, K. S., Szpunar, M. J., & Brown, E. B. (2011). Beta-adrenergic receptors (beta-AR) regulate VEGF and IL-6 production by divergent pathways in high beta-AR-expressing breast cancer cell lines. *Breast Cancer Research and Treatment, 130*(3), 747–758. https://doi.org/10.1007/s10549-011-1348-y.

Maneckjee, R., & Minna, J. D. (1990). Opioid and nicotine receptors affect growth regulation of human lung cancer cell lines. *Proceedings of the National Academy of Sciences of the United States of America, 87*(9), 3294–3298.

McEwen, B. S. (2000). The neurobiology of stress: From serendipity to clinical relevance. *Brain Research, 886*(1–2), 172–189[doi:S0006-8993(00)02950-4 [pii].

Messenger, S. W., Falkowski, M. A., & Groblewski, G. E. (2014). Ca^{2+}-regulated secretory granule exocytosis in pancreatic and parotid acinar cells. *Cell Calcium, 55*(6), 369–375. https://doi.org/10.1016/j.ceca.2014.03.003.

Meza, R., Meernik, C., Jeon, J., & Cote, M. L. (2015). Lung cancer incidence trends by gender, race and histology in the United States, 1973-2010. *PLoS One*, *10*(3). e0121323https://doi.org/10.1371/journal.pone.0121323.

Michalska, M., Katzenwadel, A., & Wolf, P. (2017). Methadone as a "tumor theralgesic" against cancer. *Frontiers in Pharmacology*, *8*, 733. https://doi.org/10.3389/fphar.2017.00733.

Michalska, M., Schultze-Seemann, S., Kuckuck, I., Katzenwadel, A., & Wolf, P. (2018). Impact of methadone on cisplatin treatment of bladder cancer cells. *Anticancer Research*, *38*(3), 1369–1375. https://doi.org/10.21873/anticanres.12360.

Moody, T. W., Ramos-Alvarez, I., & Jensen, R. T. (2018). Neuropeptide G protein-coupled receptors as oncotargets. *Frontiers in Endocrinology (Lausanne)*, *9*, 345. https://doi.org/10.3389/fendo.2018.00345.

Muller, L., Radtke, A., Decker, J., Koch, M., & Belge, G. (2017). The synthetic cannabinoid WIN 55,212-2 elicits death in human Cancer cell lines. *Anticancer Research*, *37*(11), 6341–6345. https://doi.org/10.21873/anticanres.12086.

Nabissi, M., Morelli, M. B., Amantini, C., Liberati, S., Santoni, M., Ricci-Vitiani, L., et al. (2015). Cannabidiol stimulates Aml-1a-dependent glial differentiation and inhibits glioma stem-like cells proliferation by inducing autophagy in a TRPV2-dependent manner. *International Journal of Cancer*, *137*(8), 1855–1869. https://doi.org/10.1002/ijc.29573.

Nakamura, H., & Saji, H. (2014). Worldwide trend of increasing primary adenocarcinoma of the lung. *Surgery Today*, *44*(6), 1004–1012. https://doi.org/10.1007/s00595-013-0636-z.

Nylen, E. S., Becker, K. L., Snider, R. H., Tabassian, A. R., Cassidy, M. M., & Linnoila, R. I. (1993). Cholinergic-nicotinic control of growth and secretion of cultured pulmonary neuroendocrine cells. *The Anatomical Record*, *236*(1), 129–135. https://doi.org/10.1002/ar.1092360117.

Pereira, L., Metrich, M., Fernandez-Velasco, M., Lucas, A., Leroy, J., Perrier, R., et al. (2007). The cAMP binding protein Epac modulates Ca2+ sparks by a Ca2+/calmodulin kinase signalling pathway in rat cardiac myocytes. *The Journal of Physiology*, *583*(Pt. 2), 685–694. https://doi.org/10.1113/jphysiol.2007.133066.

Powe, D. G., Voss, M. J., Zanker, K. S., Habashy, H. O., Green, A. R., Ellis, I. O., et al. (2010). Beta-blocker drug therapy reduces secondary cancer formation in breast cancer and improves cancer specific survival. *Oncotarget*, *1*(7), 628–638[doi: 101009 [pii].

Ramer, R., Bublitz, K., Freimuth, N., Merkord, J., Rohde, H., Haustein, M., et al. (2012). Cannabidiol inhibits lung cancer cell invasion and metastasis via intercellular adhesion molecule-1. *The FASEB Journal*, *26*(4), 1535–1548. https://doi.org/10.1096/fj.11-198184.

Ravi, J., Elbaz, M., Wani, N. A., Nasser, M. W., & Ganju, R. K. (2016). Cannabinoid receptor-2 agonist inhibits macrophage induced EMT in non-small cell lung cancer by downregulation of EGFR pathway. *Molecular Carcinogenesis*, *55*(12), 2063–2076. https://doi.org/10.1002/mc.22451.

Schuller, H. M. (2008). Neurotransmission and cancer: Implications for prevention and therapy. *Anti-Cancer Drugs*, *19*(7), 655–671. https://doi.org/10.1097/CAD.0b013e3283025b58 00001813-200808000-00001[pii].

Schuller, H. M. (2010). Beta-adrenergic signaling, a novel target for cancer therapy? *Oncotarget*, *1*(7), 466–469. https://doi.org/10.18632/oncotarget.101102.

Schuller, H. M. (2013). Effects of tobacco constituents and psychological stress on the beta-adrenergic regulation of non-small cell lung cancer and pancreatic cancer: Implications for intervention. *Cancer Biomarkers*, *13*(3), 133–144. https://doi.org/10.3233/CBM-130323. NT71W501W8817518 [pii].

Schuller, H. M., & Al-Wadei, H. A. (2012). Beta-adrenergic signaling in the development and progression of pulmonary and pancreatic adenocarcinoma. *Current Cancer Therapy Reviews*, *8*(2), 116–127. https://doi.org/10.2174/157339412800675351.

Schuller, H. M., Al-Wadei, H. A., & Majidi, M. (2008). Gamma-aminobutyric acid, a potential tumor suppressor for small airway-derived lung adenocarcinoma. *Carcinogenesis*, *29*(10), 1979–1985. https://doi.org/10.1093/carcin/bgn041. bgn041 [pii].

Schuller, H. M., & Orloff, M. (1998). Tobacco-specific carcinogenic nitrosamines. Ligands for nicotinic acetylcholine receptors in human lung cancer cells. *Biochemical Pharmacology*, *55*(9), 1377–1384. doi:S0006-2952(97)00651-5 [pii].

Schuller, H. M., Tithof, P. K., Williams, M., & Plummer, H., 3rd. (1999). The tobacco-specific carcinogen 4-(methylnitrosamino)-1-(3-pyridyl)-1-butanone is a beta-adrenergic agonist and stimulates DNA synthesis in lung adenocarcinoma via beta-adrenergic receptor-mediated release of arachidonic acid. *Cancer Research*, *59*(18), 4510–4515.

Sher, E., Cesare, P., Codignola, A., Clementi, F., Tarroni, P., Pollo, A., et al. (1996). Activation of delta-opioid receptors inhibits neuronal-like calcium channels and distal steps of Ca^{2+}-dependent secretion in human small-cell lung carcinoma cells. *The Journal of Neuroscience*, *16*(11), 3672–3684.

Shin, V. Y., Wu, W. K., Chu, K. M., Koo, M. W., Wong, H. P., Lam, E. K., et al. (2007). Functional role of beta-adrenergic receptors in the mitogenic action of nicotine on gastric cancer cells. *Toxicological Sciences*, *96*(1), 21–29. https://doi.org/10.1093/toxsci/kfl118. kfl118 [pii].

Singh, S., & Chellappan, S. (2014). Lung cancer stem cells: Molecular features and therapeutic targets. *Molecular Aspects of Medicine*, *39*, 50–60. https://doi.org/10.1016/j.mam.2013.08.003.

Singh, A., Jayanthan, A., Farran, A., Elwi, A. N., Kim, S. W., Farran, P., et al. (2011). Induction of apoptosis in pediatric acute lymphoblastic leukemia (ALL) cells by the therapeutic opioid methadone and effective synergy with Bcl-2 inhibition. *Leukemia Research*, *35*(12), 1649–1657. https://doi.org/10.1016/j.leukres.2011.06.035.

Tang, J., Li, Z., Lu, L., & Cho, C. H. (2013). Beta-adrenergic system, a backstage manipulator regulating tumour progression and drug target in cancer therapy. *Seminars in Cancer Biology*, *23*(6 Pt. B), 533–542. https://doi.org/10.1016/j.semcancer.2013.08.009.

Thorne Research, I. (2007). Gamma-aminobutyric acid. *Alternative Medicine Review*, *12*, 274.

Tirino, V., Desiderio, V., Paino, F., De Rosa, A., Papaccio, F., La Noce, M., et al. (2013). Cancer stem cells in solid tumors: An overview and new approaches for their isolation and characterization. *The FASEB Journal*, *27*(1), 13–24. https://doi.org/10.1096/fj.12-218222. fj.12-218222 [pii].

Travis, W. D., Brambilla, E., Nicholson, A. G., Yatabe, Y., Austin, J. H. M., Beasley, M. B., et al. (2015). The 2015 World Health Organization classification of lung tumors: Impact of genetic, clinical and radiologic advances since the 2004 classification. *Journal of Thoracic Oncology*, *10*(9), 1243–1260. https://doi.org/10.1097/JTO.0000000000000630.

Wallukat, G. (2002). The beta-adrenergic receptors. *Herz*, *27*(7), 683–690. https://doi.org/10.1007/s00059-002-2434-z.

Wong, H. P., Yu, L., Lam, E. K., Tai, E. K., Wu, W. K., & Cho, C. H. (2007). Nicotine promotes cell proliferation via alpha7-nicotinic acetylcholine receptor and catecholamine-synthesizing enzymes-mediated pathway in human colon adenocarcinoma HT-29 cells. *Toxicology and Applied Pharmacology*, *221*(3), 261–267. doi:S0041-008X(07)00163-9 [pii]. https://doi.org/10.1016/j.taap.2007.04.002.

Zhang, S., Wang, Y., Mao, J. H., Hsieh, D., Kim, I. J., Hu, L. M., et al. (2012). Inhibition of CK2alpha down-regulates hedgehog/Gli signaling leading to a reduction of a stem-like side population in human lung cancer cells. *PLoS One*, 7(6). e38996https://doi.org/10.1371/journal.pone.0038996.

Zhang, X., Zhang, R., Zheng, Y., Shen, J., Xiao, D., Li, J., et al. (2013). Expression of gamma-aminobutyric acid receptors on neoplastic growth and prediction of prognosis in non-small cell lung cancer. *Journal of Translational Medicine*, *11*, 102. https://doi.org/10.1186/1479-5876-11-102.

CHAPTER THIRTEEN

Prenatal ethanol exposure and enkephalinergic neurotransmission

Milagros Méndez[a,*], Karla Hernández-Fonseca[a], Paula Abate[b]
[a]Departamento de Neuroquímica, Subdirección de Investigaciones Clínicas, Instituto Nacional de Psiquiatría Ramón de la Fuente, Ciudad de México, México
[b]Laboratorio de Psicología Experimental, Instituto de Investigaciones Psicológicas (IIPsi - CONCET - UNC), Facultad de Psicología, Universidad Nacional de Córdoba, Córdoba, Argentina
*Corresponding author: e-mail address: ubach@imp.edu.mx

Contents

Abstract

Endogenous opioids (enkephalins, endorphins and dynorphins) are small peptides that play a main role in pain perception and analgesia, as well as in alcohol (ethanol) reinforcement and reward. Alcohol reinforcement involves the ethanol-induced activation of the endogenous opioid system, a process that may augment the hedonic value and the reinforcing properties of the drug, which in turn increases substance consumption. Changes in opioidergic transmission may contribute to alcohol intoxication and to the neuroadaptive responses produced by the long-lasting exposure to ethanol. Opioidergic transmission may be altered by ethanol at distinct levels, including the expression of precursor mRNAs, biosynthesis and release of opioid peptides, as well as ligand binding to opioid receptors. In adult rats, β-endorphinergic and enkephalinergic transmission, through activation of mu and delta opioid receptors, mediate ethanol reinforcement and high alcohol drinking behavior. Prenatal ethanol exposure (PEE) selectively modifies Methionine-enkephalin (Met-enk) content in several brain regions of infant and adolescent rats, particularly those of the reward circuits. In preweanling rats, Met-enk content is decreased in the ventral tegmental area

Vitamins and Hormones, Volume 111
ISSN 0083-6729
https://doi.org/10.1016/bs.vh.2019.05.005

but is increased in the prefrontal cortex and the nucleus accumbens and other brain areas, as a consequence of a short and moderate ethanol exposure during late gestation. PEE also increases Met-enk levels in the prefrontal cortex and other brain regions of 30-day-old adolescent rats. These findings suggest that mesocorticolimbic enkephalins are essential in ethanol reinforcement in offspring, as previously reported in adult rats.

1. Introduction

The main endogenous opioid peptides, the endorphins, enkephalins, and dynorphins, are small molecules that bind to distinct subtypes of opioid receptors with different affinities. Beta-endorphin (β-end) binds preferentially to mu opioid receptors (MOR) and to a lesser degree to delta opioid receptors (DOR). Leucine-enkephalin (Leu-enk) exhibits a higher affinity for DOR, whereas Methionine-enkephalin (Met-enk) shows similar affinity for MOR and DOR. Several dynorphins have been characterized, which bind predominantly to kappa opioid receptors (KOR): Dynorphin A (1–17) (Dyn A [1–17]), dynorphin A (1–8) (Dyn A [1–8]), dynorphin B (1–29) (Dyn B [1–29]) and dynorphin B (1 – 13) (Dyn B [1 – 13] (Berman, Juliano, & Devi, 1995; Bodnar, 2018; Borsodi & Tóth, 1995).

Opioid peptides are extensively distributed in the Central Nervous System (CNS) and peripheral tissues of several species. Beta-end (β-end) is highly concentrated in the anterior and intermediate lobes of the pituitary and is also localized in the hypothalamus, the preoptic area, the septum, the nucleus accumbens (NAcc), the bed nucleus of the stria terminalis, the periaqueductal gray, the brainstem, and the spinal cord. In general terms, Met-enk concentrations are several-fold greater than Leu-enk levels in most brain areas. Highest concentrations of Met-enk and Leu-enk are found in the globus pallidus, the caudate-putamen (CP), the hypothalamus, and the NAcc, followed by the medulla oblongata, the septum, the olfactory bulb and olfactory tubercle, and the hippocampus. The lowest levels of Met-enk and Leu-enk are detected in the cerebral cortex and the cerebellum. Highest Dyn immunoreactivities are detected in the hypothalamus and the cerebellum, followed by the pons/medulla and midbrain, the CP, the hippocampus and the cerebral cortex (Cuello, 1983; Mansour, Khachaturian, Lewis, Akil, & Watson, 1988; Schafer, Day, Watson, & Akil, 1991).

The endorphins, enkephalins, and dynorphins are coded by three different genes. These peptides are synthesized by proteolytic post-translational processing of large polypeptide precursors that code for more than one biologically active peptide. The opioid precursors (Pro-opiomelanocortin, Pro-enkephalin and Pro-dynorphin) (Pro-enk, POMC, Pro-Dyn) are nearly the same length. The sequences of the known biologically active peptides are localized in the C-terminal half of each precursor and are flanked by pairs of basic amino acids. The POMC precursor gives rise to beta-lipotropin, which in turn originates different forms of endorphins (α-, β- and γ-endorphin). The POMC precursor also gives rise to other peptides, including adrenocorticotropic hormone (ACTH) and three different forms of melanocyte-stimulating hormone (α-, β- and γ-MSH) (Civelli, Douglass, Rosen, Martens, & Herbert, 1986; Mansour et al., 1988). Processing of the POMC precursor is tissue-specific and is achieved by specific processing enzymes, which are distinctly localized in the anterior and intermediate lobes of the pituitary. This process gives rise to different fragments of the precursor, which exhibit distinct biological activities (Frederickson, 1984). The enkephalins, Leu-enk and Met-enk, are pentapeptides which only differ in the last amino acid. They are originated from the processing of Proenkephalin (Pro-enk). The Pro-enk gene contains six copies of the Met-enk sequence and one copy of Leu-enk sequence (Civelli et al., 1986; Mansour et al., 1988). The Pro-enk precursor also contains two C-terminal extended forms of Met-enk, Met-enk^{5}-Arg^{6}-Gly^{7}-Leu^{8} and Met-enk^{5}-Arg^{6}-Phe^{7} (Mansour et al., 1988). Processing of the Pro-enk precursor in brain mainly produces Leu- and Met-enk, whereas in the adrenal medulla the C-terminal extended forms of Met-enk are predominant (Gubler, Seeburg, Hoffman, Gage, & Undenfriend, 1982; Noda et al., 1982). Prodynorphin (Pro-dyn) is processed to several peptides structurally related to Leu-enk. Pro-dyn contains three sequences of Leu-enk and several opioid peptides, which are C-terminal extensions of Leu-enk (Mansour et al., 1988). These include α- and β-neoendorphin, Dyn A (1–17), Dyn A (1–8), Dyn B (1–29) and Dyn B (1–13) (Hughes et al., 1975). Differential processing of the Pro-dyn precursor originates large peptides selective for KOR (Dyn A (1–17) and α-neoendorphin). Subsequent processing yields Leu-enk, selective for DOR (Horikawa et al., 1983).

Opioid receptors are widely distributed in the CNS. All three opioid receptors belong to the seven-transmembrane G-protein coupled receptors and present extensive structural homology with each other (Bodnar, 2018; Gianoulakis, 2009; Herz, 1997). High concentrations of MOR and KOR are found in the lateral nucleus of the amygdala, the NAcc, the striatum and the nucleus tractus solitarius. Moderate levels of MOR are detected in the substantia nigra (SN) and the ventral tegmental area (VTA), whereas low densities of KOR are present in these areas. The expression of DOR in the brain is lower than that of MOR and KOR. Highest DOR densities are detected in the striatum and the NAcc, followed by the lateral nucleus of the amygdala and the olfactory tubercle. Moderate concentrations of these receptors are found in the medial nucleus of the amygdala, the hippocampus, and the bed nucleus of the stria terminalis, while low receptor densities are observed in the dentate gyrus of the hippocampus, the globus pallidus, the medial septum, and the preoptic area (Mansour et al., 1988).

Opioid peptides play a major role in pain perception and analgesia, although they are also involved in several other physiological functions and behaviors, such as regulation of food and water intake, convulsive activity, locomotion, learning and memory, sexual behavior, neuroendocrine hormone secretion, mood and motivation and alcohol addiction (Kiefer & Wiedemann, 2004; Vaccarino & Kastin, 2001). These functions are exerted by activation of specific opioid receptors. MOR, DOR and KOR are involved in analgesia and reward function. MOR and DOR mediate reinforcement, while KOR participate in drug aversion (Pasternak, 1988).

2. Alcohol reinforcement: Activation of dopaminergic mesocorticolimbic transmission

Alcohol (ethanol) is a drug with depressant actions on the CNS that displays dose-dependent biphasic effects on behavior. Low doses of alcohol induce psychomotor activation and euphoria, whereas high doses reduce locomotor activity and produce sedation in both animals and humans (Lukas & Mendelson, 1988; Martin, Earleywine, Musty, Perrine, & Swift, 1993; Méndez & Herrera, 2012). Administration of low ethanol doses may induce the progressive consumption of higher doses, a process that may lead to drug abuse and/or dependence. Activation of specific neural circuits mediates drug reinforcement and reward; gradual neuroadaptive changes may occur along this process, which may be crucial in the development of drug dependence (Nestler, Hyman, & Malenka, 2001).

Alcohol exhibits positive and negative reinforcement properties that are critical in drug seeking-behavior and dependence (Koob, 2003; Koob, Sanna, & Bloom, 1998; Wise & Bozarth, 1987). The dopaminergic (DAergic) mesocorticolombic system (VTA, prefrontal cortex [PFC] and NAcc) plays a main role in the mechanisms underlying ethanol reinforcement and reward (Wise & Bozarth, 1987). Ethanol-induced activation of DAergic transmission in this pathway may eventually lead to the development of an addictive behavior. Ethanol increases the firing rate of DAergic neurons in the VTA, as well as dopamine (DA) synthesis, release and metabolism in the NAcc (Di Chiara & Imperato, 1985; Fadda, Mosca, Colombo, & Gessa, 1989; Gessa, Muntoni, Collu, Vargiu, & Mereu, 1985) and the PFC (Di Chiara & Imperato, 1985; Fadda et al., 1989). The ethanol-induced increase in DA release from the NAcc is a critical event in the reinforcing actions of the drug.

3. Role of opioid peptides in alcohol reinforcement

Ethanol, opioid peptides and opiates (exogenous opioids) share some pharmacological properties and behavioral effects. Low doses of these substances stimulate locomotor activity through DAergic activation of the VTA, while high doses stimulate DAergic terminals in the NAcc (Joyce & Iversen, 1979; Kalivas, Widerläv, Stanley, Breese, & Prange, 1983; Wise & Bozarth, 1987). MOR agonists increase the firing rate of DAergic neurons in the VTA (Gysling & Wang, 1983) and stimulate DA release and metabolism in the NAcc and PFC, whereas DOR agonists do not have an effect (Spanagel, Herz, & Shippenberg, 1990; Wood & Rao, 1991). These findings indicate that activation of opioid receptors by endogenous opioids is critical in the regulation of DAergic transmission in the mesocorticolimbic system. Other studies show that activation of this pathway by MOR and DOR agonists induces reinforcement, while KOR stimulation is associated to dysphoria (Herz, 1997). These actions are mediated by increases or decreases in DA release from the NAcc, respectively. These findings suggest that opioid and alcohol reinforcement underly a common neurobiological mechanism that involves activation of DAergic reward circuits in the brain (Wise & Bozarth, 1987).

More recently, a fourth binding site was described, Nociceptin/orphanin (N/OFQ) and orphanin receptor like (ORL) (Bodnar, 2018). N/OFQ is the endogenous ligand of the ORL. Despite its structural homology with opioid peptides, N/OFQ does not bind to MOR, DOR or KOR opioid receptors. From a functional point of view, N/OFQ possesses anti-opioid properties. Several studies suggest that nociception reduces the rewarding properties of ethanol, as well as alcohol consumption behaviors. Some reports suggest that an ORL-1 agonist may be clinically useful for treatment of alcohol abuse (Bignan, Connolly, & Middleton, 2005;

Ciccocioppo, Economidou, Fedeli, & Massi, 2003; Ciccocioppo, Panocka, Polidori, Regoli, & Massi, 1999; Kuzmin, Kreek, Bakalkin, & Liljequist, 2007).

4. Alcohol effects on enkephalinergic and β-endorphinergic transmission

Alcohol reinforcement is partially mediated by the ethanol-induced activation of the endogenous opioid system (Froehlich, Badia-Elder, Zink, McCullough, & Portoghese, 1998; Gianoulakis, 2009; Herz, 1997; Shippenberg, LeFevour, & Chefer, 2008; Ulm, Volpicelli, & Volpicelli, 1995). This activation could increase the hedonic value and the reinforcing properties of alcohol which, in turn, would increase drug consumption. Eventually, this mechanism would be relevant in the establishment of an addictive behavior.

Pharmacological studies using opioid receptor agonists and antagonists have contributed to understand the role of opioid peptides in ethanol reinforcement and high alcohol drinking behavior. In adult rats, the pharmacological manipulation of opioidergic activity modifies alcohol reinforcement, preference and consumption (Chotro & Arias, 2007; Méndez & Herrera, 2012; Pautassi, Nizhnikov, & Spear, 2009; Spear & Molina, 2005). For instance, low doses of MOR agonists increase alcohol consumption in rats (Wild & Reid, 1990), whereas high doses decrease intake (Volpicelli, Ulm, & Hopson, 1991). The administration of MOR and DOR selective and non-selective (naloxone or naltrexone) antagonists, as well as KOR agonists, decreases alcohol consumption in different animal models (Higley & Kiefer, 2006; Hyytiä & Kiianmaa, 2001; Krishnan-Sarin, Wand, Li, Portoghese, & Froehlich, 1998; Lindholm, Werme, Brené, & Franck, 2001; Stromberg, Casale, Volpicelli, Volpicelli, & O'Brien, 1998). In alcoholic patients, naltrexone reduces alcohol craving and relapse (Volpicelli, Alterman, Hayashida, & O'Brien, 1992). Overall, these studies suggest that activation of the β-endorphinergic and enkephalinergic systems via MOR and DOR is relevant in ethanol reinforcement and high alcohol drinking behavior.

Alcohol modifies opioidergic neurotransmission at different levels (Froehlich, 1995; Gianoulakis, 2009; Méndez & Morales-Mulia, 2008a, 2008b; Ulm et al., 1995), such as the expression and release of opioid peptides, as well as ligand binding to opioid receptors (for reviews, see Méndez & Morales-Mulia, 2008a, 2008b, 2008c). For instance, ethanol increases the expression, content and release of Met-enk and β-end in brain areas of the reward circuits (Gianoulakis, 2009; Marinelli, Bai, Quirion, & Gianoulakis, 2005; Marinelli, Quirion, & Gianoulakis, 2003; Méndez, Barbosa-Luna, Pérez-Luna, Cupo, & Oikawa, 2010; Méndez & Morales-Mulia, 2006; Olive, Koenig, Nannini, & Hodge, 2001). These changes may be associated with alcohol preference and consumption in adult animals (Gianoulakis, de Waele, & Thavundayil, 1996). Moreover, a correlation has been established between the increase in β-end levels and the risk of alcoholism in humans (Gianoulakis, Krishnan, & Thavundayil, 1996). Changes in opioid neurotransmission may be relevant during ethanol intoxication, as well as in the adaptive neural responses induced by the drug.

We have previously shown that ethanol induces selective changes in enkephalinergic and β-endorphinergic systems in adult rats, particularly in mesocorticolimbic regions involved in the ethanol reinforcing actions. Ethanol selectively modifies precursor mRNA expression and peptide content, release and inactivation, as well as endogenous opioid binding to receptors (Leriche & Méndez, 2010; Méndez et al., 2010; Méndez, Leriche, & Calva, 2001; Méndez & Morales-Mulia, 2006; Méndez, Morales-Mulia, & Leriche, 2004).

4.1 Acute ethanol effects

Acute administration of high ethanol doses (2.5 g/kg) decreases the expression of Pro-enk mRNA in the VTA, the *substantia nigra pars compacta* (SNc) and the *substantia nigra pars reticulata* (SNr) but increases mRNA expression in the PFC of Wistar adult rats. This treatment also induces a sustained augmentation in Pro-enk mRNA levels in the NAcc and the CP (Méndez & Morales-Mulia, 2006; Méndez, Morales-Mulia, & Pérez-Luna, 2008). In addition, high ethanol doses (2.5 g/kg) decrease Met-enk content in the NAcc and CP but have no effect in the PFC. This treatment does not modify β-end levels in these regions, but reduces peptide concentrations in the hypothalamus (Leriche & Méndez, 2010). Intermediate-high ethanol doses stimulate Met-enk release from the NAcc, with different kinetics (Méndez et al., 2010). This effect is dose-dependent and correlates with the ascending limb of the blood ethanol curve (Barbosa-Luna, 2012; de Waele & Gianoulakis, 1993; Marinelli et al., 2005).

Overall, these studies indicate that the enkephalinergic system is an important ethanol target and Met-enk release one of the essential events in these actions. Mesolimbic enkephalins may play a key role in ethanol reinforcement.

On the other hand, ethanol administration differentially alters the binding of selective ligands to MOR and DOR with distinct kinetic patterns. A high dose of ethanol (2.5 g/kg) significantly reduces ([^{3}H]-[d-Ala^{2}, -MePhe^{4},Gly-ol^{5}]-enkephalin ([^{3}H]-DAMGO) binding to MOR in the VTA and the shell region of the NAcc, but increases binding in the PFC and has no effect in the CP (Méndez et al., 2001; Méndez, Leriche, & Calva, 2003). The same ethanol dose increases ([^{3}H]-[d-Pen^{2}, d-Pen^{5}]-enkephalin) ([^{3}H]-DPDPE) binding to DOR in the PFC, NAcc, CP and SN (Méndez et al., 2004). These studies indicate that ethanol induces selective changes in enkephalinergic and β-endorphinergic systems in adult animals, particularly in mesocorticolimbic regions involved in the reinforcing properties of the drug. These findings also suggest that opioid receptor down- and upregulation mechanisms may be involved in these actions. Differential opioid receptor sensitivity in specific brain areas may account for the impact of acute ethanol intoxication (Méndez & Morales-Mulia, 2008c).

4.2 Chronic ethanol effects

Chronic ethanol exposure selectively affects enkephalinergic and β-endorphinergic transmission as well. Ethanol (10% *v*/v, 4 weeks) increases Met-enk content in the adult rat VTA and PFC but does not affect peptide content in the SN, CP, hypothalamus and hippocampus or β-end levels in these brain areas (Barbosa-Luna, 2012; Leriche & Méndez, 2010). This treatment neither affects [^{3}H]-DAMGO nor [^{3}H]-DPDPE binding to MOR and DOR in these areas (Barbosa-Luna, 2012; Leriche & Méndez, 2010), suggesting that neuroadaptive changes in enkephalin- and β-end-containing neurons in the mesocorticolimbic system may have occurred along prolonged ethanol exposure.

4.3 Prenatal ethanol effects

Behavioral studies have allowed the analysis of alcohol-mediated mechanisms and treatment effects in vivo, as well as ethanol-related learning that can shape drug seeking and intake (Pautassi, Miranda-Morales, & Nizhnikov, 2015). Information on the impact of ethanol exposure along distinct ontogenic phases contributes to understand animal predisposition and facilitation to react toward a drug during development. Ethanol experiences throughout early ontogeny—even during prenatal periods—facilitate the posterior acceptance and consumption of the drug in animals treated with alcohol (Abate, Pueta, Spear, & Molina, 2008; Molina, Domínguez, López, Pepino, & Faas, 1999; Pautassi et al., 2009; Pueta, Abate, Haymal, Spear, & Molina, 2008). The predisposition to recognize and prefer ethanol as a function of pre- or postnatal exposure has not only been observed in animal models. Studies in humans suggest that the maternal intake of ethanol during pregnancy results in neonatal detection of ethanol odor (Faas, Spontón, Moya, & Molina, 2000) and the expression of positive facial responses to the odor of the drug (Faas, March, Moya, & Molina, 2015). These findings suggest that alcohol consumption in children is a serious risk factor that further increases the probability of drinking among these subjects (Pilatti, Godoy, Brussino, & Pautassi, 2013).

Alcohol-seeking and substance intake are modulated by the appetitive and aversive properties of the drug, as shown in animal models (Pautassi et al., 2015, 2009). These phenomena have been assessed in preweanling rats. A pattern of high acceptance of alcohol early in ontogeny seems to be associated with the pharmacological effects of the drug, rather than with its orosensory properties (Abate et al., 2008; Alati et al., 2006; Cunningham, Niehus, & Noble, 1993; Kozlov, Varlinskaya, & Spear, 2008; Pautassi et al., 2009; Spear & Molina, 2005). On the other hand, preweanling rats are sensitive to the locomotor activating effects of ethanol (Arias, Mlewski, Miller, Molina, & Spear, 2009; Arias, Mlewski, Molina, & Spear, 2009). These results suggest that infants are prone to process the stimulating effects of ethanol rather than its sedative actions.

The motivational aspects of alcohol in neonates and infant rats have been analyzed using operant approaches (Bordner, Molina, & Spear, 2008; March, Abate, Spear, & Molina, 2009; Miranda-Morales, Molina, Spear, & Abate, 2010, 2012; Miranda-Morales, Spear, Nizhnikov, Molina, & Abate, 2012; Ponce, Pautassi, Spear, & Molina, 2008). These studies strengthen the notion that the developing rat is highly sensitive to the appetitive motivational effects of ethanol, as demonstrated through rapid and robust instrumental learning.

Different components of the opioidergic system mediate some aspects of alcohol reinforcement, and therefore are involved in seeking and intake behaviors associated with reward. Experiments conducted in adult rats indicate that the motivational effects of ethanol observed as euphoric and pleasant behaviors are mediated by MOR and DOR, whereas the aversive and dysphoric states are mediated by KOR (Bodnar, 2018; Froehlich et al., 1998; Gianoulakis, 2009).

Opioid receptors follow different patterns of development across ontogeny, although MOR, DOR and KOR are functional by the second postnatal week of life (Volterra, Brunello, Restani, Galli, & Racagni, 1986). Alcohol reinforcement and acceptance in preweanling rats seem to be partially mediated by opioidergic systems, as in adult rodents (Gianoulakis, 2009). Co-administration of naloxone (non-selective opioid antagonist) with alcohol during development disrupts the facilitative effect of gestational ethanol experience on substance intake during infancy (Arias & Chotro, 2005; Chotro & Arias, 2007; Miranda-Morales et al., 2010). Moreover, the administration of opioid antagonists interferes with appetitive reinforcement toward the drug (Miranda-Morales, Molina, et al., 2012), which is modulated by MOR and KOR in newborn and infant rats (Nizhnikov, Varlinskaya, Petrov, & Spear, 2006) through inhibition of positive behaviors, such as the attachment to an artificial nipple (Nizhnikov, Varlinskaya, & Spear, 2006). During the preweanling period, ethanol intake can be reduced by non-selective or MOR and DOR selective antagonists (Arias, Molina, & Spear, 2010; Chotro, Arias, & Laviola, 2007; Hallmark & Hunt, 2004). Furthermore, ethanol-induced motor activation can also be reduced by MOR or DOR antagonists during the preweanling period (Arias et al., 2010). Overall, these findings indicate that a fully functional opioidergic system is required to promote ethanol-related responses during the second postnatal week.

Several pharmacological approaches have been implemented to assess the motivational effects of alcohol on different components of the opioidergic system during infancy (Arias et al., 2010; Miranda-Morales et al., 2010; Miranda-Morales, Molina, et al., 2012; Miranda-Morales, Spear, et al., 2012). MOR, DOR and KOR have differential developmental profiles during early ontogeny. MOR and KOR are present at birth, whereas DOR appears in the second week of life (Spain, Roth, & Coscia, 1985). In the first week of the rat, MOR and KOR activation mediates the reinforcement of alcohol (Chotro & Arias, 2007; Nizhnikov, Pautassi, Truxell, & Spear, 2009; Nizhnikov, Varlinskaya, Petrov, et al., 2006; Nizhnikov, Varlinskaya, & Spear, 2006). In preweanling rats, MOR and DOR appear to play an important role in reinforcing aspects of alcohol, enhancing consummatory behaviors and acceptance of the drug (Arias et al., 2010; Chotro & Arias, 2007). Nevertheless, the role of KOR is still controversial, since the effects of receptor activation in ethanol reward are unclear. Studies on the motivational aspects of ethanol have been assessed with pharmacological tools, such as antagonists and/or agonists of KOR.

Conflicting results were obtained depending on the agonist and antagonist doses used in these experiments. When alcohol self-administration was evaluated under the effects of a KOR agonist, a reduction of operant reinforcement was observed, particularly when employing a high dose. Pups injected with a low dose of the KOR antagonist showed a reduction in responding for ethanol, but those animals given the higher dose of the antagonist seemed unaffected (Miranda-Morales, Spear, et al., 2012). Other studies demonstrated that different antagonist doses did not affect infant alcohol intake (Arias et al., 2010). On the other hand, ORL-1 activation during infancy failed to suppress alcohol intake, as well as modulation of aversive behaviors, in contrast to the reported effects in adult rats (Miranda-Morales, Nizhnikov, Waters, & Spear, 2013; Miranda-Morales & Pautassi, 2016).

Alcohol administration may produce specific changes in opioidergic neurotransmission in different brain regions of both infant and adolescent rats. Such changes could be important in mediating some of the neurobehavioral effects of alcohol, like ethanol reinforcement, and could play a role in controlling drug consumption or other consummatory responses. Distinct effects on opioidergic transmission have been observed along the rat life span. Studies conducted in order to analyze the possible changes in protein and mRNA content of different components of opioidergic systems at distinct infancy stage in rats, have established that mRNA of precursor opioid peptides, as well as receptors are up-regulated across infancy ages. These results suggest an escalating ontogenetic expression of opioid-related genes. PEE significantly reduces opioid peptide levels, suggesting a possible mechanism by which drug exposure can affect future responsiveness toward ethanol (Bordner & Deak, 2015).

Although several studies have evaluated the effects of alcohol on opioidergic transmission in adult rats, a small number have assessed the impact of low or moderate ethanol doses during early ontogeny. Ethanol exposure (2 g/kg, i.g.) during the last period of gestation (GDs 17–20) of Wistar rats induces important changes in Met-enk content in offspring. PEE increases ethanol consumption in pups of 15 days of age (postnatal day 15; PD15) (ethanol intake test: 0%, 5% or 10%) (Abate, Hernández-Fonseca, Reyes-Guzmán, Barbosa-Luna, & Méndez, 2014). In preweanling rats, PEE produces a 45% decrease in Met-enk content in the VTA of pups that consume water (Fig. 1). This treatment increases Met-enk concentrations in the PFC (68%) and the NAcc (57%) of PD15 pups consuming water and a 10% ethanol solution, respectively (Fig. 1).

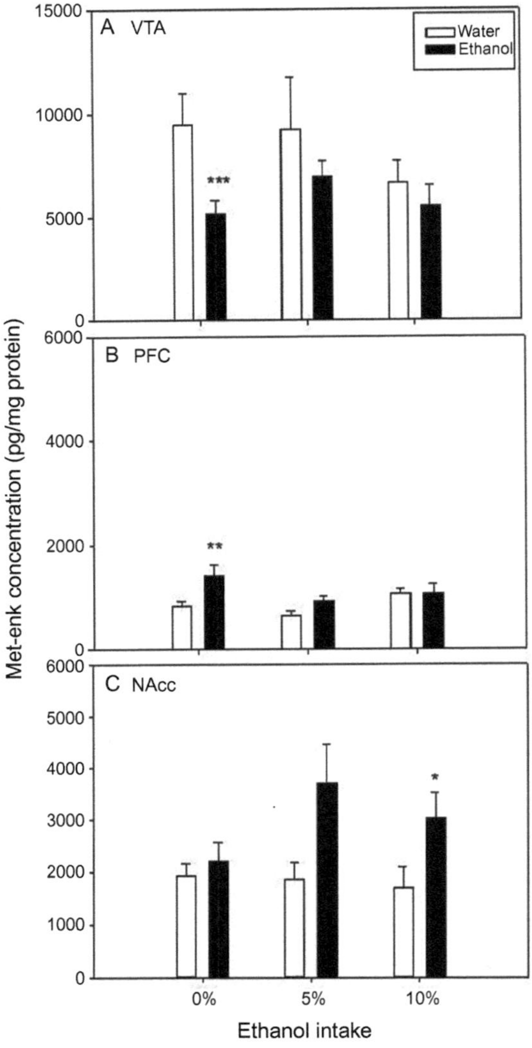

Fig. 1 Effect of prenatal ethanol treatment on Met-enk concentration in regions of the mesocorticolimbic system of PD15 pups. Wistar rat dams were daily administered with ethanol (2 g/kg body weight, i.g.) on GDs 17–20. Preweanling's intake test was performed on PDs 14 and 15. Thirty minutes after evaluation of consumption (0%-, 5%- or 10% ethanol solution), pups were sacrificed and their brains removed and frozen. Met-enk concentration was quantitated in several brain areas by radioimmunoassay (RIA). Met-enk concentration values (pg/mg protein) are shown for the VTA (A), PFC (B) and NAcc (C). Data are the mean ± SEM of 6 (prenatal water) or 7 (prenatal ethanol) pups in the VTA, 8–11 (prenatal water) or 9–11 (prenatal ethanol) in the PFC, and 10–11 (prenatal water) or 11–12 (prenatal ethanol) in the NAcc. $***P < 0.020$, $**P < 0.030$, and $*P < 0.050$ versus their own prenatal water control groups. *Taken from Abate, P., Hernández-Fonseca, K., Reyes-Guzmán, A.C., Barbosa-Luna, I.G., & Méndez, M. (2014). Prenatal ethanol exposure alters met-enkephalin expression in brain regions related with reinforcement: Possible mechanism for ethanol consumption in offspring.* Behavioural Brain Research, 274, *194–204. doi: 10.1016/j.bbr.2014.08.022.*

Met-enk content is also increased in other brain regions, such as the medial-posterior area of the CP, the hippocampus and the hypothalamus (Figs. 2 and 3) as a consequence of a short and moderate ethanol exposure during late gestation. PEE rises Met-enk levels in the mpCP of pups that consume water (57%) and a 5% ethanol solution (98%) (Fig. 2). Moreover, PEE induces important increases in Met-enk content in the hypothalamus (77%, 72% and 80%) and the hippocampus (85%, 73%, and 84%) of pups consuming water, 5%- or 10% ethanol, respectively (Fig. 3). Met-enk levels are not modified in the SN, the amCP and the amygdala by PEE (Figs. 2 and 3) (Abate et al., 2014). These findings indicate that PEE promotes ethanol consumption in infant rats and induces changes in Met-enk content in specific regions of the mesocorticolimbic and nigrostriatal pathways, as well as the hypothalamus and hippocampus. Mesocorticolimbic enkephalins may be involved in ethanol reinforcement in offspring, as has been reported in adult rats.

On the other hand, PEE produces selective changes in Met-enk content in distinct brain areas of 30-day-old adolescent rats challenged with ethanol (1.0 g/kg, i.p.) (Figs. 4–6) (Abate, Reyes-Guzmán, Hernández-Fonseca, & Méndez, 2017). PEE increases Met-enk content in the PFC (25%), the hippocampus (36%), the hypothalamus (90%) and the amCP (91%) (Figs. 4–6). The most important alterations induced by PEE are observed in the amCP and the hypothalamus. PEE did not modify Met-enk levels in the VTA and NAcc (Fig. 4), the SN and mpCP (Fig. 5) and the amygdala (Fig. 6) (Abate et al., 2017). These studies show that PEE specifically modifies Met-enk levels in the mesocorticolimbic and nigrostriatal systems, the hypothalamus and hippocampus, as has been shown in infant rats. PEE may determine sensitivity of adolescents to further ethanol exposure.

The observed effects on Met-enk content detected in infants and adolescents in response to PEE could be attributed to increases and/or decreases in Pro-enk mRNA expression, as has been shown by some authors in infant rats (Gale, Moroni, Kumakura, & Guidotti, 1979; Urwyler & Tabakoff, 1981). These changes could also be explained by alterations in Pro-enk mRNA stability and/or the post-translational processing of the precursor. Alternatively, ethanol-induced Met-enk release could also contribute to these actions. In agreement with this view, recent studies show that PEE up-regulates Pro-Dyn and KOR mRNA expression in the VTA, as well as KOR mRNA levels in the PFC of infant rats. Moreover, PEE decreases MOR gene expression in the VTA, and reduces Pro-Dyn gene levels in the NAcc (Wille-Bille et al., 2018). On the other hand, in adolescents in which

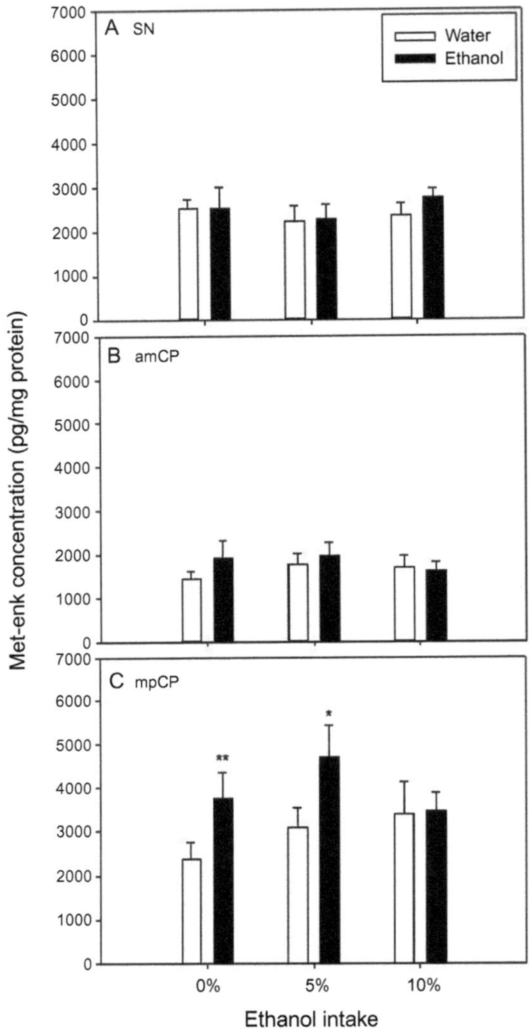

Fig. 2 Effect of prenatal ethanol treatment on Met-enk concentration in regions of the nigrostriatal pathway of PD15 pups. Animal treatment, preweanling's intake test and Met-enk quantitation were performed as described in Fig. 1. Met-enk concentration values (pg/mg protein) are shown for the SN (A), amCP (B) and mpCP (C). Data are the mean ± SEM of 5–6 pups in the SN, and 10–12 (prenatal water) or 10–13 (prenatal ethanol) in the amCP and mpCP. $^{**}P < 0.030$ and $^{*}P < 0.040$ versus their own prenatal water control groups. *Taken from Abate, P., Hernández-Fonseca, K., Reyes-Guzmán, A.C., Barbosa-Luna, I.G., & Méndez, M. (2014). Prenatal ethanol exposure alters met-enkephalin expression in brain regions related with reinforcement: Possible mechanism for ethanol consumption in offspring.* Behavioural Brain Research, 274, *194–204. doi: 10.1016/j.bbr.2014.08.022.*

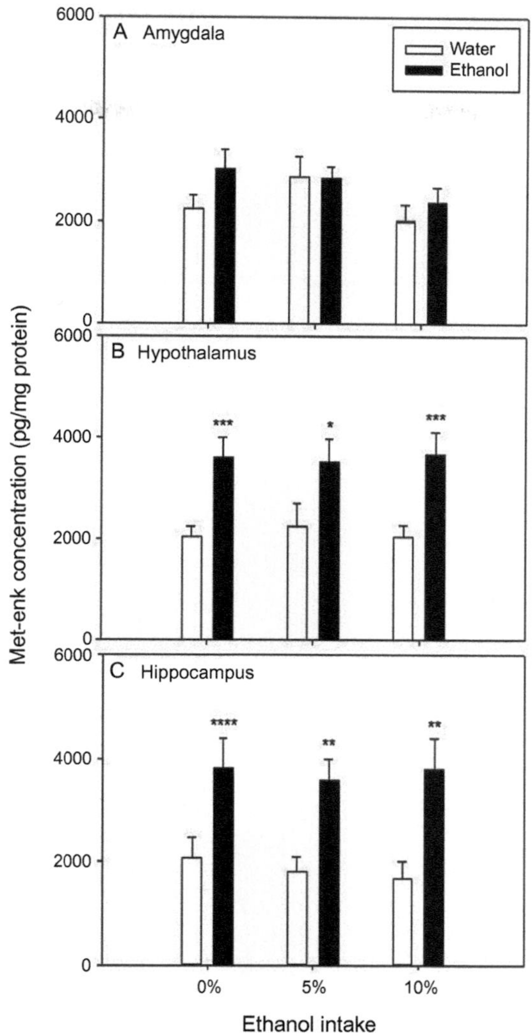

Fig. 3 Effect of prenatal ethanol treatment on Met-enk concentration in the amygdala, hypothalamus and hippocampus of PD15 pups. Animal treatment, preweanling's intake test and Met-enk quantitation were performed as described in Fig. 1. Met-enk concentration values (pg/mg protein) are shown for the amygdala (A), hypothalamus (B) and hippocampus (C). Data are the mean ± SEM of 8–10 (prenatal water) or 11–13 (prenatal ethanol) pups in the amygdala, and 12 (prenatal water) or 13 (prenatal ethanol) in the hypothalamus and hippocampus. ****$P < 0.0001$, ***$P < 0.005$, **$P < 0.010$ and *$P < 0.050$ versus their own prenatal water control groups. *Taken from Abate, P., Hernández-Fonseca, K., Reyes-Guzmán, A.C., Barbosa-Luna, I.G., & Méndez, M. (2014). Prenatal ethanol exposure alters met-enkephalin expression in brain regions related with reinforcement: Possible mechanism for ethanol consumption in offspring.* Behavioural Brain Research, 274, *194–204. doi: 10.1016/j.bbr.2014.08.022.*

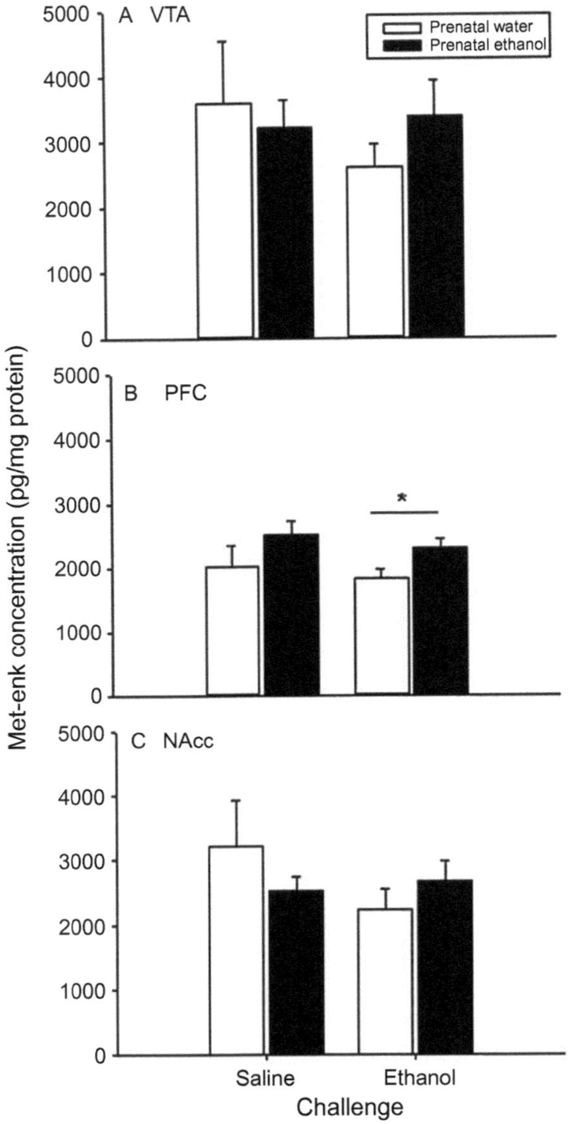

Fig. 4 Effect of prenatal ethanol treatment and drug challenge on Met-enk concentration in regions of the mesocorticolimbic system of PD30 adolescent rats. Wistar rat dams were daily administered with ethanol (2 g/kg, i.g.) or water on GDs 17–20. Animals were challenged with ethanol (1 g/kg, i.p.) or saline at PD30. Thirty minutes after ethanol or saline challenge, adolescents were sacrificed and their brains removed and frozen. Met-enk concentration was quantitated in several brain areas by RIA. Met-enk concentration values (pg/mg protein) are shown for the VTA (A), PFC (B) and NAcc (C). Data are the mean ± SEM of 6–8 rats in the PFC or 8 animals in the VTA and NAcc. *$P < 0.05$ versus prenatal water-ethanol challenge group. *Taken from Abate, P., Reyes-Guzmán, A.C., Hernández-Fonseca, K., & Méndez, M. (2017). Prenatal ethanol exposure modifies locomotor activity and induces selective changes in Met-enk expression in adolescent rats.* Neuropeptides, 62, *45–56. doi: 10.1016/j.npep.2016.11.006.*

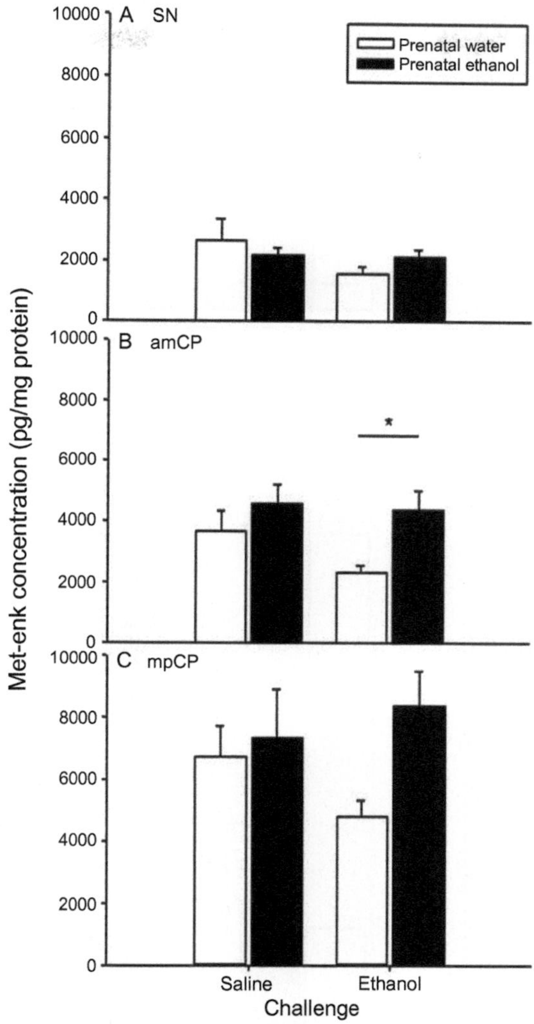

Fig. 5 Effect of prenatal ethanol treatment and drug challenge on Met-enk concentration in regions of the nigrostriatal pathway of PD30 adolescent rats. Prenatal animal treatment, drug challenge in adolescents at PD30 and Met-enk quantitation were performed as described in Fig. 4. Met-enk concentration values (pg/mg protein) are shown for the SN (A), amCP (B) and mpCP (C). Data are the mean $\pm$ SEM of 6–8 rats in the mpCP or 8 animals in the SN and amCP. $^{*}P<0.01$ versus prenatal water-ethanol challenge group. *Taken from Abate, P., Reyes-Guzmán, A.C., Hernández-Fonseca, K., & Méndez, M. (2017). Prenatal ethanol exposure modifies locomotor activity and induces selective changes in Met-enk expression in adolescent rats.* Neuropeptides, 62, *45–56. doi: 10.1016/j.npep.2016.11.006.*

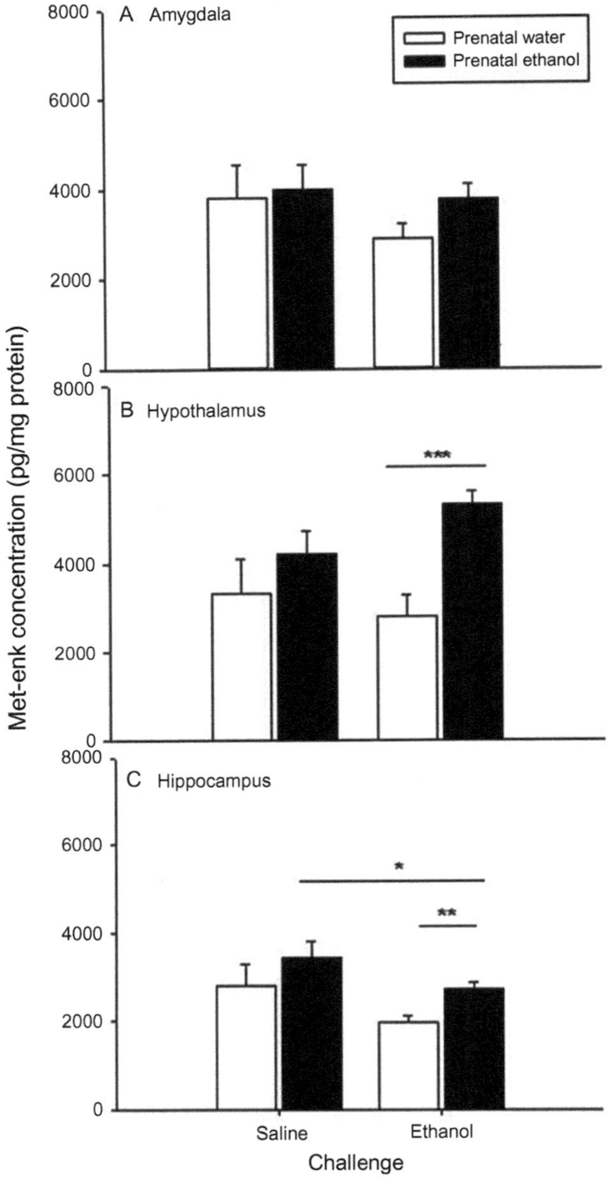

Fig. 6 Effect of prenatal ethanol treatment and drug challenge on Met-enk concentration in the amygdala, hypothalamus and hippocampus of PD30 adolescent rats. Prenatal animal treatment, drug challenge in adolescents at PD30 and Met-enk quantitation were performed as described in Fig. 4. Met-enk concentration values (pg/mg protein) are shown for the amygdala (A), hypothalamus (B) and hippocampus (C). Data are the mean ± SEM of 6–8 animals in the hippocampus or 7–8 adolescents in the amygdala and hypothalamus. ***$P < 0.001$ versus prenatal water-ethanol challenge group; **$P < 0.01$ versus prenatal water-ethanol challenge group; *$P < 0.05$ versus prenatal ethanol-saline challenge group. *Taken from Abate, P., Reyes-Guzmán, A.C., Hernández-Fonseca, K., & Méndez, M. (2017). Prenatal ethanol exposure modifies locomotor activity and induces selective changes in Met-enk expression in adolescent rats.* Neuropeptides, *62, 45–56. doi: 10.1016/j.npep.2016.11.006.*

PEE facilitates intake of the drug, increased expression of MOR mRNA has been reported in the VTA, with no effect on DOR and KOR transcripts (Fabio, Macchione, Nizhnikov, & Pautassi, 2015). These findings suggest that important differences exist in the role of opioid receptors in distinct brain areas and developmental stages.

5. Conclusions

Overall, these findings suggest a main role of the opioidergic system, particularly of enkephalinergic pathways in ethanol reinforcement and high alcohol drinking behavior in distinct developmental stages. As in adults, mesocorticolimbic enkephalins may be critical in ethanol reinforcement mechanisms in infant and adolescent rats. Ethanol-induced activation of mesocorticolimbic enkephalin-containing neurons may be relevant in infant rats. In contrast, the mesocortical enkephalinergic pathway could represent a key ethanol target in adolescents.

References

Abate, P., Hernández-Fonseca, K., Reyes-Guzmán, A. C., Barbosa-Luna, I. G., & Méndez, M. (2014). Prenatal ethanol exposure alters met-enkephalin expression in brain regions related with reinforcement: Possible mechanism for ethanol consumption in offspring. *Behavioural Brain Research*, *274*, 194–204. https://doi.org/10.1016/j.bbr.2014.08.022.

Abate, P., Pueta, M., Spear, N. E., & Molina, J. C. (2008). Fetal learning about ethanol and later ethanol responsiveness: Evidence against "safe" amounts of prenatal exposure. *Experimental Biology and Medicine (Maywood, N.J.)*, *233*(2), 139–154. https://doi.org/10.3181/0703-MR-69.

Abate, P., Reyes-Guzmán, A. C., Hernández-Fonseca, K., & Méndez, M. (2017). Prenatal ethanol exposure modifies locomotor activity and induces selective changes in Met-enk expression in adolescent rats. *Neuropeptides*, *62*, 45–56. https://doi.org/10.1016/j.npep.2016.11.006.

Alati, R., Al Mamun, A., Williams, G. M., O'Callaghan, M., Najman, J. M., & Bor, W. (2006). In utero alcohol exposure and prediction of alcohol disorders in early adulthood: A birth cohort study. *Archives of General Psychiatry*, *63*(9), 1009–1016.

Arias, C., & Chotro, M. G. (2005). Increased palatability of ethanol after prenatal ethanol exposure is mediated by the opioid system. *Pharmacology, Biochemistry, and Behavior*, *82*(3), 434–442.

Arias, C., Mlewski, E. C., Miller, S., Molina, J. C., & Spear, N. E. (2009). Novelty modulates the stimulating motor effects of ethanol in preweanling rats. *Pharmacology, Biochemistry, and Behavior*, *92*(3), 448–456. https://doi.org/10.1016/j.pbb.2009.01.012.

Arias, C., Mlewski, E. C., Molina, J. C., & Spear, N. E. (2009). Ethanol induces locomotor activating effects in preweanling Sprague-Dawley rats. *Alcohol*, *43*(1), 13–23. https://doi.org/10.1016/j.alcohol.2008.09.002.

Arias, C., Molina, J. C., & Spear, N. E. (2010). Differential role of micro, delta and kappa opioid receptors in ethanol-mediated locomotor activation and ethanol intake in preweanling rats. *Physiology & Behavior, 99*(3), 348–354. https://doi.org/10.1016/j.physbeh.2009.11.012.

Barbosa-Luna, I. G. (2012). *Participación de las encefalinas en los mecanismos de neuroadaptación al alcohol en la vía mesocorticolímbica de la rata.* Tesis de Maestría en Ciencias Biológicas (Biología Experimental)México: Facultad de Medicina, UNAM.

Berman, Y. L., Juliano, L., & Devi, L. A. (1995). Purification and characterization of a dynorphin processing endopeptidase. *The Journal of Biological Chemistry, 270*(40), 23845–23850.

Bignan, G. C., Connolly, P. J., & Middleton, S. A. (2005). Recent advances towards the discovery of ORL-1 receptor agonists and antagonists. *Expert Opinion on Therapeutic Patents, 15*(4), 357–388. https://doi.org/10.1517/13543776.15.4.357.

Bodnar, R. J. (2018). Endogenous opiates and behavior: 2016. *Peptides, 101*, 167–212. https://doi.org/10.1016/j.peptides.2018.01.011.

Bordner, K., & Deak, T. (2015). Endogenous opioids as substrates for ethanol intake in the neonatal rat: The impact of prenatal ethanol exposure on the opioid family in the early postnatal period. *Physiology & Behavior, 148*, 100–110. https://doi.org/10.1016/j.physbeh.2015.02.013.

Bordner, K. A., Molina, J. C., & Spear, N. E. (2008). Analysis of ethanol reinforcement in 1-day-old rats: Assessment through a brief and novel operant procedure. *Alcoholism, Clinical and Experimental Research, 32*(4), 580–592. https://doi.org/10.1111/j.1530-0277.2007.00609.x.

Borsodi, A., & Tóth, G. (1995). Characterization of opioid receptor types and subtypes with new ligands. *Annals of the New York Academy of Sciences, 757*, 339–352.

Chotro, M. G., & Arias, C. (2007). Ontogenetic difference in ethanol reinforcing properties: The role of the opioid system. *Behavioural Pharmacology, 18*(7), 661–666.

Chotro, M. G., Arias, C., & Laviola, G. (2007). Increased ethanol intake after prenatal ethanol exposure: Studies with animals. *Neuroscience and Biobehavioral Reviews, 31*(2), 181–191. https://doi.org/10.1016/j.neubiorev.2006.06.021.

Ciccocioppo, R., Economidou, D., Fedeli, A., & Massi, M. (2003). The nociceptin/orphanin FQ/NOP receptor system as a target for treatment of alcohol abuse: A review of recent work in alcohol-preferring rats. *Physiology & Behavior, 79*(1), 121–128.

Ciccocioppo, R., Panocka, I., Polidori, C., Regoli, D., & Massi, M. (1999). Effect of nociceptin on alcohol intake in alcohol-preferring rats. *Psychopharmacology, 141*(2), 220–224.

Civelli, O., Douglass, J., Rosen, H., Martens, G., & Herbert, E. (1986). Biosynthesis of opioid peptides. In R. S. Rapaka & R. L. Hawks (Eds.), *Opioid peptides: Molecular pharmacology, biosynthesis, and analysis* (pp. 21–42): NIDA Research Monograph.

Cuello, A. C. (1983). Central distribution of opioid peptides. *British Medical Bulletin, 39*(1), 11–16.

Cunningham, C. L., Niehus, J. S., & Noble, D. (1993). Species difference in sensitivity to ethanol's hedonic effects. *Alcohol, 10*(2), 97–102.

de Waele, J. P., & Gianoulakis, C. (1993). Effects of single and repeated exposures to ethanol on hypothalamic beta-endorphin and CRH release by the C57BL/6 and DBA/2 strains of mice. *Neuroendocrinology, 57*(4), 700–709. https://doi.org/10.1159/000126428.

Di Chiara, G., & Imperato, A. (1985). Ethanol preferentially stimulates dopamine release in the nucleus accumbens of freely moving rats. *European Journal of Pharmacology, 115*(1), 131–132.

Faas, A. E., March, S. M., Moya, P. R., & Molina, J. C. (2015). Alcohol odor elicits appetitive facial expressions in human neonates prenatally exposed to the drug. *Physiology & Behavior*, *148*, 78–86. https://doi.org/10.1016/j.physbeh.2015.02.031.

Faas, A. E., Spontón, E. D., Moya, P. R., & Molina, J. C. (2000). Differential responsiveness to alcohol odor in human neonates: Effects of maternal consumption during gestation. *Alcohol*, *22*(1), 7–17[doi: S0741832900001038].

Fabio, M. C., Macchione, A. F., Nizhnikov, M. E., & Pautassi, R. M. (2015). Prenatal ethanol increases ethanol intake throughout adolescence, alters ethanol-mediated aversive learning, and affects mu but not delta or kappa opioid receptor mRNA expression. *The European Journal of Neuroscience*, *41*(12), 1569–1579. https://doi.org/10.1111/ejn.12913.

Fadda, F., Mosca, E., Colombo, G., & Gessa, G. L. (1989). Effect of spontaneous ingestion of ethanol on brain dopamine metabolism. *Life Sciences*, *44*(4), 281–287.

Frederickson, R. C. A. (1984). Endogenous opioids and related derivatives. In M. J. Kuhar & G. W. Pasternak (Eds.), *Analgesics: Neurochemical, behavioral and clinical perspectives* (pp. 69–96). New York: Raven Press.

Froehlich, J. C. (1995). Genetic factors in alcohol self-administration. *The Journal of Clinical Psychiatry*, *56*(Suppl. 7), 15–23.

Froehlich, J. C., Badia-Elder, N. E., Zink, R. W., McCullough, D. E., & Portoghese, P. S. (1998). Contribution of the opioid system to alcohol aversion and alcohol drinking behavior. *The Journal of Pharmacology and Experimental Therapeutics*, *287*(1), 284–292.

Gale, K., Moroni, F., Kumakura, K., & Guidotti, A. (1979). Opiate-receptors in substantia nigra: Role in the regulation of striatal tyrosine hydroxylase activity. *Neuropharmacology*, *18*(4), 427–430.

Gessa, G. L., Muntoni, F., Collu, M., Vargiu, L., & Mereu, G. (1985). Low doses of ethanol activate dopaminergic neurons in the ventral tegmental area. *Brain Research*, *348*(1), 201–203.

Gianoulakis, C. (2009). Endogenous opioids and addiction to alcohol and other drugs of abuse. *Current Topics in Medicinal Chemistry*, *9*(11), 999–1015.

Gianoulakis, C., de Waele, J. P., & Thavundayil, J. (1996). Implication of the endogenous opioid system in excessive ethanol consumption. *Alcohol*, *13*(1), 19–23.

Gianoulakis, C., Krishnan, B., & Thavundayil, J. (1996). Enhanced sensitivity of pituitary beta-endorphin to ethanol in subjects at high risk of alcoholism. *Archives of General Psychiatry*, *53*(3), 250–257.

Gubler, U., Seeburg, P. H., Hoffman, B. J., Gage, L. P., & Undenfriend, S. (1982). Molecular cloning establishes proenkephalin as precursor of enkephalin-containing peptides. *Nature*, *295*(5846), 206–208.

Gysling, K., & Wang, R. Y. (1983). Morphine-induced activation of A10 dopamine neurons in the rat. *Brain Research*, *277*(1), 119–127.

Hallmark, R. A., & Hunt, P. S. (2004). Social learning about ethanol in preweanling rats: Role of endogenous opioids. *Developmental Psychobiology*, *44*(2), 132–139. https://doi.org/10.1002/dev.10163.

Herz, A. (1997). Endogenous opioid systems and alcohol addiction. *Psychopharmacology*, *129*(2), 99–111.

Higley, A. E., & Kiefer, S. W. (2006). Delta receptor antagonism, ethanol taste reactivity, and ethanol consumption in outbred male rats. *Alcohol*, *40*(3), 143–150.

Horikawa, S., Takai, T., Toyosato, M., Takahashi, H., Noda, M., Kakidani, H., et al. (1983). Isolation and structural organization of the human preproenkephalin B gene. *Nature*, *306*, 611–614.

Hughes, J., Smith, T. W., Kosterlitz, H. W., Fothergill, L. A., Morgan, B. A., & Morris, H. R. (1975). Identification of two related pentapeptides from the brain with potent opiate agonist activity. *Nature, 258*(5536), 577–580.

Hyytiä, P., & Kiianmaa, K. (2001). Suppression of ethanol responding by centrally administered CTOP and naltrindole in AA and Wistar rats. *Alcoholism, Clinical and Experimental Research, 25*(1), 25–33.

Joyce, E. M., & Iversen, S. D. (1979). The effect of morphine applied locally to mesencephalic dopamine cell bodies on spontaneous motor activity in the rat. *Neuroscience Letters, 14*(2–3), 207–212.

Kalivas, P. W., Widerläv, E., Stanley, D., Breese, G., & Prange, A. J., Jr. (1983). Enkephalin action on the mesolimbic system: A dopamine-dependent and a dopamine-independent increase in locomotor activity. *Pharmacology and Experimental Therapeutics, 227*(1), 229–237.

Kiefer, F., & Wiedemann, K. (2004). Neuroendocrine pathways of addictive behaviour. *Addiction Biology, 9*(3–4), 205–212.

Koob, G. F. (2003). Neuroadaptive mechanisms of addiction: Studies on the extended amygdala. *European Neuropsychopharmacology, 13*(6), 442–452.

Koob, G. F., Sanna, P. P., & Bloom, F. E. (1998). Neuroscience of addiction. *Neuron, 21*(3), 467–476.

Kozlov, A. P., Varlinskaya, E. I., & Spear, N. E. (2008). Ethanol, sacharine, and quinine: Early ontogeny of taste responsiveness and intake. *Alcoholism, Clinical and Experimental Research, 32*(2), 294–305. https://doi.org/10.1111/j.1530-0277.2007.00581.x.

Krishnan-Sarin, S., Wand, G. S., Li, X. W., Portoghese, P. S., & Froehlich, J. C. (1998). Effect of mu opioid receptor blockade on alcohol intake in rats bred for high alcohol drinking. *Pharmacology, Biochemistry, and Behavior, 59*(3), 627–635.

Kuzmin, A., Kreek, M. J., Bakalkin, G., & Liljequist, S. (2007). The nociceptin/orphanin FQ receptor agonist Ro 64-6198 reduces alcohol self-administration and prevents relapse-like alcohol drinking. *Neuropsychopharmacology, 32*(4), 902–910. https://doi.org/10.1038/sj.npp.1301169.

Leriche, M., & Méndez, M. (2010). Ethanol exposure selectively alters β-endorphin content but not [3H]-DAMGO binding in discrete regions of the rat brain. *Neuropeptides, 44*(1), 9–16. https://doi.org/10.1016/j.npep.2009.11.009.

Lindholm, S., Werme, M., Brené, S., & Franck, J. (2001). The selective kappa-opioid receptor agonist U50,488H attenuates voluntary ethanol intake in the rat. *Behavioural Brain Research, 120*(2), 137–146.

Lukas, S. E., & Mendelson, J. H. (1988). Electroencephalographic activity and plasma ACTH during ethanol-induced euphoria. *Biological Psychiatry, 23*(2), 141–148.

Mansour, A., Khachaturian, H., Lewis, M. E., Akil, H., & Watson, S. J. (1988). Anatomy of CNS opioid receptors. *Trends in Neurosciences, 11*(7), 308–314.

March, S. M., Abate, P., Spear, N. E., & Molina, J. C. (2009). Fetal exposure to moderate ethanol doses: Heightened operant responsiveness elicited by ethanol-related reinforcers. *Alcoholism, Clinical and Experimental Research, 33*(11), 1981–1993. https://doi.org/10.1111/j.1530-0277.2009.01037.x.

Marinelli, P. W., Bai, L., Quirion, R., & Gianoulakis, C. (2005). A microdialysis profile of Met-enkephalin release in the rat nucleus accumbens following alcohol administration. *Alcoholism, Clinical and Experimental Research, 29*(10), 1821–1828.

Marinelli, P. W., Quirion, R., & Gianoulakis, C. (2003). A microdialysis profile of b-endorphin and catecholamines in the rat nucleus accumbens following alcohol administration. *Psychopharmacology, 169*(1), 60–67. https://doi.org/10.1007/s00213-003-1490-2.

Martin, C. S., Earleywine, M., Musty, R. E., Perrine, M. W., & Swift, R. M. (1993). Development and validation of the biphasic alcohol effects scale. *Alcoholism, Clinical and Experimental Research, 17*(1), 140–146.

Méndez, M., Barbosa-Luna, I. G., Pérez-Luna, J. M., Cupo, A., & Oikawa, J. (2010). Effects of acute ethanol administration on methionine-enkephalin expression and release in regions of the rat brain. *Neuropeptides*, *44*(5), 413–420. https://doi.org/10.1016/j.npep.2010.05.001.

Méndez, M., & Herrera, S. (2012). Behavioral effects of alcohol in rodents. In S. E. Cruz-Morales & J. C. P. Arriaga-Ramírez (Eds.), *Behavioral animal models* (pp. 29–51). Kerala, India: Research Signpost.

Méndez, M., Leriche, M., & Calva, J. C. (2001). Acute ethanol administration differentially modulates μ opioid receptors in the rat meso-accumbens and mesocortical pathways. *Brain Research Molecular Brain Research*, *94*(1–2), 148–156.

Méndez, M., Leriche, M., & Calva, J. C. (2003). Acute ethanol administration transiently decreases [3H]-DAMGO binding to mu opioid receptors in the rat substantia nigra pars reticulata but not in the caudate-putamen. *Neuroscience Research*, *47*(2), 153–160.

Méndez, M., & Morales-Mulia, M. (2006). Ethanol exposure differentially alters pro-enkephalin mRNA expression in regions of the mesocorticolimbic system. *Psychopharmacology*, *189*(1), 117–124. https://doi.org/10.1007/s00213-006-0503-3.

Méndez, M., & Morales-Mulia, M. (2008a). Ethanol exposure and pro-enkephalin mRNA expression in regions of the mesocorticolimbic system. In L. Sher (Ed.), *Research on the neurobiology of alcohol use disorders* (pp. 177–202). New York: Nova Science Publishers Inc.

Méndez, M., & Morales-Mulia, M. (2008b). Role of endogenous opioid peptides in ethanol reinforcement. In M. Méndez & R. Mondragón (Eds.), *Neural mechanisms of action of drugs of abuse and natural reinforcers* (pp. 41–60). Kerala, India: Research Singpost.

Méndez, M., & Morales-Mulia, M. (2008c). Role of mu and delta opioid receptors in alcohol drinking behaviour. *Current Drug Abuse Reviews*, *1*(2), 239–252.

Méndez, M., Morales-Mulia, M., & Leriche, M. (2004). [3H]-DPDPE binding to d opioid receptors in the rat mesocorticolimbic and nigrostriatal pathways is transiently increased by acute ethanol administration. *Brain Research*, *1028*(2), 180–190.

Méndez, M., Morales-Mulia, M., & Pérez-Luna, J. M. (2008). Ethanol-induced changes in Proenkephalin mRNA expression in the rat nigrostriatal pathway. *Journal of Molecular Neuroscience*, *34*(3), 225–234. https://doi.org/10.1007/s12031-008-9039-9.

Miranda-Morales, R. S., Molina, J. C., Spear, N. E., & Abate, P. (2010). Participation of the endogenous opioid system in the acquisition of a prenatal ethanol-related memory: Effects on neonatal and preweanling responsiveness to ethanol. *Physiology & Behavior*, *101*(1), 153–160. https://doi.org/10.1016/j.physbeh.2010.04.033.

Miranda-Morales, R. S., Molina, J. C., Spear, N. E., & Abate, P. (2012). Naloxone attenuation of ethanol-reinforced operant responding in infant rats in a re-exposure paradigm. *Psychopharmacology*, *219*(1), 235–246. https://doi.org/10.1007/s00213-011-2402-5.

Miranda-Morales, R. S., Nizhnikov, M. E., Waters, D. H., & Spear, N. E. (2013). Participation of the nociceptin/orphanin FQ receptor in ethanol-mediated locomotor activation and ethanol intake in preweanling rats. *Behavioural Brain Research*, *245*, 137–144. https://doi.org/10.1016/j.bbr.2013.02.017.

Miranda-Morales, R. S., & Pautassi, R. M. (2016). Pharmacological characterization of the nociceptin/orphanin FQ receptor on ethanol-mediated motivational effects in infant and adolescent rats. *Behavioural Brain Research*, *298*(Pt. A), 88–96. https://doi.org/10.1016/j.bbr.2015.04.016.

Miranda-Morales, R. S., Spear, N. E., Nizhnikov, M. E., Molina, J. C., & Abate, P. (2012). Role of mu, delta and kappa opioid receptors in ethanol-reinforced operant responding in infant rats. *Behavioural Brain Research*, *234*(2), 267–277. https://doi.org/10.1016/j.bbr.2012.07.002.

Molina, J. C., Domínguez, H. D., López, M. F., Pepino, M. Y., & Faas, A. E. (1999). The role of fetal and infantile experience with alcohol in later recognition and acceptance patterns of the drug. In N. E. Spear, L. P. Spear, J. H. Hanningan, & C. R. Goodlett (Eds.), *Alcohol and alcoholism: Brain and development* (pp. 199–228). Hillsdale: Lawrence Erlbaum Associates.

Nestler, E. J., Hyman, S. E., & Malenka, R. C. (2001). Reinforcement and addictive disorders. In E. J. Nestler, S. E. Hyman, & R. C. Malenka (Eds.), *Molecular neuropharmacology: A foundation for clinical neuroscience* (pp. 355–382). New York: The McGraw-Hill Companies, Inc.

Nizhnikov, M. E., Pautassi, R. M., Truxell, E., & Spear, N. E. (2009). Opioid antagonists block the acquisition of ethanol-mediated conditioned tactile preference in infant rats. *Alcohol*, *43*(5), 347–358. https://doi.org/10.1016/j.alcohol.2009.06.001.

Nizhnikov, M. E., Varlinskaya, E. I., Petrov, E. S., & Spear, N. E. (2006). Reinforcing properties of ethanol in neonatal rats: Involvement of the opioid system. *Behavioral Neuroscience*, *120*(2), 267–280. https://doi.org/10.1037/0735-7044.120.2.267. 2006-05348-005 [pii].

Nizhnikov, M. E., Varlinskaya, E. I., & Spear, N. E. (2006). Reinforcing effects of central ethanol injections in newborn rat pups. *Alcoholism, Clinical and Experimental Research*, *30*(12), 2089–2096. https://doi.org/10.1111/j.1530-0277.2006.00253.x.

Noda, M., Furutani, Y., Takahashi, H., Toyosato, M., Hirose, T., Inayama, S., et al. (1982). Cloning and sequence analysis of cDNA for bovine adrenal preproenkephalin. *Nature*, *295*(5846), 202–206.

Olive, M. F., Koenig, H. N., Nannini, M. A., & Hodge, C. W. (2001). Stimulation of endorphin neurotransmission in the nucleus accumbens by ethanol, cocaine, and amphetamine. *The Journal of Neuroscience*, *21*(23), RC184.

Pasternak, G. W. (1988). Multiple morphine and enkephalin receptors and the relief of pain. *JAMA*, *259*(9), 1362–1367.

Pautassi, R. M., Miranda-Morales, R. S., & Nizhnikov, M. (2015). Operant self-administration of ethanol in infant rats. *Physiology & Behavior*, *148*, 87–99. https://doi.org/10.1016/j.physbeh.2014.08.002.

Pautassi, R. M., Nizhnikov, M. E., & Spear, N. E. (2009). Assessing appetitive, aversive, and negative ethanol-mediated reinforcement through an immature rat model. *Neuroscience and Biobehavioral Reviews*, *33*(6), 953–974. https://doi.org/10.1016/j.neubiorev.2009.03.008.

Pilatti, A., Godoy, J. C., Brussino, S., & Pautassi, R. M. (2013). Underage drinking: Prevalence and risk factors associated with drinking experiences among Argentinean children. *Alcohol*, *47*(4), 323–331. https://doi.org/10.1016/j.alcohol.2013.02.001.

Ponce, L. F., Pautassi, R. M., Spear, N. E., & Molina, J. C. (2008). Ethanol-mediated operant learning in the infant rat leads to increased ethanol intake during adolescence. *Pharmacology, Biochemistry, and Behavior*, *90*(4), 640–650. https://doi.org/10.1016/j.pbb.2008.05.007.

Pueta, M., Abate, P., Haymal, O. B., Spear, N. E., & Molina, J. C. (2008). Ethanol exposure during late gestation and nursing in the rat: Effects upon maternal care, ethanol metabolism and infantile milk intake. *Pharmacology, Biochemistry, and Behavior*, *91*(1), 21–31. https://doi.org/10.1016/j.pbb.2008.06.007.

Schafer, M., Day, K. H., Watson, S. J., & Akil, H. (1991). Distribution of opioids in brain and peripheral tissues. In T. S. Shippenberg (Ed.), *Neurobiology of opioids* (pp. 53–62). Berlin: Springer-Verlag.

Shippenberg, T. S., LeFevour, A., & Chefer, V. I. (2008). Targeting endogenous mu- and delta opioid receptor systems for the treatment of drug addiction. *CNS & Neurological Disorders Drug Targets*, 7(5), 442–453.

Spain, J. W., Roth, B. L., & Coscia, C. J. (1985). Differential ontogeny of multiple opioid receptors (mu, delta, and kappa). *The Journal of Neuroscience*, *5*(3), 584–588.

Spanagel, R., Herz, A., & Shippenberg, T. S. (1990). Identification of the opioid receptor types mediating beta-endorphin-induced alterations in dopamine release in the nucleus accumbens. *European Journal of Pharmacology*, *190*(1–2), 177–184.

Spear, N. E., & Molina, J. C. (2005). Fetal or infantil exposure to alcohol promotes ethanol ingestion in adolescence and adulthood: A theoretical review. *Alcoholism, Clinical and Experimental Research*, *25*(6), 909–929.

Stromberg, M. F., Casale, M., Volpicelli, L., Volpicelli, J. R., & O'Brien, C. P. (1998). A comparison of the effects of the opioid antagonists naltrexone, naltrindole, and β-funaltrexamine on ethanol consumption in the rat. *Alcohol*, *15*(4), 281–289.

Ulm, R. R., Volpicelli, J. R., & Volpicelli, L. A. (1995). Opiates and alcohol self-administration in animals. *The Journal of Clinical Psychiatry*, *56*(Suppl. 7), 5–14.

Urwyler, S., & Tabakoff, B. (1981). Stimulation of dopamine synthesis and release by morphine and D-ala2-D-leu5-enkephalin in the mouse striatum in vivo. *Life Sciences*, *28*(20), 2277–2286.

Vaccarino, A. L., & Kastin, A. J. (2001). Endogenous opiates: 2000. *Peptides*, *22*(12), 2257–2328. https://doi.org/10.1016/S0196-9781(01)00566-6.

Volpicelli, J. R., Alterman, A. I., Hayashida, M., & O'Brien, C. P. (1992). Naltrexone in the treatment of alcohol dependence. *Archives of General Psychiatry*, *49*(11), 876–880.

Volpicelli, J. R., Ulm, R. R., & Hopson, N. (1991). Alcohol drinking in rats during and following morphine injections. *Alcohol*, *8*(4), 289–292.

Volterra, A., Brunello, N., Restani, P., Galli, C. L., & Racagni, G. (1986). Ontogenetic studies on mu, delta and kappa opioid receptors in rat brain. *Pharmacological Research Communications*, *18*(10), 979–990.

Wild, K. D., & Reid, L. D. (1990). Modulation of ethanol-intake by morphine: Evidence for a central site of action. *Life Sciences*, *47*, PL49–PL54.

Wille-Bille, A., Miranda-Morales, R. S., Pucci, M., Bellia, F., D'Addario, C., & Pautassi, R. M. (2018). Prenatal ethanol induces an anxiety phenotype and alters expression of dynorphin & nociceptin/orphanin FQ genes. *Progress in Neuro-Psychopharmacology & Biological Psychiatry*, *85*, 77–88. https://doi.org/10.1016/j.pnpbp.2018.04.005.

Wise, R. A., & Bozarth, M. A. (1987). A psychomotor stimulant theory of addiction. *Psychological Review*, *94*(4), 469–492.

Wood, P. L., & Rao, T. S. (1991). Morphine stimulation of mesolimbic and mesocortical but not nigrostriatal dopamine release in the rat as reflected by changes in 3-methoxytyramine levels. *Neuropharmacology*, *30*(4), 399–401.

FSC
www.fsc.org
MIX
Papier | Fördert gute Waldnutzung
FSC® C083411

Kolibri 360 GmbH
Ferdinand-Jühlke-Straße 7
99095 Erfurt, Deutschland
produktsicherheit@kolibri360.de